THE MATHEMATICS OF POLITICS

SECOND EDITION

THE MATHEMATICS OF POLITICS

SECOND EDITION

E. Arthur Robinson, Jr.
George Washington University
Washington, D.C., USA

Daniel H. Ullman
George Washington University
Washington, D.C., USA

CRC Press is an imprint of the
Taylor & Francis Group, an **informa** business
A CHAPMAN & HALL BOOK

CRC Press
Taylor & Francis Group
6000 Broken Sound Parkway NW, Suite 300
Boca Raton, FL 33487-2742

Printed on acid-free paper
Version Date: 20161012

International Standard Book Number-13: 978-1-4987-9886-0 (Hardback)

Visit the Taylor & Francis Web site at
http://www.taylorandfrancis.com

and the CRC Press Web site at
http://www.crcpress.com

Contents

Preface for the Reader

It is because mathematics is so often misunderstood that people commonly believe that it has nothing to say about politics. The high school experience with mathematics, which is for so many the lasting impression they have of the subject, suggests that mathematics is the study of numbers, operations, formulas, and manipulations of symbols. Those who believe that this is the extent of mathematics will conclude that mathematics has no relevance to politics. Moreover, they will not find much in this book that they regard as mathematics.

There are some numbers here and there in the book, and they are occasionally combined with the standard operations. A few formulas make appearances, and in a few places symbols are manipulated using the basic rules of algebra. But none of that is the focus here. Instead, this book focuses on mathematical reasoning about politics. Mathematicians regard mathematics as the study of how to see whether things are true or false, whether things are possible or impossible. Applied to the political arena, in the search for ideal ways to make certain kinds of decisions, a lot of wasted effort can be averted if mathematics can determine that finding such an ideal is actually impossible in the first place.

The mathematician's notion of impossibility is absolute. To say that something is impossible, in this sense, is not merely to assert that it is highly unlikely or that it has never been done before. It is to say that it can never be done owing to the very nature of the thing that is being attempted. The aphorism that "nothing is impossible" is itself utterly false. It is impossible for $2 + 2$ to equal 5. It is impossible for a particle to travel faster than the speed of light. And in the political arena, it is impossible to create methods to vote, methods to apportion, and methods to make decisions that are satisfactory in every possible way. This is where mathematics meets politics.

In order to prove that things are impossible, mathematicians must first clearly and precisely define their terms. Indeed nothing is impossible if one allows for the possibility of twisting words into some interpretation that was not intended. For example, someone might notice that rounding all the numbers in the equation $2.3 + 2.3 = 4.6$ yields the allegedly impossible $2 + 2 = 5$. Have we achieved the impossible? No. To say that something is impossible is to say only that if the meaning of the terms is accepted, then the rules of reasoning make it irrefutably certain that the thing cannot be achieved. But nothing is impossible if you permit yourself to change the meaning of words or terms or to change the rules of reasoning.

In the first three parts of this book, we address the following three political questions:

(1) Is there a good way to choose winners of elections?

(2) Is there a good way to apportion congressional seats?

(3) Is there a good way to make decisions in situations of conflict and uncertainty?

The answer, of course, depends on what is meant by the word "good". This then becomes our focus. We carefully and precisely formulate environments within which we can discuss approaches to elections, apportionment, and decisions. We then contemplate various ways to capture a notion of goodness within each of these environments. In the end, we aim to explain why, in a certain sense of the word "good", it is impossible to find good methods for solving these problems.

In the fourth and final part of this book, we examine the Electoral College system that is used in the United States to select a president. There we bring together ideas that are introduced in each of the three earlier parts of the book.

So what makes this book mathematical is not numbers or formulas but rather reasoning. Students are expected to discern whether things are possible or impossible. Throughout the book are questions that ask whether it is always the case that something occurs. When the answer is no, the student is expected to provide a counterexample. When the answer is yes, the student is expected to provide a proof.

A proof is a persuasive argument that explains why an assertion is true. In order to be able to write a proof, it is necessary to manage technical vocabulary. The reader is advised to pay careful attention to the way that technical words are defined so that the precise technical meaning is not confused with the ordinary meaning that the word carries in natural language.

We have endeavored to keep the technical mathematical terminology to a minimum and to avoid unnecessary notation. Still, because we want to do mathematics, we do define terms precisely. Also, the cost of avoiding notation is that certain simple ideas are repeatedly spelled out in English words in what may seem like laborious detail. Our approach is intended to make the subject more accessible, although we acknowledge that it comes at the expense of conciseness.

Good writing is supposed to be both clear and precise. In mathematics, there is a curious irony that clarity and precision are not always achievable simultaneously. In order to be precise, it is sometimes necessary to provide microscopic detail in a way that obscures the big picture and makes the main idea difficult for the reader to draw out. In order to be clear, it is sometimes necessary to provide metaphors and analogies in a way that obscures the details and makes the ideas inexact. We have attempted to be as clear and precise as possible, but the reader may need to find the main idea in one passage and the exact details in another.

The book requires rather little background in mathematics or political science. The notion of a function, however, plays a pivotal role, and a feeling for numbers is useful, especially in Part II of the book. Some experience with the American system of government is assumed (e.g., that we have a Constitution, a bicameral legislature, a census each decade, an Electoral College). Still, there is little or no dependence on high school mathematics or government courses. What is really required from the reader is the ability to reason precisely within a carefully prescribed environment and to express that reasoning in clear persuasive writing. This comes easily to some readers but not to others, and this seems to be quite independent of whether those readers have been told in the past that they are good in math.

Each chapter begins with a scenario, which is a puzzle or problem meant to engage students before they read and study the text itself. The scenarios are appropriate for group study and discussion. Their purpose is to get readers thinking about the sorts of problems that will be approached in the chapter. The scenarios generally do not have a single correct answer. Readers are meant to approach them naively, trying their best to solve them just by thinking about them, without technical tools. This primes the reader to be receptive to the material in the chapter. The scenarios are meant to help build a connection between what readers already know and what is presented in the chapter.

We encourage the reader to stay in an active, rather than passive, state of mind. The text is meant to be read slowly; it may take several hours to read even one subchapter. It is best to read with paper and pencil in hand. After each sentence, the careful reader will ask: Does that make sense to me? Why is it true? Is it obvious or does it require explanation? Could I explain it to another student? What is the motivation for this? Can I think of an example? Can I work through a computation? An imprecise or vague idea of what is meant by a key concept is often insufficient. The goal is to be able to employ the concept in an exact and precise way.

Each chapter ends with a collection of exercises and problems. By an "exercise" we mean a routine application of material from the text, meant to confirm that basic notions are understood. By a "problem" we mean an open-ended challenge, where the manner of solution is not outlined for the solver. (The boundary between these two concepts is inexact.) Students should not expect that every problem is solved merely by doing some process that is outlined and explained in the text. The reader should be prepared to work through as many exercises and problems as possible. This is the key activity that develops and reinforces conceptual understanding.

There is no claim made that the four topics addressed in this book represent the only four ways or even the most important four ways that mathematics and politics intersect. But these four topics lend themselves well to mathematical analysis in a way that readily engages students without relying heavily on mathematical prerequisites. We invite the reader to undertake and embrace this study. It is fun and challenging. In the end, we are certain that the reader will have a better understanding of the rules associated with political decision-making as well as of the ways in which mathematics applies to problems of the real world.

There are many people who assisted us with the development of this book. We offer thanks to Lowell Abrams, Sudeshna Basu, Meena Bhatia, John Dougherty, Farshad Foroozan, Alex Gordon, Valentina Harizanov, Joe Herning, Chasya Hoagland, Diane Holcomb, Merilys Huhn, Ed Scheinerman, Matthew Sedlock, and Ted Turner for teaching from earlier versions of this text and offering helpful suggestions. Also thanks to Allison Pacelli, Alan Taylor, and Alex Wolitzky for valuable discussions. We appreciate as well the support of our friend John B. Conway. Last but most, we thank our wives, without whose good-humored patience this project would never have been completed.

— EAR
— DHU

Preface for the Instructor

This is a second edition of a text whose first edition was titled "A Mathematical Look at Politics". The present edition has updated examples and problems, improved exposition in selected parts, and numerous small corrections, but it is otherwise structured much like the previous edition.

The book offers an alternative to the usual mathematics courses for liberal arts students seeking to meet a general education requirement in mathematics or quantitative reasoning. Students in such a course rarely have a native interest in mathematics, and the typical course surveying the mathematical landscape rarely offers them anything that they see as useful, relevant, or engaging. By contrast, such students often have a native interest in politics, more so in recent years, as evidenced by the involvement of young people in the 2008 campaign of President Obama and the 2016 campaigns of Bernie Sanders and Donald Trump. For such students, we suggest a course that brings mathematics to their world of interests, rather than attempting to drag them against their will into our world of abstract mathematics.

The treatment here is highly mathematical, more so in some sense than many traditional mainstream mathematics courses for first-year college students. Concepts are defined precisely. Hypotheses are laid out carefully. Propositions, lemmas, theorems, and corollaries are stated clearly and proved. Counterexamples are offered to refute conjectures. Students are expected not only to make computations but also to state results, apply them to draw conclusions about specific examples, and prove them.

A course based on this book is well suited for an inquiry-based pedagogy, and each chapter begins with a "scenario" for student discussion in small groups. We find it useful, especially with highly engaged students, to invite students to engage with the problems of voting, apportionment, and games before we present to them the theories that address these issues. Students are naturally much more receptive to learning a theorem when they see that it solves a problem that interests them. At a minimum, the scenarios that open Chapters 2, 7, and 13 should be investigated at some length before the exposition of the theory is begun.

As in any mathematics course, the exercises and problems are of central importance, and we offer a large number of them at a variety of levels of difficulty. Students will learn a great deal from doing many of them, and yet assigning too many problems will make the course overwhelming to some students. Complete answers to odd-numbered problems appear in an appendix.

There is much more material here than can be covered in a single semester. The four parts of the book (on voting, apportionment, conflict, and the Electoral College, respectively) are essentially independent, although the fourth part relates to all three of the earlier parts. This affords the instructor a number of options for designing a semester-long course. One could very easily choose any two of the first three parts of the book and find much more to cover than time allows. This is especially true if one allows plenty of class time for exploration of the scenarios.

On the other hand, we recommend in a one-semester course touching on Parts I, II, and III of the book. Unless students are very able, this will require skipping certain topics and even entire chapters. Within each part, the chapters must be covered in order. (Chapter 12 is an exception to this rule.) Generally, it is best to cover the early chapters in each part with care, and then cover only briefly or skip entirely some later chapters. In Part I, the proof of Arrow's Theorem in Chapter 5 can be skipped, as can all of Chapter 6. In Part II, Chapters 10 and 11 can be skipped, while Chapter 12 can be left as reading for the students. In Part III, the first few sections of Chapter 14 may be skipped or covered rapidly if students are familiar with the first concepts of probability; Chapter 17 is more technical and may also be skipped.

Part IV makes a nice closing unit for the course, but saving time to cover it may require even further cuts to earlier parts. The material there is certainly worth a couple of weeks of study, but it is possible also to tell the story of these two chapters in one breezy lecture. In the fall semesters of years congruent to 0 modulo 4, the unit will be of special significance. Some instructors will skip Part IV entirely, with no dire consequences.

The Notes section that concludes each part of the book directs students toward related topics and relevant references. Students will find therein many ideas for independent projects and investigations.

A technical point that persists throughout the book is dealing with ties in various contexts. When an electorate is exactly divided between two candidates, no theory we present can help decide who should win. When a census finds two states with equal populations, we do not attempt to resolve the battle between the two states for a subsequent seat in the House. When two strategies in a game yield equivalent payoffs, we won't be able to say which one is preferable. In some parts of the book, we ignore the possibility of ties. In other parts, we explicitly include a careful treatment of ties, in the name of precision but at the expense of simplicity of exposition.

We find that the course is more deeply engaging for students than is a traditional general education course touring the mathematical landscape. At the same time, it is also more challenging for the students, and many of the problems are hard enough to stymie and then frustrate even strong students. By contrast with a course in, say, precalculus, students bring no prior experience or knowledge of the subject to the course, and they must master a substantial vocabulary of new words and concepts to succeed. A prudent approach is to proceed slowly, assess progress regularly, and assign difficult problems in moderation.

The joy in teaching this course is in the interplay of theory and application. The concept of independence in the theory of voting is brought to life by a discussion of Ralph Nader's role in the 2000 presidential election. The definition of Dean's method in the theory of apportionment is brought to life by a discussion of Montana's 1992 lawsuit challenging the constitutionality of the apportionment process. The explanation of the prisoner's dilemma in the theory of games is brought to life by a discussion of the arms race. When students realize that mathematical thinking can address these sorts of pressing concerns of the political world — of the human world — it naturally sparks their interest in the underlying mathematics.

We hope you enjoy teaching students about the connections between mathematics and politics as much as we have.

— EAR
— DHU

Part I

Voting

Introduction to Part I

There can be no democracy without elections. In order for political power to reside with the people, the people must have the means to express that power. They do this by going to the polls and communicating their personal views through ballots. Then, based on these ballots, a single decision is rendered on behalf of the collective voters. This is how political leaders emerge in democratic societies, and it is what restrains them from despotic rule.

In establishing rules governing an election, a key issue is to determine precisely how voters will be permitted to express themselves on the ballots. Will they be asked which candidate they most favor, or to what degree they favor each candidate, or whether they approve of each candidate, or the order in which they rank the candidates, or perhaps yet something else? Once that is settled, a new and unexpected difficulty arises. It isn't always obvious from the ballots which candidate best represents the collective choice of the voters. A decision process is needed that names a candidate the winner. Certainly if the electorate is unanimous in its preference (or even nearly so) the decision will be obvious. But we must be alert to the possibility of a split electorate. In close elections, different decision procedures may yield different winners.

When people think of rules governing elections, they tend to think of such notions as suffrage, voter registration procedures, poll taxes, campaign financing laws, and term limits. We do not consider these issues here. We ignore rules governing how candidates are allowed to persuade voters, rules about who is permitted to vote, and rules about who may run for office. Rather, our interest is in the process used to decide on a winner *after* all the votes have been cast. So we begin our study where others finish. The slate of candidates has been determined, the electorate has registered to vote, and minds are made up. Even the election itself has proceeded without a hitch. Yet a critical and controversial part of the process makes its appearance at this final stage: How do we decide who wins?

In Part I, we lay out a context for studying methods for deciding elections. We discover that there are many such methods, and that no one of them is the clear and obvious choice. This study will lead us to a proof of a famous theorem of Kenneth Arrow, which asserts (roughly) that it is impossible to find a method that is desirable from every point of view.

1

Two Candidates

"We always want the best man to win an election. Unfortunately, he never runs." — Will Rogers

1.0 Scenario

A student organization needs to elect a new president, but unfortunately the bylaws do not specify how this is to be done. One candidate, called candidate A, is the incumbent president and a senior. Running against candidate A is a sophomore, candidate B. The organization has 57 members, all of whom attend a meeting called to elect a new president.

Candidate A presides over the meeting, and opens the floor to suggestions. Someone responds "We should just take a vote, and whoever gets the most votes should win." Everyone agrees this is a good idea. Assuming everyone votes, what is the minimum number of votes a candidate needs to win? What might go wrong if some members abstain? What do the words "majority" and "plurality" mean?

Membership records show that in this organization there are 8 seniors and 22 juniors. It is a given that all the seniors will vote for candidate A, while all the freshmen and sophomores will vote for candidate B. However, the votes of the juniors are evenly divided between the two. If the vote were taken right away, who would win?

Just before the vote, one of the seniors speaks. "In the past," he recalls, "seniors each got three votes, and juniors each got two." Without thinking about it, everyone agrees to adopt this new "weighted" voting system. If the vote were taken at this point, who would win? Where in the real world does weighted voting seem appropriate?

Just before the ballots are distributed, a certain junior who supports candidate A says "It is unfair to discriminate against freshmen and sophomores. 'One person one vote' is a solid American tradition." He points out, however, that in the past "we always required 2/3 of the members to vote against a sitting president to unseat him." Under this new system, are there enough votes to unseat candidate A? Where in the real world does it seem appropriate to give one alternative an advantage over another in an election?

The vote is finally ready to be taken, but just before that happens, a representative from the Student Association arrives to oversee the election. "This year," he says, "all student organizations must use the 'parity method' in which a candidate can be elected only if he or she receives an even number of votes." Who would win the election under this new method? What features of this method would you mention if your goal was to highlight the absurdity of this Student Association mandate? In particular, what would occur if candidate A, in frustration with campus politics, decided to vote for the opposing candidate B?

1.1 Two-Candidate Methods

The theory of elections becomes especially rich and complicated when there are at least three candidates running for office. Before we take on these complications, however, we begin with the simpler situation in which there are just two candidates. (We can safely ignore elections with just one candidate.)

Society uses voting for many purposes besides deciding who will hold public office. Often voting is required to decide between two alternatives. For example, when a legislature votes on a bill to enact a new law, the two alternatives are simply to "pass" or to "defeat" the bill. In a vote about which beverage to serve at a picnic, the alternatives might be "beer" or "wine". And in a vote in a law firm on whether to promote an associate, the alternatives may be to "favor" or "oppose" the promotion.

We will generally focus on the political setting, describing every situation that requires voting as an "election", and referring to the alternatives as "candidates". In the two-candidate case, we will usually call the two candidates A and B. Still, it is worth remembering that the model is widely applicable and that candidates need not be people running for office (or people at all).

There are several reasons why the two-candidate case is simpler than the general case. First, there is not much doubt about how to design a ballot for this situation. The ballot merely needs to inquire whether a voter prefers candidate A or candidate B. The purpose of a ballot is to divide the voters into two classes: those who prefer candidate A and those who prefer candidate B. Provided this is accomplished, we do not concern ourselves with practical matters such as whether the ballot is electronic or paper, or whether the name of candidate A comes before the name of candidate B. We assume that each voter expresses a preference and that the intentions of the voters have been properly registered by the balloting process.

Another reason that the two-candidate case is simple is that it is clear what decision needs to be rendered in the end: Either candidate A wins or candidate B does. Actually, it will be useful to allow for a third possibility — that no decision is rendered. In this case, the result of the election is considered to be a **tie**. Ties may appear to be undesirable outcomes, but we will soon see that election procedures that always resolve ties necessarily suffer from other problems.

A third reason that restricting to two candidates simplifies the story is that it prevents an anomaly that can occur when there are more than two candidates. A candidate who gets more votes than any other candidate is said to have a **plurality** of the votes. A candidate who gets more than half the votes is said to have a **majority** of the votes. When there are three or more candidates, it is possible — and even familiar — for the candidate with a plurality of the votes not to have a majority of the votes. In the case of two candidates, however, the distinction between plurality and majority disappears. With just two candidates, a candidate who gets more votes than the opponent necessarily gets more than half of all the votes cast.

We call the decision procedure that is used to render the result of an election a **social choice function** or **voting method**. A **function** is a rule that assigns to every input from one set (the **domain**) an output chosen from another set (called the **codomain**). No coin-tossing or other extraneous information is permitted to be used in the process. A function must be repeatable in the sense that the output is always the same whenever the input is repeated. We do not allow decision procedures that work one way on weekdays but a different way on weekends. Finally, a function must produce an output from every logically possible input. In other words, the procedure must give a result no matter how the public votes. In the context of two-candidate elections, a social choice function is a function whose domain is the set of all possible preferences of the voters and whose codomain is the set {"A wins", "B wins", "It's a tie"}.

The two-candidate case may seem unusually simple, and the reader may be baffled at all the fuss. After all, there seems to be an obvious method for deciding elections with two candidates: Just count the votes, and whichever candidate gets the most votes is the winner. If the number of voters is odd, this completely specifies a social choice function, because one of the two candidates will always achieve a majority, while the other will fail to achieve a majority. But if the number of voters is even, the vote can be evenly split. In this case it seems clear that we should declare the result to be a tie.

This leads to the following **definition**, producing our first example of a social choice function. In mathematical discourse, a definition is meant to establish a word or phrase — identified in boldface — as a technical term, and to highlight that the word or phrase will have a restricted and limited usage. It is important to use such terms only in the technical sense and not to confuse them with the informal meaning they may have acquired from natural language.

Definition 1.1. The **simple majority method** is the social choice function that, in a two-candidate election, selects as the winner the candidate who gets more than half of all the votes cast. If each candidate gets exactly half of the votes, then the result is a tie.

Example 1.2. Suppose that there are 10 voters in a certain election that is to be decided by the simple majority method. Suppose their votes are A, B, A, B, A, A, A, A, B, and B, respectively. This voting data can be summarized in a diagram we call a **profile** of the electorate.

A	B	A	B	A	A	A	A	B	B

To apply the simple majority method, we need to count votes. We see that six voters prefer A, while four voters prefer B. We can summarize this data in a table we refer to as a **tabulated profile**:

6	4
A	B

Since there are 10 voters, a majority requires 6 or more votes. Clearly candidate A has achieved this, so candidate A wins by the simple majority method.

Example 1.3. Suppose that at the last minute one of the A voters in the previous example, say voter 8, changes her vote from A to B. The new profile is

A	B	A	B	A	A	A	B	B	B

and the new tabulated profile is

5	5
A	B

and now the result according to the simple majority method is a tie.

1.2 Supermajority and Status Quo

The simple majority method may seem like the natural, obvious choice to use in conducting elections with two candidates. Yet there are many other methods, and in certain circumstances these methods may be not only reasonable but actually more appropriate than the simple majority method.

Sometimes a bare majority may not be regarded as sufficiently strong to be decisive. For example, to stop a filibuster, the rules of the United States Senate require 3/5 of all the senators to vote to stop it. If the 3/5 vote is not achieved, the filibuster can continue. There are actually two separate ideas in this example, and we consider them one at a time.

A method that requires more than a simple majority is called a supermajority method. A supermajority method depends on the choice of a **parameter** p, which is a number specified in advance.

Definition 1.4. Let p be a number satisfying $1/2 < p \leq 1$. The **supermajority method** with parameter p is the social choice function that selects as the winner the candidate who gets a fraction p or more of all the votes. In particular, if there are t voters, a candidate must get at least pt votes to win. If no candidate gets pt votes, then the result is a tie.

The minimum number q of votes that a candidate needs to win is called the **quota**. Supermajority methods are sometimes called **quota methods**. The quota q is given by the smallest whole number greater than or equal to pt. The mathematical notation for this is $q = \lceil pt \rceil$ (read "the ceiling of pt"), where the expression $\lceil x \rceil$ denotes the smallest integer greater than or equal to x.

If we allowed the case $p = 1/2$, the supermajority method would reduce to the simple majority method, provided that, when the vote is evenly split and both candidates claim victory, we regard the election as a tie. As p increases toward 1, the standard for winning becomes increasingly difficult to meet. In the ultimate case, $p = 1$, the supermajority method requires **unanimity** (or **consensus**) to produce a winner.

So far, the supermajority method — as we have stated it — is not exactly what goes on in the Senate to end a filibuster. If those senators trying to end a filibuster cannot muster the necessary 3/5 majority, the result is not a tie, but rather the filibuster continues. The continuation of the filibuster, which is to say, the status quo, is given privileged status. We can add the idea of a status quo to any other method as follows.

Definition 1.5. Start with a social choice function (e.g., the simple majority method or a supermajority method), and call it the "base method". The following methods are called **status quo methods**. Designate one of the two candidates as the status quo and the other as the challenger. If either candidate wins under that base method, then that candidate wins under that status quo method as well. However, if there is a tie under the base method, then the status quo method names the status quo candidate as the winner.

Example 1.6. Suppose in Example 1.3 we are using a status quo method based on the supermajority method with $p = 2/3$ and candidate A has been designated as the challenger. Since the quota is $q = \lceil \frac{2}{3} \cdot 10 \rceil = \lceil 6.666... \rceil = 7$, a two-thirds majority requires 7 votes. Thus the challenger is one vote short of winning, and the status quo B prevails.

This illustrates that the outcome of an election depends critically on the choice of the social choice function. In Examples 1.2 and 1.6, there is no difference in the information collected from the electorate, yet the election yields different results in the two cases.

In the Senate, with 100 senators all told, $q = \lceil \frac{3}{5} \cdot 100 \rceil = 60$ votes are needed to end a filibuster. There are many other circumstances where it is useful to combine a status quo with a supermajority method. For example, when a jury in a criminal trial votes on whether a defendant is guilty, a fundamental principle of American jurisprudence is that defendants are innocent until proven guilty, which means, in effect, that innocence is the status quo. Moreover, in jury voting for criminal trials, the parameter is generally taken to be $p = 1$, which means a jury vote must be unanimous to convict.

It seems appropriate to use a status quo method in situations where a decision between two alternatives is required but where one of the alternatives is preferred. For example, procedures to amend constitutions and bylaws often use a status quo method with a supermajority. On the other hand, we tend to avoid status quo methods in elections for public office because they favor one of the two candidates. In particular, status quo methods do not treat the candidates equally.

1.3 Weighted Voting and Other Methods

Just as we tend to believe elections for public office should treat all the candidates equally, we also tend to believe that they should treat all the voters equally. This belief is captured in the familiar principle of "one person, one vote". Nevertheless, in some situations it may be more appropriate not to treat all the voters equally.

One convenient way to give different voters different voices in an election is to give them different numbers of votes. For example, in a corporate election, each shareholder may be accorded a number of votes proportional to the number of shares that she owns. This leads to another type of social choice function.

Definition 1.7. A **weighted voting method** works as follows. Suppose there are n voters: $1, 2, \ldots, n$, and, for each i from 1 to n, voter i is assigned a positive number w_i of votes (called the **weight** of voter i). Let $t = w_1 + w_2 + \cdots + w_n$ be the total number of votes (the sum of the weights). A candidate who gets more than half of all the votes cast (i.e., more than $t/2$ votes) is the winner. If no candidate gets more than half of the votes, then the result is a tie.

Another possibility for conducting two-candidate elections is to use a method that is a hybrid of the supermajority, status quo, and weighted voting methods, as follows. Suppose there are n voters with nonnegative weights $w_1, w_2, \ldots w_n$, and two candidates, one designated as the status quo and the other designated as the challenger, and a parameter p satisfying $1/2 \leq p \leq 1$. Let $t = w_1 + w_2 + \cdots + w_n$ be the sum of the weights (the total numbers of votes). If the challenger gets a fraction p or more of all the votes (i.e., at least pt votes), then the challenger is the winner. Otherwise, the status quo candidate is the winner. (The threshold $q = pt$ is another example of a quota.) We reconsider weighted voting in Chapter 19, where we denote this system as $V(q; w_1, w_2, \ldots, w_n)$.

Example 1.8. A small business is governed by a board consisting of six partners, who according to the partnership agreement get $10, 5, 5, 3, 2,$ and 1 votes, respectively, in business meetings. A decision to amend the partnership agreement requires a $p = 3/5$ supermajority to pass. Suppose that a certain amendment is proposed, that alternative A is to pass the amendment, and that alternative B, to reject the amendment, is the status quo alternative. Suppose that the amendment is put to a vote, and the results are given by the following profile.

B	A	A	A	B	A

We assume this profile lists the partners in the same order as the list of voting strength. The number of votes is $t = 10+5+5+3+2+1 = 26$. The number of votes in favor is $5+5+3+1 = 14$, which is a simple majority of the total votes, but less than the quota $q = \lceil \frac{3}{5} \cdot 26 \rceil = \lceil 15.6 \rceil = 16$. Thus the status quo B prevails. Now suppose that the fifth voter changes his vote to A, as follows.

B	A	A	A	A	A

This time the vote in favor is $5 + 5 + 3 + 2 + 1 = 16$, so the amendment passes.

Another potential twist in election procedures is what one finds in the case of the election for the American president. In this approach, each state conducts a popular election, and then sends a delegation of electors (to the Electoral College) with instructions (in most cases) to cast all of its votes for the winner of the state's popular election. The size of each state's delegation to the Electoral College is equal to the size of its congressional delegation: its number of representatives plus two (for its senators). The next definition describes a type of voting method that follows a similar plan.

Definition 1.9. The following method is called the **bloc voting method**. First, the electorate is partitioned into n blocs (every voter is in exactly one bloc), and, for each i from 1 to n, bloc i is assigned a positive number w_i of votes. Each bloc conducts a "popular vote" election using the simple majority method (resolving any ties by some method chosen by that bloc). Then the bloc casts all of its votes in the "national" election for the candidate that won its simple majority election. The winner is the candidate receiving the most votes in the national election.

The term "social choice function" does not incorporate any value judgements about whether the function is fair or reasonable or appropriate for elections in a democratic system. At this point, we want to stretch our imagination and contemplate all kinds of methods for deciding elections. Thus we are led to consider some less democratic methods.

Definition 1.10. In the **monarchy method**, one of the candidates is a **monarch**. That candidate wins no matter how anybody votes.

Definition 1.11. In the **dictatorship method**, one of the voters is the **dictator**. Whoever the dictator prefers is the winner.

Example 1.12. Consider a weighted voting method based on simple majority with 4 voters having $6, 2, 2, 1$ votes, respectively. Then the first voter is (in effect) a dictator. This means that, if we compute the election by tallying votes, we obtain the same winner as if we had instead simply reported the first voter's choice as the winner.

Certain other methods may seem ridiculous or just plain unfair but nonetheless need to be considered if we want to fathom the limits of possibility. Here are two such methods.

Definition 1.13. In the **parity method**, if just one candidate gets an even number of votes, then that candidate wins. If both candidates get an odd number of votes or if both candidates get an even number of votes, then the result is a tie.

Definition 1.14. In the **all-ties method**, the election is a tie, no matter how the electorate votes.

Our goal in the end will be to rule out the monarchy and dictatorship methods, because they are undemocratic, the parity method, because it is absurd, and the all-ties method, because it is useless. For the time being, however, they remain in the mix as reminders of how much variety is possible in methods to decide elections.

1.4 Criteria

In the remainder of this chapter, we develop a theory that identifies, for two-candidate elections, one method of choice that is, under certain assumptions, superior to the others. We do this by identifying **criteria** (we may also call them "conditions", "rules", or "properties") that we may impose on our election methods. We want these criteria to formalize our notions of fairness, appropriateness, or reasonableness. This is a preview of the approach we will adopt throughout the book.

In imposing a criterion, one establishes a certain philosophical position about voting. One difficulty is that it may be overly restrictive to limit the search for an optimal method to methods that satisfy certain criteria. For example, we may wish to impose the criterion that the method never ends in a tie. It is hard to find fault with this criterion until one realizes that even the simple majority method fails to satisfy it. So we must be cautious about imposing criteria hastily. We nevertheless proceed to enunciate criteria that attempt to capture some principles of fairness that we will consider imposing on our decision procedures. We then look for procedures that satisfy as many of these criteria as possible. Ultimately, we prove a theorem (due to mathematician and historian Kenneth May) that identifies criteria that single out the simple majority method as the "best" method for elections with two candidates.

Think of the criteria as tests that apply to election decision procedures. We want to be able to answer a question of the form "Does method X satisfy criterion Y?" with a "yes" or "no". We devise our criteria so that a "yes" reflects favorably on the method. Some criteria are particularly compelling, meaning that people tend to regard methods that violate them to be inappropriate or offensive. Other criteria may be less compelling, establishing requirements that people may regard as perhaps desirable but certainly not essential, or desirable in some situations but not others. We may occasionally consider criteria that do not seem desirable at all to some people. Whether a criterion is compelling is ultimately a question of political philosophy. One can argue at length about what "fairness" is. We tend to avoid engaging in such debates. Nonetheless, all of our criteria are offered as measures of some sort of appropriateness or fairness of social choice functions.

Let us begin with a criterion that, in particular, eliminates from consideration the dictatorships.

Definition 1.15. A method satisfies the **anonymity criterion** (or, more simply, is **anonymous**) if it treats all the voters equally.

A more complicated, but more precise, way to phrase this is that a method is anonymous if the outcome of the method (either "A", "B", or "tie") does not change if the voters exchange ballots among themselves. In other words, if I were to present your ballot and you were to present mine, this twist should not impact the election outcome.

Anonymity is the property of a social choice function that allows elections to be conducted via secret ballots. Up to now, we have not assumed ballot secrecy, and one may imagine that voters are expected to write their names on their ballots. In the absence of anonymity, the result of the election can change if the names on the ballots are permuted. In the presence of anonymity, however, the names on the ballots can be ignored, and therefore the ballots themselves need not even ask for the names of voters.

Clearly a dictatorship favors one voter (the dictator!) over the others, and therefore cannot operate unless it is possible to discern which ballot was contributed by the dictator. Thus a dictatorship does not satisfy the anonymity criterion. We express this informally via the slogan "Dictatorships violate anonymity."

Weighted voting methods distinguish between voters, making some voters more powerful than others. Weighted voting therefore also violates anonymity. While this may be acceptable in certain situations, it violates the fundamental principle of "one person, one vote" for democratic elections to public office.

It is easy to see that the simple majority method and the supermajority methods *do* satisfy the anonymity criterion, as do any status quo methods that are based on these methods. This is because it does not matter *which* voters vote for a certain alternative; all that matters is *how many* voters vote for a certain alternative.

We formalize this observation by displaying it as a proposition.

Proposition 1.16. *A method is anonymous if and only if its outcomes depend only on the tabulated profile.*

Throughout the book, when we want to highlight a fact or result of special significance, we generally present it in this way as a theorem, a proposition, a lemma, or a corollary, and then follow it with a proof. A **theorem** is generally a deep result of major importance. Other results of less importance are given the name **proposition**. A **lemma** is a preliminary result that may not be of independent interest but whose purpose is to help explain why another result is true. A **corollary** is a result that follows readily from a theorem or proposition and often helps explain why the theorem or proposition is of value. In this book, we follow the age-old tradition of mathematical discourse by displaying these types of results on the page in a way that brings them emphasis and facilitates reference.

Following the presentation of a result, we usually present a **proof**. A proof is nothing more than a careful explanation of why the result is true. Many of our results make the claim that something always occurs; the associated proof therefore is a demonstration of why the thing does indeed always have to occur. Usually there are infinitely many possibilities, so to deal with the "always", one must consider an arbitrary possibility and show why the thing occurs in this general case. A single example does not prove a general rule. On the other hand, sometimes our results take the opposite form: "It is not the case that something always occurs." In this case, all that is required for a proof is a single example in which the thing does not occur. We often call such an example a **counterexample**. No general argument is appropriate when just a counterexample will do.

We use the symbol □ to indicate the end of a proof. We also use the symbol □ at the end of statement of a result (e.g., a theorem) if we do not give a proof.

Proof of Proposition 1.16. The phrase "if and only if" indicates that two separate claims must be justified. We must show that the condition before this phrase implies the one after it and separately that the condition after it implies the one before.

Let us first suppose that a particular method under consideration is anonymous. Suppose that two profiles (for the same electorate) give the same tabulated profile. We must show that they result in the same candidate being elected. Since the two profiles give the same tabulated profile, the same number of voters prefer candidates A and B in both of them. Thus we can rearrange (permute) the voters in the first profile to change it into the second. Since the method is anonymous, such a permutation does not change the outcome. So the outcome depends only on the tabulated profile.

Conversely, suppose that the outcome of a certain method depends only on the tabulated profile. Starting with one profile, we may obtain different profiles by permuting the voters. Since all these permuted profiles give the same tabulated profile, we know by our assumption that the method yields the same outcome for all of them. Thus the method does not change under any permutation of voters, which implies that it is anonymous. □

Even though status quo methods that come from anonymous base methods are themselves anonymous, they violate another important criterion called neutrality.

Definition 1.17. A method satisfies the **neutrality criterion** (or is **neutral**) if it treats both candidates equally.

To be more precise, a method is neutral if whenever all the voters for A change their votes to B, and all the voters for B change their votes to A, the winner changes from A to B or from B to A (or the outcome remains a tie if it was a tie before the change).

Clearly status quo methods are not neutral (they favor the status quo candidate). Similarly, monarchy is not neutral (in much the same way that dictatorship is not anonymous). In elections for public office, at least, fairness and impartiality almost always require neutrality. Neither candidate should have an advantage built into the decision procedure.

So, any reasonable method for choosing candidates for public office should be anonymous and neutral, but is this enough? The parity method is anonymous and neutral. Nevertheless it still seems to be an abomination as a way to make decisions. This is because it violates another compelling criterion, known as monotonicity.

Definition 1.18. A method satisfies the **monotonicity criterion** (or is **monotone**) if the following holds: Suppose the votes are cast and the method selects one candidate as the winner. Then suppose that the method is employed again after one or more voters change their votes from the losing candidate to the winning candidate. The candidate who was the winner before the change must remain the winner after the change.

In other words, if candidate A wins when facing one profile but then some voters switch their votes from B to A, then A must still win. This criterion requires the consideration of two profiles, the first before some voters change their minds, the second after. The voters who change their minds all do so in the same direction, deciding to vote for A where they had voted for B before. It would be disturbing if a social choice method gave the election to A before the change, but failed to give the election to A after the change. After all, if such a method were used in practice, voters would be afraid to cast their ballots in an honest way, aware that a vote for A could harm the cause of electing A. We therefore impose the monotonicity criterion in most situations. Another way to phrase the monotonicity criterion informally is: "A vote for a candidate can never harm that candidate."

Proposition 1.19. *The parity method violates monotonicity.*

Proof. We are going to construct an example (a "counterexample") of a profile in which A wins, but then, after one additional voter votes for A, in the new profile B wins and A loses.

Consider the profile

B	B	B	B	A	B	B	A	B

Recall that under the parity method, a candidate wins if she receives an even number of votes and her opponent receives an odd number. Since A receives 2 votes in this profile (an even number) and B receives 7, A wins the election conducted via the parity method. Now suppose the second voter changes his vote from the loser B to the winner A.

B	A	B	B	A	B	B	A	B

Now A has 3 votes, and since B now has 6 (an even number), B wins. □

It does not seem desirable for a change like the one in the preceding proof to cause candidate A to go down to defeat — after all more voters voted for A in the second profile! For this reason it seems essential to always require our decision procedures to satisfy monotonicity. Monotonicity is a compelling criterion. Non-monotone methods, like the parity method, seem like especially inappropriate ways to run elections.

Whenever possible we would like elections between two candidates to end in a decision for one candidate over the other. Therefore, we might hope to choose a method that never results in a tie.

Definition 1.20. A method satisfies the **decisiveness criterion** (or is **decisive**) if it always chooses a winner (i.e., it never ends in a tie).

As disappointing as a tie may be, however, we sometimes need to settle for a method that allows for the possibility of a tie. For example, if we want to use the simple majority method, and the number of voters is even, ties will always be a possibility. On the other hand, we might try to impose a criterion that eliminates ties, except when absolutely necessary.

Definition 1.21. A method satisfies the **near decisiveness criterion** (or is **nearly decisive**) if the only situation in which a tie can occur is if both candidates receive exactly the same number of votes.

Here again, some care is necessary with the way we use language. Near decisiveness is meant to be inclusive in the sense that any method that is decisive must also be nearly decisive. The converse of this statement, however, is false. A nearly decisive method need not be decisive. For example, the simple majority method with an even number of voters is nearly decisive without being decisive.

Many methods fail to satisfy near decisiveness. For example, every super-majority method (without status quo) fails to be nearly decisive, because with a nearly 50:50 split in the electorate, neither candidate will have enough votes to win. The worst violator of the near decisiveness criterion is, of course, the all-ties method, which ends in a tie when presented with any profile whatsoever.

1.5 May's Theorem

At this point we have discussed various methods for deciding two candidate elections. Many more methods exist, but we will not be concerned with them for now. We have also identified five criteria for evaluating methods. So this begs the question: Which methods satisfy which criteria? Table 1.1 answers this question. The answer "yes" or "no" in a box in this chart indicates whether the method to the left satisfies the criterion above.

The diligent reader will want to verify that the answers are as we claim. Generally, our criteria ask if a social choice function behaves a certain way always. A "yes" answer begs for a proof. On the other hand, a "no" answer begs for a counterexample, as in Proposition 1.19.

As an illustration, let us demonstrate that the "yes" that appears in the upper right entry of Table 1.1 is correct.

Proposition 1.22. *The simple majority method is nearly decisive.*

Proof. Suppose first that the number t of voters is odd. Then no candidate can receive exactly half of the votes (because $t/2$ is not a whole number), so one must receive more than half. This will be a majority and that candidate will be the winner. Now suppose t is even, so that $t/2$ is a

Table 1.1 Two-candidate methods and the criteria that they satisfy — "status quo" refers to the status quo method based on the simple majority method

	Anonymous	Neutral	Monotone	Decisive	Nearly decisive
Simple majority	Yes	Yes	Yes	No	Yes
Super-majority	Yes	Yes	Yes	No	No
Status quo	Yes	No	Yes	Yes	Yes
Weighted	No	Yes	Yes	No	No
Parity	Yes	Yes	No	No	No
Dictator	No	Yes	Yes	Yes	Yes
Monarchy	Yes	No	Yes	Yes	Yes
All-ties	Yes	Yes	Yes	No	No

whole number. If both candidates get $t/2$ votes, they tie. Otherwise, one candidate must get more than $t/2$ votes for a majority (and the other gets less). The candidate with the majority is the winner. So the only tie occurs when each candidate gets exactly half the votes. □

In Table 1.1, a row with the word "Yes" throughout would point to a method that simultaneously meets all of our desired criteria. That would be ideal. Unfortunately, we do not see any such method listed, so it appears that our search for an ideal method must continue. It turns out, however, that no such method exists. The theorem of Kenneth May, proved in 1952, identifies the simple majority method as the only method, not merely among the methods described above but among all possible methods, that satisfies anonymity, neutrality, monotonicity, and near decisiveness. But even that method fails to be decisive.

Theorem 1.23. (***May's Theorem***) *In an election with two candidates, the only voting method that is anonymous, neutral, monotone, and nearly decisive is the simple majority method.*

Proof. Suppose that a method is anonymous, neutral, monotone, and nearly decisive. Since the method is anonymous, Proposition 1.16 tells us that we need only consider tabulated profiles. So let a be the number of voters who prefer candidate A, and let b be the number of voters who prefer candidate B. Then $t = a + b$ is the total number of voters. Our goal is to show that the method we are imagining must in fact be the simple majority method.

First consider the case when t is even. If $a = b = t/2$, then each candidate has exactly half the votes. By the neutrality criterion the result must be a tie, because the method must treat the candidates in the same way. This is consistent with the simple majority method.

Next suppose that candidate A has a majority. That is, suppose that $a \geq t/2+1$. We need to show that candidate A must win, and for this it is enough to show that the result cannot be a tie or a win for candidate B. Candidate B cannot win, because if candidate B were to win with $b = t - a \leq t - (t/2 + 1) = t/2 - 1 < t/2$ votes, monotonicity would imply that candidate B must also win with exactly $t/2$ votes. But we have already remarked that neutrality requires that the election must be a tie if candidates A and B both received exactly $t/2$ votes. This contradiction shows that candidate B cannot win. Moreover, the near decisiveness criterion implies that the result cannot be a tie. Therefore candidate A must win. In summary, if candidate A gets a majority of the votes, then candidate A must win the election. By neutrality, the same is true of B. Hence the method is the simple majority method in the case when t is even.

Now suppose t is odd. In this case, neither candidate can get $t/2$ votes, because $t/2$ is not a whole number. Thus by the near decisiveness criterion, there can be no ties. Now let $a = t/2+1/2$, the smallest whole number greater than $t/2$. The number a is the smallest number of votes that delivers a majority for candidate A. If candidate A gets a votes, then candidate B will receive $b = t - a = t - (t/2 + 1/2) = t/2 - 1/2 = a - 1$ votes. We claim that candidate B cannot be the winner. For if B wins with $b = a - 1$ votes, then, by neutrality, A would also win with $a - 1$ votes. But by monotonicity A would then also win with a votes, which is a contradiction. Since B cannot be the winner, and since the election cannot be a tie, A must be the winner with $a = t/2+1/2$ votes. By monotonicity, A is the unique winner whenever A receives at least $t/2 + 1/2$ votes. But by neutrality, it is also true that B is the unique winner whenever B receives at least $t/2 + 1/2$. So, any candidate who has more votes than the other is the unique winner. Again, the method must be the simple majority method. □

We leave as an exercise the proof of the following extension of May's Theorem.

Theorem 1.24. *In an election with two candidates, a voting method that is anonymous, neutral, and monotone and that isn't the all-ties method must be either the simple majority method or a supermajority method.*

Note that this theorem does not merely say that quota methods are the only methods we know *so far* that satisfy these criteria. It says that no other method will ever be found, no matter how clever or how imaginative the searcher. It says that it is impossible — mathematically impossible — to find such a method that is not a quota method. It closes the book forever on the search for better methods for two-candidate elections.

On the other hand, Table 1.1 shows that there are other methods worthy of consideration if we can tolerate the absence of any one of the three criteria anonymity, neutrality, or monotonicity, or if we can live with the method that always gives ties.

Since it can end in a tie, the simple majority method is not a perfect method. On the other hand, Theorem 1.24 assures us that it is impossible to do better. We sometimes phrase this inference in the following way.

Corollary 1.25. *It is impossible for a voting method with two candidates to satisfy anonymity, neutrality, monotonicity, and decisiveness.* □

With regard to elections for public office, decisiveness is generally regarded as less compelling than anonymity, neutrality, and monotonicity. In any case, near decisiveness is a reasonable substitute for the decisiveness, and the simple majority method is nearly decisive. This, then, is how the simple majority emerges as the clear choice for most elections with two candidates.

In the next chapter we move on to consider elections with more than two candidates. But May's Theorem continues to play an important role, because we may need to hold a run-off election between two candidates or we may want to ask how one candidate would fare against one other if all the remaining candidates were to drop out of the race. When we reduce the slate of candidates in such a way to just two, May's Theorem tells us how we should render our decision, and it does so by reaffirming what may have seemed obvious to begin with: The candidate who garners the most votes should be declared the winner.

1.6 Exercises and Problems

1.1. Consider the profile

A	A	B	A	B	A	A	B	B	B	B	B	B	A	B	B	B

Who wins under:

(a) the simple majority method?
(b) the super-majority method with $p = 2/3$?
(c) the super-majority method with $p = 2/3$ but with A as the status quo?

(d) the weighted simple majority method where the first 7 voters have 3 votes and the rest have 1 vote?

1.2. Consider the following two-candidate profile:

A	A	B	A	A	A	B	B	B	B	B	B	B	B	B	A	B	B

Who wins under:
(a) the simple majority method?
(b) the 2/3 super-majority method?
(c) the parity method?
(d) the weighted voting system where the first 6 voters get 3 votes, the next 6 voters get 2 votes, and the rest get 1?

1.3. Suppose you favor one of two alternatives but only 10% of the electorate agrees with your position. Is there a voting method that leads to victory for your position that is
(a) anonymous?
(b) neutral?
(c) anonymous and neutral?
(d) monotone?
(e) anonymous, neutral, and monotone?

1.4. Suppose you are against one of two alternatives but 90% of the electorate disagrees with your position and favors that option. Is there a voting method that is anonymous, neutral, and monotone that prevents that option from being selected as the winning alternative?

1.5. Explain why the simple majority method satisfies:
(a) anonymity,
(b) neutrality, and
(c) monotonicity.

1.6. Explain why a dictatorship is monotone and neutral, but not anonymous.

1.7. For each of the following methods for settling elections with two candidates (A and B), determine whether the method is (1) anonymous, (2) neutral, and (3) monotone.
(a) Candidate A wins if she has a prime number of votes and B doesn't. Candidate B wins if he has a prime number of votes and A doesn't. Otherwise it is a tie.
(b) Whichever candidate has the most votes from male voters is the winner. If the number of males voting for A equals the number of males voting for B, then the election is a tie.
(c) Candidate A wins no matter what.

1.8. Devise an example to show that:
(a) the 2/3 super-majority method is not nearly decisive.

(b) a status quo voting method is not neutral.
(c) the monarchy method is not neutral.
(d) weighted voting is not anonymous.

1.9. Consider a hybrid method that works as follows. The voters register their weight in ounces with the election board. Candidate A wins if the total weight (literally) of the voters that support him is at least 70% of the total weight of all voters. Otherwise candidate B wins. Does this hybrid method satisfy
(a) the anonymity criterion?
(b) the neutrality criterion?
(c) the monotonicity criterion?
(d) the decisiveness criterion?
(e) the near decisiveness criterion?

1.10. The US Federal System is the voting method for passing a bill into law. It involves 100 members of the Senate, 435 members of the House of Representatives, the vice president, and the president, for a total of 537 voters. For a bill to pass, it must have a majority in the House, a majority in the Senate, with the possibility of the vice president breaking a tie, and the signature of the President; or if the president opposes the bill, it must have a 2/3 super-majority in both the House and Senate to override the veto. Otherwise the bill is defeated.

Think of this as a social choice function where the alternatives are A, to pass the bill, or B, to defeat the bill. Does the US federal system satisfy:
(a) the anonymity criterion?
(b) the neutrality criterion?
(c) the monotonicity criterion?
(d) the decisiveness criterion?
(e) the near decisiveness criterion?

1.11. Because ties are possible in the United States Senate, the Constitution provides for the vice president to cast a tie-breaking vote. One can therefore think of the Senate as a body with 101 voters, 100 of whom are senators and one of whom is the vice president. The rules dictate that the vice president votes only to break ties and not otherwise. Is this system anonymous?

1.12. Consider a weighted voting method with 5 voters assigned weights 17, 15, 14, 12, and 7, respectively, with a simple majority of the weighted votes sufficient for victory. Explain why this method is in effect the (unweighted) simple majority method.

1.13. Consider a method that we call the simple minority method: Whichever candidate receives the fewest (!) votes is declared the winner, and if the two candidates receive the same number of votes, then the method declares a tie. Determine whether this method is anonymous, neutral, monotone, and nearly decisive.

1.14. Consider a variation on the idea of a weighted voting system, in which one or more voters is given a negative weight. Show that such a system is not monotone. If a voter knows that the method assigns a negative weight to her vote, how will she be inclined to cast her ballot?

1.15. Consider the Supreme Court as an electorate of 9 members that votes yes-or-no on a variety of issues. It is customary for this electorate to use the simple majority method, and since 9 is an odd number, no ties can result when no justices are absent. Imagine the following alternative method for determining Supreme Court decisions. The decision is affirmative whenever at least 3 of the first 5 justices vote to affirm. Otherwise the decision is negative. This is just the bloc voting method, where the justices form blocs of size 5, 1, 1, 1, and 1. (This method emulates what would occur if the first 5 justices were to agree to always vote as a bloc and to decide among themselves how the bloc should rule.)
(a) Is this method anonymous?
(b) Is this method neutral?
(c) Is this method monotone?
(d) If you were one of the 4 justices not among the first 5, why would you complain about this voting method?

1.16. Imitate the proof of Theorem 1.23 to prove Theorem 1.24.

1.17. Jones and Smith are running for public office. There are 99 voters.
(a) Suppose the rules say that Jones wins if she receives 50 or more votes; otherwise Smith wins. Is this method neutral?
(b) Suppose the rules say that a candidate wins if he or she receives a number of votes that is between 60 and 89. (If neither candidate gets a winning number of votes, then the election is a tie.) Is this method monotone?
(c) Suppose the rules say that a candidate wins if he or she receives an odd number of votes. (If neither candidate gets a winning number of votes, then the election is a tie.) Is this method decisive?
(d) Suppose that the rules say that Smith wins the election no matter what. Is this method anonymous?

2

Social Choice Functions

"The best argument against democracy is a five-minute conversation with the average voter." — Winston Churchill

2.0 Scenario

Three candidates, denoted A, B, and C, are running for public office. Each of the 99 voters contributes a ranking of the candidates. Here are the results of the voting.

- 18 voters prefer A to B and prefer B to C;
- 15 voters prefer A to C and prefer C to B;
- 24 voters prefer B to A and prefer A to C;
- 8 voters prefer B to C and prefer C to A;
- 16 voters prefer C to A and prefer A to B; and
- 18 voters prefer C to B and prefer B to A.

Decide who should be declared the winner and why.

In particular, you may also wish to consider the follow questions. Who should be in second place? What if some candidate drops out of the race? What other ways of collecting information from voters — more or less informative than asking for preference orders — might be useful? How many reasonable methods can you come up with for deciding elections with three or more candidates?

2.1 Ballots

In this chapter we consider elections with more than two candidates. This situation features a richer set of possible voting methods than we encountered in the previous chapter, and a number of curious examples emerge that challenge intuition. To prepare for this, we lay out with some care the vocabulary and assumptions that we impose on the problem.

We begin with a set of candidates or alternatives (the **slate**) and with a set of voters (the **electorate**). Both of these sets are finite and nonempty. We often designate the candidates with the capital letters A, B, C,..., since we don't want to inflame passions, say, by naming them Cruz, Sanders, and Trump. We designate the voters, when necessary, by $1, 2, 3, \ldots$.

We assume that each voter submits a **preference ballot**, on which she expresses her preferences by listing the candidates in descending order of preference. This means that each voter indicates which candidate is her first choice, which candidate is her second choice, and so forth until the entire slate is ranked from top to bottom. This is called a voter's **preference order**.

Every voter must vote. No voter may express an equal preference for two candidates. Every voter must continue to rank candidates all the way to the end of the slate, no matter how many candidates there are. So, for example, if the set of candidates is $\{\mathrm{A}, \mathrm{B}, \mathrm{C}, \mathrm{D}\}$, then the preference order for a voter might be: "B is my first choice, D is my second choice, C is my third choice, and A is my last choice." We might denote this preference order by B $>$ D $>$ C $>$ A. Often we illustrate it schematically with a vertical column that lists the candidates from top to bottom, like this:

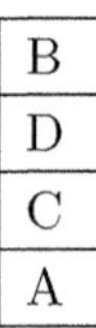

Our electorate is assumed to be composed of **rational** voters, in the following sense. From the preference B $>$ D $>$ C $>$ A we deduce that the voter favors not only B to D, D to C, and C to A, but also B to C, B to A, and D to A. Put in a more mathematical way, we assume that voters hold **transitive** preferences.

All these assumptions are part of the modeling process. The assumption that voters express preference orders may strike some as unorthodox, because it differs from the way elections are usually conducted. On the usual **vote-for-one ballot**, voters indicate only their top choice. By contrast, we insist that our voters report their entire preference order. At the same time, we forbid them from expressing information beyond their preference order, such as the degree to which one candidate is preferred over another or how many candidates they deem acceptable. We make our assumptions because they seem to be broad enough to be useful but narrow enough to lead to a tractable mathematical theory.

It may seem that our assumptions about the setting for the election are unrealistic. For example, in real elections not all eligible voters go to the polls. But in our setting, we may as well define "voter" to mean someone who has *already* voted, so the assumption that every voter goes to the polls is not unnecessarily restrictive. It is also true that, in practice, not all voters would wish to rank every single candidate. Some might prefer to express "B is my first choice, D is my second choice, but I don't care after that." The preference ballot forbids this. Still, the preference ballot gives rich information about a voter's sentiments. In particular, if the election is somehow reduced to a subset of the original slate that contains just two candidates, then we are able to discern for any voter which of these two candidates is preferred. This is a convenience if there is any prospect of a run-off election, since the voters have already expressed their wishes and so don't need to be dragged back to the polling places. Finally, it is also true that real elections often do not have a fixed slate of candidates but rather allow write-in votes. In our theory, this is forbidden, since the slate of candidates on which the voters express preferences is fixed and every voter ranks every candidate. With a write-in candidate, we would not have complete information about the electorate's view of the candidate. In the end, it is probably pointless to defend our assumptions, since they are imposed not because they best capture reality but for the convenience of our theory. We will consider other kinds of ballots in Chapter 6.

When every voter in the electorate expresses a preference order (i.e., every voter casts a ballot), we obtain what we call a **profile**: a set of preference orders, one for each voter. After all ballots are cast and tallied, the profile contains all the information about the preferences of the voters. As an example with 3 candidates and 7 voters, we may have 3 voters (voters 1, 3, and 5, say) who vote A > B > C, 2 voters (2 and 6, say) who vote B > C > A, and 2 voters (4 and 7, say) who vote C > A > B. Our usual schematic for representing this profile will be:

A	B	A	C	A	B	C
B	C	B	A	B	C	A
C	A	C	B	C	A	B

Profile 2.1

In this profile, each column represents the preference order of one voter. In many cases, however, it will be enough to *count* how many voters have each particular preference order. If we do that in the case of Profile 2.1, we get the **tabulated profile** shown in Profile 2.2.

3	2	2
A	B	C
B	C	A
C	A	B

Profile 2.2

The numbers in the top row of a tabulated profile represent the number of voters who contribute the preference order indicated below that number in the column. From a tabulated profile, it is not possible to discern which voter contributed which preference list. But that information is often of no importance.

2.2 Social Choice Functions

In setting out rules for elections, the ideal would be a rule that assigns, to every profile, a candidate to be declared the winner. Such a rule would be a function whose domain is the set of all profiles and whose range is the set of candidates. Unfortunately, this is asking for a bit too much, because sometimes the profile describes an electorate that is evenly split, and in such cases we want to allow the possibility of a tie between two or more of the candidates. Bearing this in mind, we define a **social choice function** (or, informally, a **voting method**) to be a function that assigns, to every profile, a nonempty subset of the slate, and we refer to the candidates in this subset as the **winners** of the election. We input a profile into the method and the method outputs a set of winners. Naturally, we call candidates who don't win **losers**. From the practical point of view, we will be pleased if the set of winners has just a single candidate in it, but we recognize the need to declare some elections to be ties. A tie in this context refers to an election with more than one winner.

The **domain** of a social choice function is the set of all profiles. The **codomain** (the set of all allowed outputs) is the set of nonempty subsets of the slate. To say "function" here is to forbid random elements such as coin tossing and to insist that the method does not change with the weather or other irrelevancies. To say that the domain of the function is the set of *all* profiles is to mandate that a decision of some sort must be rendered, no matter how the electorate votes. A function must produce an output on *every* possible input. This requirement is sometimes referred to as the hypotheses of **universal domain**. To say that the codomain consists only of nonempty subsets assures that no election can end without any winners. (We do, however, allow for the possibility that *all* candidates could be declared winners, which in practice is equivalent to and just as useless as declaring all the candidates to be losers.)

We might imagine a social choice function as an enormous rulebook for declaring a winner or winners in an election. This rulebook lists every possible profile, and after each profile it identifies the set of winners associated with that profile. To conduct an election, you send the voters to the polls, obtain the profile, look it up in the book, and read off the winners. There is no requirement at this point that the social choice function be logical or reasonable. We will eventually look among the social choice functions in search of desirable ones for conducting elections, but at first we leave ourselves open to the possibility that a social choice function in practice might be utterly chaotic or absurd.

Even with just 3 candidates and 4 voters there are an amazingly huge number of social choice functions. In fact, we can say exactly how many. Each of the 4 voters has a choice of 6 possible preference orders to submit. There are therefore $6^4 = 1296$ possible profiles all told. For each of the 1296 profiles, a social choice function must choose a non-empty subset of the slate. There are 7 possible such choices (A wins; B wins; C wins; A and B tie; A and C tie; B and C tie; and A, B, and C tie). Hence, to construct a social choice function, we must choose one of these 7 outputs for each of the 1296 inputs. The number of ways to do this is 7^{1296}, a number whose decimal expansion contains 1096 digits. That's a lot! And that's just for 3 candidates and 4 voters. With 100,000,000 voters and dozens of candidates, the number becomes truly incomprehensible. But most of these myriad social choice functions would not serve very well as a way of conducting fair and reasonable elections. Only a small handful of them have a sensible kind of logic to them.

The first example of a reasonable social choice function is one that is clearly inspired by the success of the simple majority method in the two-candidate case.

Definition 2.1. The **plurality method** is the social choice function that selects as the winner the candidate who is ranked as the first choice of the most voters. In the event of a tie, the plurality method selects all the candidates who tie for the most first-place votes.

Example 2.2. Let's apply the plurality method to Profile 2.1. We must count the first-place votes for each candidate. To do this, it is sufficient to look at the tabulated profile Profile 2.2. Since candidate A garners 3 first-place votes, whereas each other candidate garners just 2, the plurality method selects A as the unique winner.

The plurality method is the most familiar social choice function. You simply tally the first-place votes for each candidate, and the candidate with the most first-place votes wins. Such a candidate is said to have a **plurality** of the votes, that is, more votes than anyone else. This is how elections are customarily run in most situations. Note that we do not require a **majority**, because doing so would sometimes result in no winner being selected. To determine the plurality winner of an election, one need know only the first-choice preference of each voter. In other words, this method permits us to ignore all the information in the profile about second, third, and further preferences of the voters. In many elections, voters are given the chance to submit only a vote-for-one ballot and we do not have any further information about the voters' preferences. It is still possible to compute the plurality winner in vote-for-one elections, because the lost information from the profile about lower-ranking preferences is not used. But most of the other methods we will consider require more complete knowledge of the profile. As a result, most of the methods we will study cannot be implemented in an election conducted with a vote-for-one ballot. Hence the plurality method often becomes the default method in vote-for-one elections. We investigate vote-for-one ballots in more detail in Chapter 6.

2.3 Alternatives to Plurality

Since we assume that voters contribute preference ballots, the plurality method is no longer the only option. But why would anybody ever want a different method? To see why the plurality method might raise concerns, we consider an example.

Example 2.3. Consider an election that results in the tabulated Profile 2.3. Here candidate A wins the plurality election with 5 votes, more than any of the other four candidates. However, candidate A is considered to be the worst choice by a whopping 75% of the electorate, all of whom seem to prefer candidate B by a wide margin. This is not to say that candidate A definitely ought to lose the election. It is merely to say that it is not overwhelmingly clear that A is the only rational, reasonable choice. Rather, we might wish to entertain a method that would select candidate B when facing inputs like Profile 2.3.

5	4	4	4	3
A	B	C	D	E
B	C	B	B	D
C	E	D	E	B
E	D	E	C	C
D	A	A	A	A

Profile 2.3

One such method is named after the French military engineer Jean Charles de Borda (1733–1799).

Definition 2.4. The **Borda count method** is the social choice function that works as follows. If there are n candidates in the slate, then assign $n-1$ points to a candidate for every voter who ranks that candidate first, $n-2$ points for every voter who ranks that candidate second, $n-3$ points for every voter who ranks that candidate third, and so forth until you assign 1 point for every voter who ranks that candidate in position $n-1$ (i.e., second-to-last). A candidate receives no points for any last-place votes they get. Tally the points, and the winner is the candidate who gets the most points, or, in the event of a tie, the winners are the candidates who tie for the most points.

If $n = 5$, as in Profile 2.3, then the Borda count method gives a candidate 4, 3, 2, 1, or 0 points for every voter who ranks them first, second, third, fourth, or fifth (last), respectively. Candidate A in Profile 2.3 gets 4 points from each of the 5 voters represented by the first column but gets no points at all from the other voters, so candidate A tallies 20 points. By contrast, candidate B obtains 3 points from each of the 5 voters represented by the first column, but 4 additional points from each of the next 4 voters, 3 points from the next 4, 3 from the subsequent 4, and 2 points from the final 3, for a total of $3\times5+4\times4+3\times4+3\times4+2\times3 = 61$, easily defeating candidate A. The reader can check that the Borda count tallies for candidates C, D, and E are 45, 37, and 37, so it is something of a landslide for candidate B in a Borda count election, with plurality winner A in distant last place.

The Borda count method is attractive because it uses all the rich information that the profile provides. Still, some may be unhappy about a method that allows a candidate to win when that candidate wouldn't register even a single vote in a vote-for-one election. The next example shows that this is unfortunately the case for the Borda count.

Example 2.5. In Profile 2.4, Candidate A scores 6 points in the Borda count tally, while the other three candidates each score 4 points. Thus candidate A wins the Borda count election, and yet not a single voter ranks candidate A first. With vote-for-one ballots, this election would be a three-way tie between candidates B, C, and D.

B	C	D
A	A	A
D	B	C
C	D	B

Profile 2.4

Example 2.6. A more troubling feature of the Borda count method is illustrated in the following profile:

5	4	4	4	3
B	C	A	D	E
C	A	B	A	A
E	B	E	B	B
D	E	D	E	D
A	D	C	C	C

Profile 2.5

The Borda count totals for the candidates A, B, C, D, and E are 49, 54, 31, 28, and 38, respectively, so candidate B wins the Borda count election. Notice, however, that 75% of the electorate rates candidate A ahead of candidate B. Those voters may be perplexed about why their preference of A over B does not seem to have mattered. This example suggests that we should investigate a method in which candidate A would have a chance to challenge candidate B to a head-to-head run-off.

The next method addresses this concern. It was originally suggested by the English barrister Thomas Hare (1806–1891) and is often called the Hare method. It is also known by a variety of other names, such as the instant run-off method, the single transferable vote system, or plurality with elimination.

Definition 2.7. Hare's method is the social choice function that operates as follows. Unless all the candidates have the same number of first-place votes, identify the candidate who has (or candidates who have) the fewest first-place votes, and eliminate them from consideration. When a candidate is eliminated, allow all voters to remove that candidate from their preference list and move up all the candidates that were ranked lower. This produces a new profile applicable to the remaining candidates. Then repeat the process, eliminating from contention the candidate or candidates with the fewest first-place votes in this new profile. Continue the process until either one candidate remains or until all remaining candidates have the same number of first-place votes. The candidate or candidates who survive all the elimination rounds are declared to be winners.

Let's follow the method as it computes a winner in Profile 2.5. In the first round, Candidate E has the fewest first-place votes (just 3) and is eliminated. After eliminating candidate E from the slate, the profile becomes:

5	4	4	4	3
B	C	A	D	A
C	A	B	A	B
D	B	D	B	D
A	D	C	C	C

In the second round, the 3 voters who originally had candidate E as their top choice now throw their support behind their second-choice candidate A. So the tally now has 7 first-place votes for A, 5 for B, 4 for C, and 4 for D. At this stage, candidates C and D are eliminated simultaneously for having the fewest first-place votes. This leaves only candidates A and B remaining, and the profile now simplifies to:

5	4	4	4	3
B	A	A	A	A
A	B	B	B	B

or what amounts to the same thing:

5	15
B	A
A	B

The 5 voters who originally placed candidate B as their top preference continue to do so. But the remaining 15 voters all prefer A to B and so candidate B is eliminated in this final round, and candidate A becomes the Hare winner of this election.

Note that the algorithm for computing the Hare winner of an election can be terminated if at any point one candidate has a majority of the votes among the remaining candidates. That is because that candidate will retain his votes even as other candidates are eliminated and votes are transferred. Put another way, a majority is a robust plurality, so strong that it persists as a majority even as votes for other candidates are potentially consolidated behind a single opponent. Thus, once a candidate has a majority of votes in our elimination scheme, that candidate can be declared the unique winner, without having to carry out subsequent elimination rounds.

Although we have described the Hare method as consisting of rounds, it is important to recognize that the original profile contains all the information needed to conduct every round, and so the process of elimination and iteration occurs only in the vote-counting machine that in a fraction of a second tabulates the profile and computes the winner. A significant advantage of the preference ballot is that one can compute methods like this one without a need to send all the voters back to the ballot box to recast their votes. We know from their preference ballot how they will transfer their votes when candidates are eliminated.

Many organizations use the Hare method. It is used widely for elections in Australia and Ireland, as well as for municipal elections in San Francisco, California, and Takoma Park, Maryland. It protects voters who support less popular candidates from the concern about wasting their votes, since their votes are effectively transferred if their candidate is eliminated early on. This reduces the tendency toward strategic voting, inviting voters to be honest about their preferences. It therefore reduces the instability associated with third-party candidates. But we will see later that the Hare method has flaws of its own.

The following method is a variation on the Hare method. It takes its name from the American psychologist Clyde Coombs (1912-1988).

Definition 2.8. Coombs's method is the social choice function that operates as follows. Unless all the candidates have the same number of last-place votes, identify the candidate who has (or candidates who have) the most last-place votes, and eliminate them from consideration. When a candidate is eliminated, allow all voters to remove that candidate from their preference list and move up all the candidates that were ranked lower. This produces a new profile applicable to the remaining candidates. Then repeat the process, eliminating from contention the candidate or candidates with the most last-place votes in this new profile. Continue the process until either one candidate remains or until all remaining candidates have the same number of last-place votes. The candidate or candidates who survive all the elimination rounds are declared to be winners.

The Coombs method has some of the same advantages as the Hare method. Coombs's idea is to eliminate not the candidates who are least loved, as with Hare's method, but rather the candidates who are most hated. Bland candidates who do not inflame passions at either end of the spectrum will have few strong supporters or detractors; they may be able to survive in the Coombs method but are likely to be quickly eliminated by the Hare method.

One concern with the methods of Hare and Coombs is that the elimination of a candidate early on in the process is irrevocable. Let us subject Profile 2.3 to the Hare method. The first candidate eliminated is E, with only 3 votes. Those three voters transfer their support to candidate D, and candidates B and C are eliminated in the next round with just 4 first-place votes. That leaves candidates A and D for the final run-off, and D wins with 15 of the 20 votes. Candidate B, the Borda count winner, would have been able to defeat candidate D in a run-off had candidate B survived the second elimination round. But unfortunately he was knocked out.

This brings us to a method that focuses entirely on head-to-head matches without eliminating any candidates in early rounds.

Definition 2.9. Copeland's method is the social choice function in which every candidate earns one point for every other candidate whom she can beat in a head-to-head match-up (using the simple majority method). A candidate earns half a point for every other candidate that she ties in a head-to-head match-up. The candidate with the most points becomes the winner. If there is a tie for most points, then all candidates with the most points become winners.

This method gets its name from A. H. Copeland (1910–1990), who was a professor of mathematics at the University of Michigan. Let us explore how it works on the following profile:

A	E	A	D	B
C	C	B	E	D
B	A	C	C	C
D	B	D	A	E
E	D	E	B	A

Profile 2.6

For any pair of candidates, we can determine how many voters prefer one to the other. For example, looking at candidates A and D, we see that the first three voters prefer A to D, while the last two voters prefer D to A. We record this as:

3	2
A	D

It is as if we are momentarily imagining that candidates B, C, and E have dropped out of the race and that the resulting two-candidate election is executed according to the simple majority method. This victory of A over D is worth one point for candidate A in the Copeland method. The other nine head-to-head match-ups are shown in Profile 2.7. Altogether,

4	1
A	B

2	3
A	C

2	3
A	E

2	3
B	C

4	1
B	D

3	2
B	E

3	2
C	D

3	2
C	E

4	1
D	E

Profile 2.7

A wins two matches, B wins two, C wins four, and D and E each win one. So candidate C is the Copeland winner of this election.

The Copeland method is like a round-robin tournament in the world of sports. Each team plays every other team, and the team with the best record at the end is the winner. Tie matches award "half a victory" to each team. On the other hand, sports tournaments often use run-offs at the end to make certain that a single victor is identified; the Copeland method permits multiple winners, since several candidates may tie for the most victories.

There is another common type of sports tournament, which involves the idea of an elimination grid. A team keeps playing in such a tournament as long as it keeps winning its individual contests, and once it loses it is eliminated for good. The team that wins its final contest is the tournament winner. In the theory of elections, the schedule of head-to-head match-ups is known as the **agenda**. One possibility for an agenda comes from a list of the candidates in a particular order. In the first round, the first two candidates on the list compete in a head-to-head match. Then the winner of this first round challenges the next candidate on the list, and the winner of that match challenges the next, etc. The process stops once the last candidate on the list has participated in a match. An agenda of this type is called a **sequential agenda**.

Definition 2.10. Suppose a sequential agenda is given. The **agenda**

method based on this agenda is the social choice function in which rounds of head-to-head contests are conducted between the candidates following the order indicated by the agenda. In the first round, the first two candidates on the agenda face off. A candidate who loses in a certain round is eliminated. A candidate who wins (or ties) goes on to the next round and faces the next candidate on the agenda. A candidate who wins (or ties) in the last round is a winner of the election.

Example 2.11. Suppose to make an agenda we list the candidates in Profile 2.6 in alphabetical order. In this contest, A wins the first round against B and takes on C. But C beats A, goes on to beat D and E, and wins.

It is easy to see that since C wins all her head-to-head contests in Profile 2.6, she will win no matter what the agenda is. In fact, for the same reason, C also wins the Copeland method. Later, we shall see that if no candidates win all their head-to-head matches, changing the agenda can change the winner.

Of course more complicated kinds of agendas can also be used in agenda methods. Consider, for example, the kind of "binary" elimination grid used in the NCAA March Madness basketball tournament. With agendas more complicated than a sequential agenda, however, some care must be taken in the way that ties are handled. (Fortunately, a basketball game cannot end in a tie.)

The following class of methods generalizes the idea behind the Borda count method.

Definition 2.12. Assume that there are n candidates and that $a_1, a_2, a_3, \dots, a_{\mathrm{n}}$ is a sequence of real numbers satisfying $a_1 \geq a_2 \geq a_3 \geq \cdots \geq a_n$. The **positional method** associated with this sequence assigns a_1 points to every first-place vote, a_2 points for every second-place vote, and so forth, until it assigns a_n points for every last-place vote. Points are tallied, and the candidate with the most points wins.

We denote this social choice function by the symbol $P(a_1, a_2, a_3, \dots, a_n)$. For example $P(3, 1, 0)$ is the social choice function that assigns each of the three candidate 3 points for every first-place vote and 1 point for every second-place vote. The Borda count method with n candidates is the prototype of a positional method and is denoted $P(n-1, n-2, n-3, \dots, 2, 1, 0)$. The plurality method is the positional method $P(1, 0, 0, \dots, 0, 0)$.

The positional method $P(1, 1, 1, \dots, 1, 0)$ is known as the **antiplurality method**. Candidates earn a point for every voter that does not place them in last place. In effect, then, the antiplurality winner is the candidate with the fewest last-place votes.

The positional method $P(1,1,0,\ldots,0,0)$ is called the **vote-for-two method**.

We examine a number of other interesting social choice functions in the exercises. Most of these are variants on the methods explored in this section.

2.4 Some Methods on the Edge

The following methods mark the outlying areas of the theory. They are all familiar from the two-candidate case.

Definition 2.13. Let v be one voter. The **dictatorship method** with voter v as dictator is the social choice function in which the candidate ranked first by v is the unique winner, regardless of how anyone else votes.

For each voter in the electorate, there is one social choice function that makes that voter the dictator. Certainly these dictatorships are a preposterous way to run elections in a democratic society. But dictatorships do have something to recommend them: A dictatorship election can never end in a tie. On the other hand, it is disturbing that all but one of the preference orders in the profile is ignored.

In fact, it is possible to consider a social choice function in which *every* preference order in the profile is ignored.

Definition 2.14. Let A be one of the candidates. The **monarchy method** with A as monarch is the social choice function in which A is the unique winner regardless of how anyone votes.

While this certainly does not sound like a method that we should favor in conducting elections, history tells us that this method has often been implemented.

There is even a method that not only ignores the voters but also ignores the candidates:

Definition 2.15. The **all-ties method** is the social choice function that selects all candidates to be winners, no matter how any of the voters vote.

The previous two methods are examples of what mathematicians call **constant functions**. In each of those cases the output of the method (the set of winner) does not depend on the input (the ballots). Such methods are absolutely unresponsive to the will of the electorate.

To emphasize that we do not insist *a priori* on reasonableness for a social choice function, let us conjure up a few truly absurd voting methods. In this vein, we have: the function that selects the candidate with the most last-place votes; or the function that selects the candidate with the most first-place votes from the first three voters who vote; or the function that selects the candidate whose name shares the most letters with the names of the voters. These are all social choice functions. Our aim is to develop some criteria that rule out those social choice functions that are not consistent with the goals of democracy. For now, they and all possible social choice functions remain under consideration.

2.5 Exercises and Problems

2.1. Consider the following profile involving 5 candidates and 11 voters:

4	3	2	1	1
A	B	C	D	E
C	E	B	B	D
D	D	D	E	B
B	C	A	C	A
E	A	E	A	C

(a) Who wins the election if the plurality method is used?
(b) Who wins the election if the Borda count method is used?
(c) Who wins the election if the Copeland method is used?
(d) Who wins the election if the Coombs method is used?
(e) By what method might candidate E win?

2.2. For the profile

4	3	3	2	1	2	1	2
A	B	C	D	D	E	E	F
B	E	F	F	F	D	D	C
D	C	D	B	E	B	F	B
E	D	B	E	C	C	B	A
F	F	E	A	A	A	C	E
C	A	A	C	B	F	A	D

who wins an election conducted via the
(a) plurality method?

(b) Borda count method?
(c) Hare method?
(d) Copeland method?
(e) antiplurality method?
(f) vote-for-two method?

2.3. For the profile

7	3	8	4	1	4	3
A	A	B	C	B	B	C
B	C	A	A	D	C	B
D	D	D	D	A	D	D
C	B	C	B	C	A	A

who wins an election conducted via the
(a) plurality method?
(b) Borda count method?
(c) Hare method?
(d) Copeland method?
(e) Coombs method?

2.4. For the profile

A	C	B	C	A
B	A	C	B	C
C	B	A	A	B

who wins an election conducted via the
(a) plurality method?
(b) Borda count method?
(c) Hare method?
(d) Copeland method?
(e) Coombs method?

2.5.
(a) Show by example that the Hare method and the Coombs method need not always produce the same winner.
(b) Show by example that the plurality winner can be the first eliminated by the Coombs method.
(c) Show by example that the plurality winner can come in last place in the Borda count.
(d) Show by example that the Borda count winner need not have any first-place votes.

2.6. Create a profile for an election with 4 candidates such that, for each of the 4 candidates, there is a positional voting method that selects that candidate as the unique winner.

2.7. Create a profile in which candidate A wins the antiplurality election, candidate B wins the Borda count election, while candidate C wins the Copeland method.

2.8. Give an example of a profile in which the Hare winner is the first eliminated by the Coombs method and in which the Coombs winner is the first eliminated by the Hare method.

2.9. The following informal descriptions of voting methods do not represent social choice functions as we have defined them. Explain why not.

(a) (Mini-quota) Any candidate with at least 25% of the votes is declared a winner.

(b) (Approval voting) Every voter answers a yes-or-no question about each candidate: "Is she acceptable?" Whichever candidate is acceptable to the most voters is declared the winner.

(c) (Copeland without ties) The winners of the Copeland method become our finalists. If there is more than one finalist, the tie is resolved by drawing straws.

(d) (Condorcet candidate) Pick the candidate who defeats each of the other candidates in a head-to-head election.

2.10. With 4 candidates, the Borda count method is the positional voting method that we denote $P(3, 2, 1, 0)$.

(a) Consider instead the method $P(4, 2, 1, 0)$. Show by example that this need not give the same outcome as the Borda count method.

(b) Consider now the method $P(4, 3, 2, 1)$. Explain why this always gives the same outcome as the Borda count method.

(c) Consider the method $P(8, 6, 4, 2)$. Explain why this always gives the same outcome as the Borda count method.

(d) Characterize those sequences (a, b, c, d) with the following property: The positional voting method $P(a, b, c, d)$ always selects the same winner as the Borda count method.

2.11. Explain why no positional method can make candidate D the unique winner when facing the following profile:

B	D	D	B	C	B
A	C	B	D	B	D
C	A	C	C	A	A
D	B	A	A	D	C

2.12. Consider the social choice function that assigns points to each candidate in the following way: Candidate A gets one point for every pair

(v, C) such that v is a voter, C is a candidate, and voter v ranks candidate C below candidate A. Then tally points and declare the candidate(s) with the most points to be the winner. Show that this social choice function is identical to the Borda count method.

2.13. Devise an example of a profile involving 5 candidates that results in a tie among all candidates no matter what positional voting method is used and yet where candidate A is the unique winner of the Copeland election.

2.14. Which of the methods introduced in this chapter reduce to the simple majority method in the special case where there are only two candidates?

2.15. Assume that the slate consists of precisely 3 candidates and the electorate consists of an odd number of voters. Consider the social choice function that operates as follows: Candidates A and B face off in a head-to-head election, and the winner of this match faces off in a head-to-head final round with candidate C. The winner of the final round is the winner of the election. Create an example of a profile in which candidate C wins but if every voter switches candidates A and C on their preference ballots, then candidate A does not win.

2.16. A social choice function known as iterated plurality works as follows: If the plurality method gives a unique winner, then that candidate is also the unique winner of the iterated plurality method. Otherwise, the candidates who tie for victory with the plurality method survive for another round and the other candidates are eliminated. The process is then repeated until a unique winner is obtained or until all remaining candidates tie in a plurality election.

(a) Devise an example of a profile in which the iterated plurality method lasts three rounds.

(b) Devise an example of a profile in which the iterated plurality method, the Hare method, and the Coombs method give three different unique winners.

2.17. The **nomination method** is a social choice function that names a candidate to be a winner if the candidate receives at least one first-place vote. (In practice, this method is likely to produce a tie among many candidates.) Is it possible for the set of winners via the nomination method to have no candidates in common with the winners of the

(a) plurality method?

(b) Borda count method?

(c) Hare method?

(d) Coombs method?

2.18. Consider a social choice function that works as follows. Any candidate who defeats every other candidate in a head-to-head match-up is a winner. If no candidate defeats every other candidate in a head-to-head match-up, then the election is a tie among all candidates.

(a) Show that if this method produces a unique winner, then that candidate is also the unique winner of the Copeland method.

(b) Show that this method produces two winners only if it produces a tie among all candidates.

(c) Provide an example of a profile in which this method produces a tie among all candidates.

2.19. A variation on the Coombs method is obtained if one terminates the process as soon as any one candidate has a majority of the first-place votes, declaring that candidate to be the unique winner. It is unclear whether Coombs himself intended this variation. Show that the variation is not identical to the Coombs method as it is defined in this chapter.

2.20. Midway through the presidential campaign of 2008 but before the party conventions, John McCain, Barack Obama, and Hillary Clinton were the three remaining viable candidates for the presidency. For each of the six possible ways of ranking these three candidates, make an estimate of what percentage of American voters held this ranking as their preference order. Determine the winner of a presidential election according to several reasonable social choice functions, assuming that only these three candidates were running and that profile was determined by your estimates.

2.21. Gather data about the nationwide popular vote in the presidential elections from the year 2000. Assume that only George W. Bush, Al Gore, and Ralph Nader were candidates. Make some assumptions about voter preferences. (Election data can tell us only about voters' first choices, so make an educated guess about voters' second choices.) Now analyze your data via the various social choice procedures we've discussed. Can you envision any reasonable voting procedure via which Nader, according to your assumptions, would have been declared the winner? (Remark: a dictatorship doesn't qualify as "reasonable".)

2.22. In early 2016, the field of candidates for the presidential elections was being winnowed to Hillary Clinton, Ted Cruz, John Kasich, Bernie Sanders, and Donald Trump. Make an educated guess as to the preference ballots were all five of these candidates to be offered to the full American electorate. (This forces one to ignore the two-party system, in which voters in one party do not express their preferences among the candidates in the other party.) Express your guess as a tabulated profile. Is there a Condorcet candidate? Who would win the plurality election? Who would win the Borda count election?

3

Criteria for Social Choice

"A democracy is nothing more than mob rule, where fifty-one percent of the people may take away the rights of the other forty-nine."
— Thomas Jefferson

3.0 Scenario

Four candidates, denoted A, B, C, and D, are interested in public office. Seven voters rate the candidates according to the profile:

3	2	2
C	B	A
B	A	D
A	D	C
D	C	B

Assume that the preference orders of the voters do not change with time and that candidate A is the incumbent. Imagine that, one year, candidate B challenges A in a two-candidate election. The next year, in a second two-candidate election, candidate C challenges the winner of the first election. Finally, a year after that, candidate D challenges the winner of the second election.

Who wins the last of these elections? Has the will of the electorate been implemented through these three elections? Which of the four candidates, in your opinion, has the best claim to the office, given the preferences of the voters?

In general, given three alternatives A, B, and C, is it possible for a populace to prefer A to B, B to C, and C to A? Does it make sense for an individual voter to prefer A to B, B to C, and C to A? In track and field, is it reasonable for someone to expect A to beat B, B to beat C, and C to beat A? What about in soccer? In golf? In boxing?

Suppose that every voter in an election prefers A to D? Would it be possible for candidate D to win the election?

What should we mean by calling a social choice function "reasonable" or "appropriate" or "fair"? What assumptions (or axioms) should we insist upon? Or, to put it in negative terms, what flaws in a social choice function should lead us to brand it "unreasonable"?

3.1 Weakness and Strength

Among the many social choice functions, how are we to decide which ones are appropriate for use in democratic elections? That is to say, how can we decide which voting methods are good and which are bad? As we have already seen when studying the two-candidate case, the answer depends on what we mean by good and bad. Let us attempt to describe the properties of a social choice function that we would expect of a reasonable and fair voting system. There are many aspects to fairness, and we will have many properties to describe.

As in Chapter 1, we want to enunciate yes-no criteria (we may sometimes call them properties, conditions, principles, or rules) by which to measure the appropriateness of social choice functions. The criteria that we discussed there in the context of two-candidate elections also pertain to elections with more than two candidates, but we will also introduce many new criteria that emerge only when there are more than two candidates. Before we move on to any of these, however, we want to discuss the question of what characteristics the criteria themselves should have.

First, we want the criteria to be "not too weak". This means that we want to make sure they are not too lenient — that they are not satisfied by every conceivable social choice function. A criterion that is too weak does not help us discriminate between good and bad methods. Second, we want the criteria to be "not too strong". This means that we want to make sure they are not too strict — that they are satisfied by at least some reasonable social choice functions. A criterion that is too strong also does not help us discriminate between good and bad methods. Third, we want the criteria to be "compelling" in the sense that we want them to represent clearly desirable qualities. A compelling criterion is one that people readily agree represents a desirable attribute for a method to possess.

Let us examine via examples what we mean by "not too weak", "not too strong", and "compelling". We illustrate these ideas in a contrarian fashion by giving some peculiar criteria that fail to exhibit these characteristics. An example of a criterion that we might regard as being too weak is the following.

Definition 3.1. A method satisfies the **unanimity criterion** if, whenever all voters place the same candidate at the top of their preference orders, that candidate must be the unique winner.

This criterion is certainly compelling. It is easy to agree that we should insist that our social choice function meet this condition. But it turns out that every reasonable social choice function that we can imagine satisfies this criterion. The only method that is even remotely reasonable that violates this criterion is monarchy. Consequently it cannot very well serve as an important way to discriminate between the best and the worst alternatives.

An example of a condition that we might regard as too strong is one we have already seen:

Definition 3.2. A method is **decisive** (or satisfies the **decisiveness criterion**) if it always selects a unique winner.

There is no doubt that this criterion is compelling. It would be nice if our social choice functions could avoid ties. However, just as in the two-candidate case, we may regret imposing this condition. To see why, consider what happens when a decisive social choice function faces the following profile:

A	B	C
B	C	A
C	A	B

Since we prohibit a tie in this situation, we force the method to declare one of the three candidates to be the unique winner. This in turn forces us to impose a certain bias, either on the voters or on the candidates. We will formalize this observation in Proposition 3.12. We note that plurality, Borda count, Hare, Coombs, Copeland, and all the positional methods declare a tie when facing this profile. Therefore all of these methods violate the decisiveness criterion. Since we don't want to dismiss all these methods too abruptly — after all, their use is widespread — we are led to conclude that it may be too much to ask that a social choice function is decisive.

A criterion is most certainly too strong if it is satisfied by no method whatsoever. Sometimes, however, even a criterion such as this can sound quite compelling. For example, one might wish to impose the requirement that in order to be named as a winner, a candidate must have a majority of the first-place votes. Why is this condition vacuous? Because any social choice function that satisfies this condition cannot name a winner when faced with a profile in which no candidate has a majority. We know that such profiles exist when there are more than two candidates. But a social choice function is, by definition, a function and, as such, must name at least one winner no matter what profile is input. In other words, this criterion is impossible to satisfy.

A small but significant rewording of this criterion leads to the following criterion, which proves to be quite valuable.

Definition 3.3. A method satisfies the **majority criterion** if, whenever some candidate receives a majority of the first-place votes, that candidate must be the unique winner.

This criterion is not vacuous and, in fact, is quite interesting to analyze. The point of formulating criteria is to discriminate between different voting methods. An effective criterion should be satisfied by some but not all methods. But that is not enough to make a criterion appropriate. We still must ask if it is compelling, that is, whether satisfying it corresponds to some measure of the appropriateness of the method. A compelling criterion should capture some sense of fairness or impartiality or stability or democracy.

Our goal in this chapter is to establish a set of principles by which we can assess our social choice functions. In our effort to be dispassionate, we should be careful not to enunciate principles that might bias our assessment. So for instance we have no reason to consider what one might call the "is-the-Borda-count" criterion: A social choice function satisfies this criterion if and only if it is the Borda count method. An advocate for the Borda count method may lobby for this principle, but unless you believe in advance that the Borda count method is the only one worth considering, it is hard to call this criterion compelling or unbiased.

3.2 Some Familiar Criteria

Even if we restrict ourselves to compelling criteria that are neither too strong nor too weak, we still have no shortage of criteria to consider. We consider several such criteria now. We begin by restating, in the context of three or more candidates, the three main criteria that arose in our study of elections with two candidates.

Definition 3.4. A method is **anonymous** (or satisfies the **anonymity criterion**) if the outcome is unchanged whenever any two voters exchange their ballots.

Imagine two profiles that are identical except that the first two voters have swapped their preference lists. The principle of one-person-one-vote demands that the method should assign the same output to these two inputs. In the presence of the anonymity criterion, the only information that one needs from a profile is how many voters — but not which voters — select each of the possible preference orders. We note that Lemma 1.16 also holds for elections with more than two candidates.

Lemma 3.5. *A social choice function is anonymous if and only if it depends only on the tabulated profile.* □

The second criterion is the following.

Definition 3.6. A method is **neutral** (or satisfies the **neutrality criterion**) if the following condition holds: Imagine a profile for which the method names a certain candidate A to be a winner. For some other candidate B, suppose that all the voters alter their preference orders by interchanging the positions of candidates A and B. Then the method must name candidate B a winner when facing the altered profile.

The point is this: If the electorate feels a certain way about candidate A and that is enough to get candidate A elected, then the same rules should apply to any other candidate. If the electorate were to feel just the same way about candidate B, then candidate B would have to get elected. Neutrality, also called indifference or impartiality, is a fairness condition that requires that the rules apply equally to all candidates. By contrast, anonymity is a fairness condition that requires that the rules apply equally to all voters.

The third criterion from Chapter 2 that we revisit now is the following.

Definition 3.7. A method is **monotone** (or satisfies the **monotonicity criterion**) if the following condition holds: Suppose under a certain profile that some candidate A wins and that some voter v puts a different candidate, say B, immediately ahead of A on her preference list. Imagine a second profile that is identical to the first except that voter v moves A up one place and moves B down one place on her preference list. Then the method must still declare candidate A to be a winner when facing the second profile.

Suppose that some monotone method declares candidate A to be a winner when facing the profile here.

A	C	B	C	C	A	B
B	A	C	B	B	B	A
C	B	A	A	A	C	C

Then it must also declare candidate A to be a winner when facing the profile here.

A	C	B	C	C	A	B
B	A	C	B	A	B	A
C	B	A	A	B	C	C

This is because this profile is obtained from the preceding one by moving candidate A up on the preference list of one voter (namely voter 5). It further follows by the same sort of reasoning that candidate A must win when facing the profile here.

A	C	B	C	A	A	B
B	A	C	B	C	B	A
C	B	A	A	B	C	C

The point of monotonicity is that a candidate cannot be harmed by being more favored by a voter or voters. From the voter's point of view, the point is that raising a candidate up on the voter's preference order ought not harm the candidate. In the absence of monotonicity, voters find themselves paralyzed at the polls, not certain if they should admit to their honest preferences for fear that placing their favorite candidate at the top of their list may in fact doom their favorite candidate to defeat. This conjures up thoughts of tactical or strategic voting and leads to election instability. Monotonicity is compelling.

3.3 Some New Criteria

We now move on to the consideration of some criteria whose significance only emerges now that more than two candidates are in the picture. One such particularly compelling criterion is the following.

Definition 3.8. A method satisfies the **Pareto criterion** (or is **Pareto**) if, whenever every voter prefers one candidate to another — say that every voter prefers candidate X to candidate Y — the method does not select candidate Y as a winner.

This criterion is named after the Italian economist Vilfredo Pareto (1848–1923). If the electorate unanimously agrees that candidate A is preferable to candidate B, then the choice of candidate A to win the election seems to be strictly better than the choice of candidate B. Thus, it seems inappropriate to select candidate B as a winner.

In economics, an alternative is called **Pareto optimal** if no other alternative is preferable in every respect. In our context, we call a candidate Pareto optimal if there is no other candidate who is preferred by *all* the voters. The Pareto criterion demands that winners of elections are Pareto optimal candidates.

A common misunderstanding is the belief that the Pareto criterion demands that, whenever every voter places candidate A over candidate B, candidate A must be a winner. The Pareto criterion says no such thing. After all, in the profile

7	6
D	C
C	D
A	A
B	B

every voter rates candidate A ahead of candidate B. Yet few would argue that A should be declared the winner here. In fact, since every voter places C ahead of A, the Pareto criterion demands that A is *not* a winner. The Pareto criterion demands that neither A nor B can be a winner of this election, since neither is a Pareto optimal choice. It is a contest between candidates C and D.

The next two criteria invoke the name of the Marquis de Condorcet (1743–1794), a French mathematician and philosopher, sometimes credited with introducing mathematics into the social sciences. Given a profile, a candidate is called a **Condorcet candidate** if that candidate can beat every other candidate in a head-to-head match-up. (In view of May's Theorem 1.23, all head-to-head elections are conducted via the simple majority method.) Similarly, an **anti-Condorcet candidate** is one who loses to every other candidate in a head-to-head match-up. A profile can have at most one Condorcet candidate and at most one anti-Condorcet candidate, because if there were two, then each defeats the other in a head-to-head election, which is clearly an impossibility.

A Condorcet candidate is sometimes referred to as a Condorcet winner. The latter term can be misleading, however, because it suggests that there is a "Condorcet method" for running elections that selects the Condorcet candidate as the winner. But this is not a "method" at all in our sense of the word, because not every profile has a Condorcet candidate (or an anti-Condorcet candidate). For example, the simple symmetric Profile 3.1 involving just three voters and just three candidates has A beating B head-to-head, B beating C head-to-head, and C beating A head-to-head, so there is no Condorcet candidate or anti-Condorcet candidate. Thus deciding elections by choosing the Condorcet candidate does not constitute a social choice function at all.

Even though there may not be a Condorcet candidate, if there is one, then that candidate can stake a certain claim to the electorate's

A	C	B
B	A	C
C	B	A

Profile 3.1

preference. No single other candidate can defeat a Condorcet candidate. It is tempting to insist that a social choice function, if it is reasonable, must name as a winner the Condorcet candidate if there is one.

Definition 3.9. A method satisfies the **Condorcet criterion** if, whenever a profile has a Condorcet candidate, the method must choose this candidate to be the unique winner of the election.

A distinct but related condition is the following.

Definition 3.10. A method satisfies the **anti-Condorcet criterion** if, whenever a profile has an anti-Condorcet candidate, the method must not choose this candidate to be a winner of the election.

These criteria represent the viewpoint of an election observer who asks, after the election is complete, whether a different candidate would have been preferred by the electorate to the actual winner. A Condorcet candidate is preferred to any single alternative. Hence the electorate will feel dissatisfied when a social choice function selects some alternative if a Condorcet candidate was available. It is a compelling viewpoint.

There are, however, limitations to this viewpoint. For example, we must not impose a criterion that demands that any candidate who is beaten by some other candidate head-to-head cannot be a winner. Profile 3.1 above shows that this criterion is vacuous, because a profile can have every candidate losing to some opponent. The universal domain assumption does not allow us to rule every candidate ineligible for victory.

Most elections for public office involve more than two candidates. A familiar but unfortunate election anomaly is when a candidate with little chance to win influences the outcome of an election in unexpected ways. For example, in the 2000 US presidential race in Florida, there was a close contest between George W. Bush and Al Gore. Why did Bush beat Gore? One curious and disturbing answer is "Because Ralph Nader was more popular then Pat Buchanan." Nader was widely regarded as outflanking Gore on the left wing, while Buchanan was seen as outflanking Bush on the right wing. We therefore imagine that most of Nader's supporters preferred Gore to Bush, while most of Buchanan's supporters preferred Bush to Gore. Thus Nader and Buchanan had the effect of taking votes away from Gore and Bush, respectively. But since the "real" contest was between Bush and Gore, it seems peculiar that this contest was affected at all by the introduction of these two other candidates who themselves were not able to gain significant support. One may even imagine that Nader was a Bush operative introduced strategically to steal votes from Gore, while Buchanan was a Gore operative introduced to steal votes from Bush. It is troubling that our method of conducting elections allows for the potential of this kind of sabotage candidacy. To rule it out, we establish a criterion that goes back to Kenneth Arrow in 1950. He coined the term "independence of irrelevant alternatives" to describe this criterion; it is sometimes called "independence of infeasible alternatives" and is often abbreviated "IIA". We prefer to call it simply "independence".

Definition 3.11. A social choice function satisfies the **independence criterion** (or is **independent**) if the following condition holds. Suppose that there are two profiles between which no voter changes his mind about whether candidate A is preferred to candidate B. (In other words, if a voter puts A ahead of B in one profile, then he puts A ahead of B in the other, and vice versa.) Suppose also that when facing the first profile, the method chooses candidate A but not candidate B as a winner. Then the method must not choose candidate B as a winner when facing the second profile.

Like monotonicity, independence requires the consideration of two profiles, the second related to the first in some critical way. In addressing monotonicity, we first imagine one profile where candidate A wins, and then we imagine a second profile where one voter raises candidate A up. In addressing independence, we first imagine one profile where candidate A defeats candidate B, and then we imagine a second profile where no voters change the relative position of candidates A and B on their preference lists compared to their preference list in the first profile. In going from the first profile to the second, all voters may alter their views about all other candidates, but the one proviso is that no voter may rank A ahead of B in one profile and B ahead of A in the other. These changes should not cause candidate B to become a winner.

In particular, the independence criterion asserts that if A defeats B, then a change in a voter's view about C should not allow B to become a winner. After all, were B able to benefit from voters' better impression of C, then supporters of B might spend their time strategically campaigning for C even though they actually want candidate B to win. In this setting, C is the "irrelevant alternative", since A defeats B without the extra support for C. It isn't sensible that the fortune of B depends on the irrelevant alternative, since the voters had A defeating B before the change, and no voters changed their opinion about the relative positions of candidates A and B on their preference lists. If the electorate chose A but not B before the change, then it would be inconsistent of them to choose B after the change, given that their feelings about A versus B have not changed.

Suppose that an independent method declares candidate A to be the unique winner when facing the profile here.

A	C	B	C	C	A	B
B	A	C	B	B	B	A
C	B	A	A	A	C	C

Then it must not declare candidate B to be a winner when facing the profile here.

A	C	B	C	C	A	B
B	A	C	B	B	B	C
C	B	A	A	A	C	A

The only difference between the first profile and the second is with the last voter, who moves candidate C up from a third-place ranking to a second-place ranking. There is no saying that candidate A will still win in the second profile. After all, candidate C is looking somewhat better now relative to candidate A, and one might imagine that a reasonable method could now favor C. But should candidate B be able to win now? Candidate B couldn't win before candidate C moved up, so why should she look better now? Independence forbids candidate B from benefiting from a new viewpoint about candidate C. In the presence of the independence criterion — assuming still that candidate A won and candidate B lost when facing the first profile — the method cannot name candidate B as a winner when facing the second profile.

These then are the principal criteria that we want to investigate. We would love to discover a method that satisfies all of them. Unfortunately, that isn't possible. In fact, the following proposition shows that three of our criteria are mutually contradictory. This result involves criteria that were all introduced in Chapter 1 and applies even when there are just two candidates. It strengthens Corollary 1.25.

Proposition 3.12. *Any social choice function that satisfies anonymity and neutrality must violate decisiveness.*

Proof. Suppose that we have a method that is anonymous and neutral. For the sake of argument, suppose that there are just two candidates, A and B. Suppose further that there are $2n$ voters and that they present the profile

n	n
A	B
B	A

Some candidate has to be named a winner here, so let us suppose, without loss of generality, that candidate B is a winner. Now suppose that the first n voters change places with the second n voters. The resulting profile is

n	n
B	A
A	B

By anonymity, candidate B must still be a winner. But the second profile can be obtained from the first by another procedure: interchanging candidate A and candidate B. By neutrality, candidate A must now be a winner. Hence the method must declare a tie when facing this profile.

A similar argument can be made for any number of candidates. Hence a neutral and anonymous method must allow ties. □

The previous proposition may be phrased as follows: It is impossible to find a social choice function that is neutral, anonymous, and decisive. These three criteria are incompatible. This is the first of several impossibility results that we will encounter. An impossibility theorem is an indication that we are forced to choose among the criteria. As much as we may desire anonymity, neutrality, and decisiveness, we simply cannot have all three.

Another impossibility theorem is the following.

Proposition 3.13. (Taylor) *No social choice function involving at least three candidates satisfies both the independence criterion and the Condorcet criterion.*

Proof. Suppose, in order to obtain a contradiction, that we have an independent method that satisfies the Condorcet criterion. Imagine that there are just three candidates, A, B, and C, and imagine that there are just three voters, who present the profile

A	C	B
B	A	C
C	B	A

Profile 3.2

We argue that candidate A cannot be named a winner here. To see why, consider the profile

A	C	C
B	A	B
C	B	A

Profile 3.3

In Profile 3.3, candidate C is the Condorcet candidate, and so, by the Condorcet criterion, candidate C must be the unique winner. Hence candidate A is not a winner facing Profile 3.3. But all the voters in Profile 3.2 place candidates A and C in the same relative position as they do in Profile 3.3. The only difference between them is that the third voter, in going from Profile 3.3 to Profile 3.2, has repositioned candidate B higher up on her preference list. By independence, candidate A cannot be a winner in Profile 3.2.

So the method cannot name candidate A as a winner in Profile 3.2. But an identical argument (whose details we omit) shows that candidate B cannot be named a winner there either. Neither can candidate C, by the same argument once again. So the social choice function gets stuck on Profile 3.2, unable to name any of the candidates a winner, contradicting the universal domain condition that we impose on all social choice functions.

A slight modification of this argument works with more than three candidates and more than three voters. □

The moral of this story is that we must make a choice between the Condorcet criterion and independence. As we look over the possible social choice functions, we may be impressed that certain ones satisfy the Condorcet criterion and other ones satisfy independence. Proposition 3.13 tells us that, unfortunately, it is impossible to find a method that satisfies both. If we want to debate which social choice functions are best, we have to establish which criteria should be imposed to measure them. A decision to impose the Condorcet criterion, say, is a philosophical one; it may derive from an investigation into the fundamental principles of fairness and democracy. It is not a decision about which we will argue. All we can guarantee is that, however tempting it may be to impose both the Condorcet criterion and independence, a choice must be made. We can't have it both ways.

Sometimes one criterion implies another.

Proposition 3.14. *If a method satisfies the Condorcet criterion, then it satisfies the majority criterion.*

Proof. Notice that a candidate who has a majority of the first-place votes must be a Condorcet candidate. This is because that same majority will favor this candidate in any head-to-head match-up. Suppose that a method satisfies the Condorcet criterion, and imagine that this method faces a profile that has one candidate receiving a majority of the first-place votes. Since that candidate is a Condorcet candidate, the method must choose that candidate as the unique winner. Hence it satisfies the majority criterion. □

One way to express Proposition 3.14 is to say that the Condorcet criterion is stronger than the majority criterion. While we should be pleased if a method satisfies the Condorcet criterion, we have already seen that this involves a trade-off. (It would preclude satisfaction of the independence criterion, for example.) We could decide to relax the requirement and demand instead the weaker version, the majority criterion. In so doing, according to Proposition 3.14, we continue to encompass all the methods that satisfy the Condorcet criterion. In other words, it is impossible for a method to satisfy the Condorcet criterion yet fail to satisfy the majority criterion.

The king of all impossibility theorems in this area is the celebrated theorem of Kenneth Arrow, winner of the 1952 Nobel Prize in economics. The proof of Arrow's Theorem is more involved than the proofs of the propositions above, and we will devote an entire chapter to an explication of the result and its proof. First we examine which of the voting methods introduced in Chapter 2 satisfy which of the criteria introduced here in Chapter 3.

3.4 Exercises and Problems

3.1. Determine if the following profile has a Condorcet candidate, and if so, identify that candidate.

A	D	E	C	D
B	B	B	A	E
C	E	A	B	C
D	A	C	E	B
E	C	D	D	A

3.2. Determine if the following profile has a Condorcet candidate, and if so, identify that candidate.

2	2	1
A	B	C
F	A	B
E	D	A
C	F	E
B	E	D
D	C	F

3.3. Give an example of a profile with an anti-Condorcet candidate but no Condorcet candidate, or explain why it is impossible to do so.

3.4. Give an example of a profile in which the plurality winner has fewer Borda count points than any other candidate, or explain why it is impossible to do so.

3.5. Assume that a method satisfies the Pareto criterion. Explain why it also satisfies the unanimity criterion.

3.6. Explain why Lemma 3.5 holds by arguing that the information that is lost when tabulating a profile can affect the outcome of an election if and only if the method violates the anonymity criterion.

3.7. Answer each of the following questions with a response of the form "X must win" or "Y must lose", or write "No conclusion can be drawn".

(a) A social choice function satisfies the Condorcet criterion and the profile is:

A	A	B	B	C
C	D	D	D	D
D	B	C	C	B
B	C	A	A	A

What can you conclude?

(b) A social choice function satisfies the neutrality criterion and candidate A loses in the "before" profile here:

A	B	B
C	A	C
B	C	A

before

C	B	B
A	C	A
B	A	C

after

What can you conclude about the "after" profile?

(c) A social choice function satisfies the Pareto criterion and the profile is:

A	A	B	B	C
B	C	D	D	B
D	B	C	A	A
C	D	A	C	D

What can you conclude?

(d) A social choice function satisfies the monotonicity criterion and candidate A wins in the "before" profile here:

A	B	B	A
C	A	A	B
B	C	C	C

before

A	B	B	A
C	A	C	B
B	C	A	C

after

What can you conclude about the "after" profile?

(e) A social choice function satisfies the independence criterion and candidate B wins and candidate C loses in the "before" profile here:

B	B	C	A	C
C	C	A	B	D
D	A	B	D	B
A	D	D	C	A

before

C	B	C	A	C
D	C	A	B	D
B	A	B	D	B
A	D	D	C	A

after

What can you conclude about the "after" profile?

3.8. Answer each of the following questions with a response of the form "X must win" or "Y must lose", or write "No conclusion can be drawn".

(a) A social choice function satisfies the anti-Condorcet criterion and the profile is:

A	A	B	B	C
C	D	D	D	D
D	B	C	C	B
B	C	A	A	A

What can you conclude?

(b) A social choice function satisfies the neutrality criterion and candidate A wins in the "before" profile here:

A	B	B
C	A	C
B	C	A

before

C	B	B
A	C	A
B	A	C

after

What can you conclude about the "after" profile?

(c) A social choice function satisfies the Pareto criterion and the profile is:

A	A	A	B	D
D	B	D	D	C
B	D	C	C	B
C	C	B	A	A

What can you conclude?

(d) A social choice function satisfies the monotonicity criterion and candidate A loses in the "before" profile here:

A	B	B	A
C	A	A	B
B	C	C	C

before

A	B	B	A
C	A	C	B
B	C	A	C

after

What can you conclude about the "after" profile?

(e) A social choice function satisfies the independence criterion and candidate B wins and candidate C loses in the "before" profile here:

B	B	C	A	C
A	D	A	B	D
D	A	B	C	B
C	C	D	D	A

before

B	B	C	A	C
C	C	B	B	A
D	A	A	D	D
A	D	D	C	B

after

What can you conclude about the "after" profile?

3.9. Consider the two profiles:

7	5	4	1
A	C	B	B
B	A	C	A
C	B	A	C

Profile P

8	5	4
A	C	B
B	A	C
C	B	A

Profile Q

(a) Suppose that a method chooses candidate A as the unique winner in Profile P but candidate B as the unique winner in Profile Q. What can you say about such a method?
(b) Suppose that a method chooses candidate A as the unique winner in Profile P but candidate C as the unique winner in Profile Q. What can you say about such a method?

3.10. Consider the two profiles:

A	B	C
B	D	D
C	A	A
D	C	B

Profile P

A	B	D
D	D	C
B	A	A
C	C	B

Profile Q

(a) Suppose that a method chooses candidate D as the unique winner in Profile P but candidate C as the unique winner in Profile Q. What can you say about such a method?
(b) Suppose that a method chooses candidate B as the unique winner in Profile P but candidate A as the unique winner in Profile Q. What can you say about such a method?

3.11. What is a simple way of describing the Pareto criterion as it applies to social choice functions for two candidates?

3.12. Give a complete description of the social choice functions for two candidates that satisfy the Condorcet criterion.

3.13. Give a complete description of the social choice functions for two candidates that satisfy the independence criterion.

3.14. A candidate with a majority of the last-place votes is called the antimajority candidate. A voting method is said to satisfy the **anti-majority criterion** if the antimajority candidate, whenever there is one, is always a loser of the election.
(a) Give an example of a profile where there is no antimajority candidate.
(b) Prove that if there is an antimajority candidate, that candidate must be the anti-Condorcet candidate as well.
(c) Explain why any method that satisfies the anti-Condorcet criterion must satisfy the antimajority criterion.

3.15. A social choice function is said to be **stable** if, whenever a majority of voters prefer one candidate (call him A) to another (call her B), the other candidate (B) must lose. (A voting method violating this criterion may seem to be unstable, because were candidate B chosen as a winner in this situation, a majority of the electorate would want candidate A instead.) Explain why no social choice function is stable.

3.16. A method is called **unrestricted** if, for every candidate, there is some profile that results in that candidate becoming the unique winner of the election. Show that any method that satisfies the Pareto criterion or the majority criterion or the unanimity criterion must be unrestricted.

3.17. Consider the social choice function that selects as a winner the candidate (or candidates) with the most second-place votes.
(a) Show that this method violates monotonicity.
(b) Show that this method violates the Pareto criterion.
(c) Show that this method is anonymous and neutral.

3.18. Consider the social choice function that selects as a winner the candidate with the fewest first-place votes. (This should not be confused with the much more reasonable antiplurality method.)
(a) Show that this method violates monotonicity.
(b) Show that this method violates the Pareto criterion.
(c) Show that this method is anonymous and neutral.

4

Which Methods Are Good?

"It's not the voting that's democracy, it's the counting."
— Tom Stoppard

4.0 Scenario

Suppose you are in the mood to eat a piece of fruit. Which kind of fruit should you choose? Imagine that you attempt to answer this question by listing a number of popular fruits and identifying a number of criteria for assessing them. The fruits you consider by no means exhaust all possibilities. In fact, you might consider only the apple, the banana, the cherry, the date, and the elderberry.

Here are seven yes-no criteria for assessing fruit:

(1) Does the fruit taste good?

(2) Is the fruit healthy?

(3) Is the fruit inexpensive?

(4) Is the fruit nonpoisonous?

(5) Does the common English name of the fruit start with a vowel?

(6) Is the fruit yellow and shaped like a banana?

Create a 5-by-7 matrix whose rows are indexed by the five fruits and whose columns are indexed by the 7 criteria. Put the word "yes" or "no" in each cell of the matrix according to whether the fruit indexing that row has the property indexing that column. (Resist the temptation to answer equivocally. Understand that there is no wrong answer, since, as they say, there's no accounting for taste. Naturally, some of the answers are a matter of opinion.) When the answer is yes, we regard the fruit as meeting the criterion.

Which of the criteria are compelling in the sense that you would be pleased to sample a fruit that meets the criterion? Which of the criteria are too weak? Which are too strong? Which seem inappropriately biased? After removing any criteria that you regard as biased or inappropriate or unimportant or too strong, do any fruits meet all the remaining criteria? If not, which fruits meet more criteria than others?

4.1 Methods and Criteria

We have introduced a number of social choice functions and a number of criteria by which to measure them. For each social choice function X and each criterion Y, we want to answer the question "Does method X satisfy criterion Y?" The methods that satisfy many criteria, with special emphasis on the criteria that we regard as most compelling, are the methods that we may want to recommend. In this chapter, we investigate 10 of our favorite criteria: anonymity, neutrality, unanimity, decisiveness, the majority criterion, the Condorcet criterion, the anti-Condorcet criterion, monotonicity, Pareto, and independence, all of which were discussed in Chapter 3. We also investigate 12 of our favorite social choice functions.

Among the 12 methods are plurality, antiplurality, Borda count, and the methods of Hare, Coombs, and Copeland, methods described in Chapter 3. In addition, we consider a method named after Scottish political scientist Duncan Black (1908–1991).

Most of the methods we've considered up to now are anonymous, so by Lemma 3.5 all of them depend only on tabulated profiles. We also consider in this chapter two methods that are not anonymous. The first is the method of bloc voting, and the second is dictatorship.

Most of the methods considered so far are also neutral. This is to say that each method delineates rules by which a candidate — any candidate — wins the election. We consider in this chapter just two methods that are not neutral. The first is the agenda method with a sequential agenda. The second is the monarchy method, the method that declares the election to be a victory for one particular candidate no matter how the electorate votes.

The final method that we investigate here is the all-ties method. Methods such as all-ties, monarchy, and dictatorship are clearly useless from the point of view of implementing the ideals of democracy. They are considered in this chapter nevertheless, in the spirit of what a scientist might call controls.

Most of these 12 methods satisfy the unanimity criterion. To test a method for unanimity, imagine how the method would work facing a unanimous electorate, in which every voter places the same candidate in first place. The all-ties method fails to name this candidate as the unique winner, of course, as does the monarchy method, unless the monarch just happens to be the candidate ranked first. All of the other 11 methods, however, even dictatorships, always name such a unanimously favored candidate as the unique winner. In this sense, unanimity is a weak criterion and not so useful in discriminating among proposed methods. The decisiveness criterion, by contrast, is satisfied by rather few of the methods that we are considering. The only exceptions are dictatorships and monarchies. In this sense, decisiveness is too strong a condition to be useful, ruling out as it does all the familiar reasonable methods. This demonstrates why permitting ties in elections is an essential element of our model.

In the next section, we test each of our 12 methods against each of our 10 criteria. It would be nice to discover a method that satisfies all the criteria, but we won't find any, because impossibility theorems lurk. Instead, we shall be content to discover a method that satisfies many criteria.

4.2 Proofs and Counterexamples

The propositions that follow address the question "Does method X satisfy criterion Y?" for the main examples that we are considering. Since most criteria demand that something always happens, it is insufficient to provide just one example in arguing that a method does satisfy a criterion. Instead what is needed is an explanation of why the thing does indeed always happen. On the other hand, to demonstrate that a method does not satisfy a criterion, one needs to provide just one counterexample where the thing does not happen. In the proofs below, one sees explanations where the answer is "yes" and counterexamples where the answer is "no".

In what follows, our exposition occasionally adopts an abbreviated style. For example, "The Hare method violates Condorcet" or even "Hare violates Condorcet" might take the place of the more formal "The Hare method does not satisfy the Condorcet criterion." The advantages of this manner of writing are that it permits clear and efficient communication, that basic words are not repeated incessantly, and that the language points directly to the central issue. The disadvantages are that the writing becomes more like jargon, more like slang, and that comprehension depends somewhat more heavily on a savvy reader.

Proposition 4.1. *The plurality method satisfies majority, monotonicity, and Pareto but not Condorcet, anti-Condorcet, or independence.*

Proof. The plurality method satisfies the majority criterion because a majority is always a plurality. It satisfies monotonicity because raising candidate A up on some preference lists can never decrease the number of first-place votes that A receives, nor can it increase the number of first-place votes that any other candidate receives. Hence, if A wins the election before such a change, she wins afterwards as well. The plurality method satisfies Pareto because if candidate A is above candidate B on every preference list, then B will have no first-place votes at all. Without first-place votes, B cannot win a plurality election.

In Profile 4.1, candidate A defeats candidate B in a head-to-head match-up by a score of 4-to-3. Candidate A defeats candidate C in a head-to-head match-up by a score of 5-to-2, and candidate C defeats candidate B in a head-to-head match-up by a score of 4-to-3. Therefore A is the Condorcet candidate and B is the anti-Condorcet candidate. Yet B wins the plurality election.

2	3	2
A	B	C
C	A	A
B	C	B

Profile 4.1

This demonstrates that the plurality method fails both the Condorcet and the anti-Condorcet criteria.

Profiles 4.2 and 4.3 provide an example of a before-and-after scenario. In the "before" profile, A wins the plurality election and B loses. In the "after" profile, B has become the plurality winner.

A	A	A	A	B	B	B
B	B	C	C	A	A	A
C	C	B	B	C	C	C

Profile 4.2 "before"

A	A	C	C	B	B	B
B	B	A	A	A	A	A
C	C	B	B	C	C	C

Profile 4.3 "after"

Yet no voter changes the relative placement of candidates A and B in the two profiles; the only change that has occurred between the two profiles is that candidate C has moved on the preference lists of two voters. This demonstrates that plurality violates independence. □

The failure of the plurality method to satisfy independence is a familiar concern in American politics. Third-party candidates, even if relatively unpopular, can impact the outcome of an election. In Profiles 4.2 and 4.3, it seems clear that there is a race between candidates A and B, with candidate C the least well-liked of the three. Comparing these two profiles, one notes that no voters change their minds about their relative placement of A and B. (The first four voters prefer A while the last three voters prefer B.) Yet the emergence of minor candidate C in Profiles 4.2 and 4.3 hands the plurality election over to B. Candidate C has stolen critical votes from A and has indeed altered the outcome of the election. An electorate that selects A "before" but B "after" is inconsistent about whether it favors A or B.

Proposition 4.2. *The antiplurality method satisfies monotonicity but not majority, Condorcet, anti-Condorcet, Pareto, or independence.*

Proof. The antiplurality method satisfies the monotonicity criterion because raising candidate A up on some preference lists can never increase the number of last-place votes that A receives, nor can it decrease the number of last-place votes that any other candidate receives. Hence if A could win the antiplurality election before such a change, she wins afterwards as well.

Profile 4.4 shows that a candidate with a majority of first-place votes can also have a plurality of the last-place votes and therefore lose the antiplurality election.

C	C	B	B	B
A	A	C	A	A
B	B	A	C	C

Profile 4.4

In fact, B is the Condorcet candidate while A is the anti-Condorcet candidate. Yet A wins the antiplurality election. Hence antiplurality violates the majority, Condorcet, and anti-Condorcet criteria.

In Profile 4.5, candidate A is rated higher than candidate B by all voters.

C	A	A	A	D
A	C	B	D	A
B	B	D	B	B
D	D	C	C	C

Profile 4.5

Yet B is a winner in the antiplurality election (tying with A). Hence antiplurality violates the Pareto criterion. Finally, consider the before-and-after Profiles 4.6 and 4.7.

C	A	A	B	B
A	B	B	A	A
B	C	C	C	C

Profile 4.6 "before"

C	A	A	B	B
A	B	B	C	C
B	C	C	A	A

Profile 4.7 "after"

Candidate A wins and candidate B loses the antiplurality election in Profile 4.6. But after two voters move candidate C up on their preference lists — keeping A and B in the same relative position — we obtain Profile 4.7 and B emerges as the antiplurality winner. This violates independence. □

Proposition 4.3. *The Borda count method satisfies monotonicity, anti-Condorcet, and Pareto but not majority, Condorcet, or independence.*

Proof. The Borda count method satisfies the monotonicity criterion because raising candidate A up on some preference lists can never decrease the Borda count score that candidate A receives, nor can it increase the Borda count score that any other candidate receives. Hence if candidate A could win the Borda count election before such a change, she wins afterwards as well.

We next address why the Borda count satisfies the anti-Condorcet criterion. This argument is somewhat more delicate than most of the others in this chapter. Assume that candidate A is the anti-Condorcet candidate. We must show that A cannot win the Borda count election. Note that in the Borda count, candidate A gets one point for each pair (X, v) consisting of a candidate X and a voter v such that v ranks candidate A ahead of candidate X. For example, in Profile 4.8, A gets one point from the first voter for being above C, and A gets another point from the first voter for being above D.

B	A	C	D	D
A	B	D	C	C
C	C	B	B	A
D	D	A	A	B

Profile 4.8

The fact that A is the anti-Condorcet candidate means that for every other candidate Y, A is below Y more often than A is above Y. In other words, A gets less than half the maximum number of Borda points it is possible for any candidate to get. For example, in Profile 4.8, A gets 2 (out of a possible 5) points from B, 2 (out of 5) points from C and 2 (out of 5) points from D. So altogether A gets 6 out of a possible 15 Borda points.

Now suppose there are n candidates and m voters. If all m voters put candidate Y first, Y will receive $m(n-1)$ Borda points. This is the maximal number of Borda points, so the Condorcet loser, if there is one, must receive less than $m(n-1)/2$ Borda points. On the other hand, we leave it as an exercise for the reader to show that each voter gives out $n(n-1)/2$ Borda points. It follows that the sum of all the Borda points given out by all the voters to all the candidates is $mn(n-1)/2$. Thus, since there are n candidates, the average Borda score for each is $m(n-1)/2$ votes. Now since the anti-Condorcet candidate A receives fewer points than this average, some other candidate Y must receive more. Thus, A cannot win.

To see why Borda count satisfies the Pareto criterion, imagine a profile in which candidate A is rated higher than candidate B by every voter. Then certainly A will obtain a higher Borda count score than B, and so B does not win such an election.

In Profile 4.9, candidate A is the Condorcet candidate. In fact, A has a majority of the first-place votes. Yet A loses the Borda count election to B, who outscores A by a 7-to-6 margin.

A	A	A	B	B
B	B	B	C	C
C	C	C	A	A

Profile 4.9

Hence Borda count violates both the majority criterion and the Condorcet criterion.

The "before" and "after" Profiles 4.6 and 4.7 also serve to demonstrate that the Borda count violates independence. In the "before" profile, the Borda count scores for candidates A, B, and C are, respectively, 7, 6, and 2, so candidate A wins. The "after" profile is obtained when two voters move candidate C up on their lists. The Borda count scores for candidates A, B, and C become 5, 6, and 4, respectively, making candidate B the winner "after". This violates independence, since the voters have maintained their views about which of candidates A and B they prefer. □

The proof that the Borda count method satisfies the anti-Condorcet criterion is one of the subtler arguments found in this chapter. Another subtle argument, this time in the form of a counterexample, is the part of the following proof that demonstrates the failure of the Hare method to satisfy monotonicity. It is striking that this widely used and popular method fails to satisfy what is arguably one of the most compelling of the criteria.

Proposition 4.4. *Hare's method satisfies majority and Pareto but not monotonicity, Condorcet, anti-Condorcet, or independence.*

Proof. A candidate at the top of a majority of the preference lists wins Hare's method, because they remain at the top of a majority of the preference lists even when candidates are eliminated and therefore never face elimination themselves. Hence Hare's method satisfies majority. At the other extreme, a candidate at the top of no preference lists is always eliminated in the first round of the Hare method. Now if candidate A is ahead of candidate B on every preference list, candidate B has no first-place votes and therefore suffers elimination in the first round. Hence Hare's method satisfies Pareto.

Profiles 4.10 and 4.11 illustrate the surprising fact that Hare's method violates monotonicity.

6	5	4	2
A	C	B	B
B	A	C	A
C	B	A	C

Profile 4.10 "before"

6	5	4	2
A	C	B	A
B	A	C	B
C	B	A	C

Profile 4.11 "after"

Facing the "before" profile, the Hare method eliminates candidate C, who has only 5 first-place votes, in the first round, and candidate A beats candidate B in the final head-to-head match-up, 11 to 6. Now imagine that the final two voters move A up on their preference lists to the top position. The result is the "after" profile, in which B is eliminated in the first round, and C subsequently defeats A in the final round, 9 to 8. Hence moving A up on some lists has caused A to lose the Hare election.

Profile 4.1 shows that Hare's method violates the Condorcet and anti-Condorcet criteria. Candidate B (the anti-Condorcet candidate) is the Hare winner there, as candidate A (the Condorcet candidate) and candidate C are eliminated in the first round.

To see that the Hare method violates independence, consider the before-and-after tabulated Profiles 4.12 and 4.13.

2	2	1
B	A	A
A	C	B
C	B	C

Profile 4.12 "before"

2	2	1
B	C	A
A	A	B
C	B	C

Profile 4.13 "after"

In Profile 4.12, Hare's method eliminates candidate C in the first round and candidate B in the second round, leaving candidate A as the winner. In fact, A has a majority of the first-place votes in Profile 4.12, so is guaranteed to be the unique winner of the Hare election. Note that B loses in Profile 4.12. In Profile 4.13, every voter expresses the same relative preference regarding candidates A and B that they expressed in Profile 4.12. Facing Profile 4.13, the Hare method eliminates A in the first round, and candidate B wins the second round over C. Thus B is promoted to winner by the movement of C. This violates independence. □

Proposition 4.5. *Coombs's method satisfies Pareto but not majority, monotonicity, Condorcet, anti-Condorcet, or independence.*

Proof. If candidate A is ahead of candidate B on every preference list, then A will never acquire last-place votes until B is eliminated. Hence as long as B remains in the mix, A will survive each round while at least one other candidate is eliminated. Therefore the Coombs election cannot come to an end until B is eliminated. This shows that Coombs's method satisfies the Pareto criterion.

Profile 4.4 shows that a candidate with the majority of first-place votes can lose an antiplurality election, hence getting eliminated in the first round of Coombs's method. Hence Coombs violates majority and therefore the Condorcet criterion.

We leave as an exercise (**4.13**) the argument that Coombs's method violates monotonicity.

Profile 4.14 demonstrates that Coombs's method violates the anti-Condorcet criterion. There, candidates B and C are eliminated simultaneously in the first round, and so the anti-Condorcet candidate A leapfrogs to victory.

C	C	C	B	B
B	A	A	A	A
A	B	B	C	C

Profile 4.14

Finally, we leave as an exercise (**4.14**) the argument that Coombs's method violates independence. □

There is a phony argument that purports to show that Coombs's method does in fact satisfy the anti-Condorcet criterion. Think about how an anti-Condorcet candidate might fare in a Coombs election. Such a candidate may stave off elimination for several rounds. But even if she manages to survive until the last round, she will certainly be defeated in the final head-to-head match-up, since anti-Condorcet candidates are defeated in all head-to-head match-ups. So it appears that an anti-Condorcet candidate cannot win a Coombs election. The flaw in this argument is revealed in Profile 4.14. Coombs elections need not have a final run-off between two candidates, and an anti-Condorcet candidate can indeed eke out a Coombs victory if other candidates are eliminated simultaneously in the final round.

Copeland's method, relying as it does on information about head-to-head match-ups, is quite different from the positional and elimination methods.

Proposition 4.6. *Copeland's method satisfies majority, Condorcet, anti-Condorcet, monotonicity, and Pareto, but not independence.*

Proof. In the Copeland system, a Condorcet candidate obtains a perfect score and the other candidates do not, since the Condorcet candidate defeats each of them in their head-to-head match-up. Hence Copeland's method satisfies Condorcet and therefore majority. An anti-Condorcet candidate gets no points at all in Copeland's method and therefore loses. Hence Copeland's method satisfies anti-Condorcet.

To see that Copeland's method satisfies monotonicity, note that moving candidate A up on a preference list can never harm candidate A's standing in any head-to-head match-up. Such a change has no effect on the head-to-head match-ups not involving A. Thus such a change can never decrease A's Copeland score, nor increase any other candidate's Copeland score. If A was the Copeland winner before such a change, she remains the Copeland winner afterwards as well.

To see that Copeland's method satisfies Pareto, consider a profile in which candidate A is above candidate B on every preference list. In such a situation, if B can defeat another candidate in a head-to-head election, so can A. In other words, A gets a Copeland point (or half-point) for every Copeland point (or half-point) that B gets. Moreover, A defeats B head-to-head, so A gets one point that B does not get. Hence A has a Copeland score that is at least one more than the Copeland score of B. Candidate B therefore cannot win the Copeland election.

To see that Copeland's method violates independence, it is enough to quote Proposition 3.13, which asserts that any method satisfying Condorcet must violate independence. For those who would like to see an example of before-and-after profiles that illustrate the dependence on an irrelevant alternative, we offer Profiles 4.15 and 4.16.

B	A	A
C	B	C
A	C	B

Profile 4.15 "before"

B	A	C
C	B	A
A	C	B

Profile 4.16 "after"

In Profile 4.15, the Copeland scores of the candidates A, B, and C are 2, 1, and 0, respectively, and A is the unique Copeland winner. In changing to Profile 4.16, no voter changes their relative ranking of A and B. Facing that profile, candidates A, B, and C all obtain Copeland scores of 1 and there is a three-way tie in a Copeland election. Thus B becomes a Copeland winner as C becomes more popular. This violates independence. □

We now come to the method of Duncan Black. **Black's method** is the method that selects the Condorcet candidate when there is one and selects the Borda count winner when there isn't. It is one of many hybrid methods that one can invent.

Proposition 4.7. *Black's method satisfies majority, Condorcet, anti-Condorcet, monotonicity, and Pareto, but not independence.*

Proof. By its definition, Black's method selects a Condorcet candidate if there is one as its unique winner. Therefore Black's method satisfies the Condorcet and the majority criteria. To see that Black's method also satisfies anti-Condorcet, note that an anti-Condorcet candidate cannot be the Condorcet candidate, nor can he be a Borda count winner, because the Borda count method satisfies anti-Condorcet. So the Black method inherits the anti-Condorcet property from the Borda count method. To see that the Black method is monotone, consider the impact on a Black winner from a change in the profile that moves this Black winner up on some preference lists. If the Black winner was a Condorcet candidate in the first place, she remains one after the change, and she remains the Black winner because of this. If the Black winner was, on the other hand, not a Condorcet candidate, then she must have been the Borda count winner. After the change, she remains the Borda count winner, since the Borda count method is monotone. Moreover, no other candidate could have become the Condorcet candidate via this change, because the improved standing of the Black winner on some preference lists benefits no other candidate in any head-to-head match-ups. Therefore Black's method is monotone.

To see that Black's method is Pareto as well, imagine a profile in which candidate A is favored over candidate B by all voters. Candidate B cannot be the Condorcet candidate in this situation, since she loses to A. Moreover, she cannot be the Borda count winner, because A must have more Borda count points than B. So B has no prospects for winning a Black election.

Finally, Black's method violates independence, because it satisfies Condorcet. (Profiles 4.15 and 4.16 provide the concrete example.) □

It may surprise some readers to learn that dictatorships satisfy a number of our criteria. In fact, they satisfy the elusive decisiveness criterion that few of the other methods satisfy. It is also one of relatively few methods that we have seen that satisfies independence. This should not be understood to mean that dictatorship is a reasonable way to conduct democratic elections. One must remember that dictatorships violate anonymity, and anonymity seems fundamental to modern notions of fairness.

Proposition 4.8. *Dictatorship methods satisfy monotonicity, Pareto, and independence but not Condorcet, anti-Condorcet, or majority.*

Proof. If candidate A is at the top of the dictator's preference list, then moving A up on some preference lists cannot displace him from the top spot on the dictator's list. So dictatorships are monotone. Moreover, if candidate A is higher on every preference list than candidate B, then in particular A is higher on the dictator's preference list than B. Hence B cannot hold the top spot on the dictator's list and so cannot win the dictatorship election. This shows that dictatorships satisfy Pareto. If candidate A is at the top of the dictator's preference list and if a second profile has every voter placing A and B in the same relative position, then that second profile cannot have B at the top of the dictator's preference list. So B cannot win the dictatorship election facing the second profile. It follows that dictatorships satisfy independence.

In Profile 4.17, assume that the third voter is a dictator. Candidate A has a majority of first-place votes and is the Condorcet candidate, while Candidate B is the anti-Condorcet candidate. Yet B is named the winner by the dictator.

A	A	B
B	B	A

Profile 4.17

This shows that dictatorships violate the Condorcet, the anti- Condorcet, and the majority criteria. □

Proposition 4.9. *The all-ties method and the monarchy method satisfy monotonicity and independence but not Condorcet, anti-Condorcet, majority, or Pareto.*

Proof. The all-ties method and the monarchy method are what are known as constant functions, meaning that they provide output that pays no attention whatsoever to the input. The winner of the election does not depend on how the electorate votes. Such methods satisfy monotonicity and independence, because no candidate can win facing one profile while losing facing another. However they clearly violate Condorcet, anti-Condorcet, majority, and Pareto, because their output ignores the profile, which may well have a Condorcet candidate, an anti-Condorcet candidate, a candidate holding a majority of the first-place votes, or one candidate preferred to another by every voter. □

We leave as exercises (**4.15** and **4.16**) the analysis, in the spirit of the propositions above, of the bloc voting method and of the agenda method.

4.3 Summarizing the Results

The results from this chapter are summarized in Table 4.1. Included there also are results about the bloc voting method, mentioned in Chapter 1 but easily extended to elections with more than two candidates, and the agenda method, defined in Chapter 2.

Table 4.1 This table provides the answers to 120 questions of the form "Does method X satisfy criterion Y?" The rows are indexed by 12 methods. The columns are indexed by 10 criteria: Anonymity, Neutrality, Unanimity, Decisiveness, Majority, Condorcet, Anti-Condorcet, Monotonicity, Pareto, and Independence.

	Ano	Neu	Una	Dec	Maj	Con	AC	Mon	Par	Ind
Plurality	Yes	Yes	Yes	No	Yes	No	No	Yes	Yes	No
Antiplur.	Yes	Yes	Yes	No	No	No	No	Yes	No	No
Borda	Yes	Yes	Yes	No	No	No	Yes	Yes	Yes	No
Hare	Yes	Yes	Yes	No	Yes	No	No	No	Yes	No
Coombs	Yes	Yes	Yes	No	No	No	No	No	Yes	No
Copeland	Yes	Yes	Yes	No	Yes	Yes	Yes	Yes	Yes	No
Black	Yes	Yes	Yes	No	Yes	Yes	Yes	Yes	Yes	No
Dictator	No	Yes	Yes	Yes	No	No	No	Yes	Yes	Yes
All-ties	Yes	Yes	No	No	No	No	No	Yes	No	Yes
Monarch	Yes	No	No	Yes	No	No	No	Yes	No	Yes
Bloc	No	Yes	Yes	No	No	No	No	Yes	Yes	No
Agenda	Yes	No	Yes	No	Yes	Yes	Yes	Yes	No	No

What does Table 4.1 illustrate? First, notice that no method satisfies all 10 of the criteria. This should not surprise us, because Proposition 3.12 shows that no method can simultaneously satisfy anonymity, neutrality, and decisiveness. If you insist on decisiveness, you must tolerate a violation of anonymity (like a dictatorship) or a violation of neutrality (like a monarchy). Most people are willing to give up decisiveness in favor of the very compelling notions of anonymity and neutrality.

Even if we are willing to dispense with decisiveness, we still face a difficulty because Proposition 3.13 shows that no method can satisfy the Condorcet and independence criteria simultaneously. Table 4.1 offers three methods that satisfy the Condorcet criterion and three methods that satisfy independence. Perhaps surprisingly, all of the first five methods in Table 4.1, which are familiar, natural, and in wide use, violate both Condorcet and independence. If we were to agree to dispense with the Condorcet criterion, we might hope for a desirable method that satisfies independence. In particular, we might hope for a method that would satisfy independence along with the fundamental criteria of anonymity and Pareto. But even this turns out to be more than we can hope for. The impossibility theorem of Kenneth Arrow asserts that no such method exists.

4.4 Exercises and Problems

4.1. The **vote-for-two method** works as follows: Candidates get a point whenever a voter ranks them first or second. The candidate with the most points is declared to be the winner (or if several candidates tie for the most points, they are all declared to be winners).
(a) Does the vote-for-two method satisfy the Condorcet criterion?
(b) Does the vote-for-two method satisfy the anti-Condorcet criterion?
(c) Does the vote-for-two method satisfy the Pareto property?
(d) Is the vote-for-two method independent?

4.2. The following social choice function is called the **COP method**: If there is a Condorcet candidate, then that candidate is declared to be the winner. Otherwise the COP method selects the plurality winner(s). (COP is an acronym for Condorcet Otherwise Plurality.)
(a) Does the COP method satisfy the Condorcet criterion?
(b) Does the COP method satisfy the anti-Condorcet criterion?
(c) Is the COP method independent?

4.3. The following social choice function is called the **French method** because it models the way the president is elected in France: From the profile, extract the two candidates who have the greatest and second greatest number of first-place votes. The winner of the French method is the winner of a head-to-head contest — call it the run-off — between these two candidates. (If there is a tie for second greatest number of first-place votes, allow all these candidates to go on to the second round and conduct the run-off via the Copeland method.)
(a) Does the French method satisfy the Condorcet criterion?
(b) Does the French method satisfy the anti-Condorcet criterion?
(c) Is the French method monotone?

4.4. The **anti-French method** works as follows: From the profile, extract the two candidates who have the least and second least number

of last-place votes. The winner of the anti-French method is the winner of a head-to-head contest — call it the run-off — between these two candidates. (If there is a tie for second least number of last-place votes, allow all these candidate to go on to the second round and conduct the run-off via the Copeland method.)
(a) Does the anti-French method satisfy the Condorcet criterion?
(b) Does the anti-French method satisfy the anti-Condorcet criterion?
(c) Is the anti-French method monotone?

4.5. The **single elimination tournament** (SET) social choice function works as follows, assuming that there are precisely 4 candidates A, B, C, and D, and an odd number of voters: In the first round, determine the winner of a head-to-head match-up between A and B and the winner of a head-to-head match-up between C and D. In the second round, have these two winners face each other in a head-to-head match-up, the winner of which is elected.
(a) Does SET satisfy the Condorcet criterion?
(b) Does SET satisfy the neutrality criterion?

4.6. An **anti-dictatorship** is a social choice function that identifies a particular voter (called the anti-dictator) and chooses as the unique winner the candidate on the bottom of the preference list of the anti-dictator. When a member of a committee is so provocative and irritating that the other committee members retaliate by favoring everything he opposes, the effect is that the provocateur becomes an anti-dictator.
(a) Is an anti-dictatorship monotone?
(b) Is an anti-dictatorship independent?
(c) Does an anti-dictatorship satisfy the Pareto criterion?
(d) Does an anti-dictatorship satisfy the Condorcet criterion?

4.7. The **instant recall** method works as follows. If there is a candidate with the majority of the first-place votes, that candidate is declared the unique winner. Otherwise, the candidate with the plurality of the first-place votes is eliminated from the slate and the process is repeated among the remaining candidates. (Ignore the possibility that there may be a tie for plurality winner.) The method is inspired by the recall elections that are permitted for California gubernatorial elections, which in effect leave a system where plurality winners lacking majority support are recallable, and so the ultimate holder of the governor's office, if the electorate so chooses, is the first candidate able to withstand a recall (the first candidate to have majority support) after prior plurality winners are eliminated (i.e., recalled). Arnold Schwarzenegger took office in California via a recall election.
(a) Does instant recall satisfy anonymity?
(b) Does instant recall satisfy neutrality?

(c) Does instant recall satisfy the Condorcet criterion?
(d) Does instant recall satisfy the anti-Condorcet criterion?
(e) Does instant recall satisfy monotonicity?
(f) Does instant recall satisfy Pareto?
(g) Does instant recall satisfy the majority criterion?
(h) Does instant recall satisfy the independence criterion?
(i) Does instant recall satisfy the top criterion (from Problem **4.11** below)?

4.8. The **Pareto method** is a social choice function that names a candidate a winner if there is no other candidate preferred by every voter. In other words, a candidate is a winner if and only if the candidate is a Pareto optimal candidate.
(a) Does the Pareto method satisfy the Pareto criterion?
(b) Is the Pareto method monotone?
(c) Does the Pareto method satisfy the Condorcet criterion?

4.9. Call a social choice function **robust** if the addition of a new candidate ranked last by all voters would not alter the outcome of an election.
(a) Is the plurality method robust?
(b) Is the antiplurality method robust?
(c) Is the Borda count method robust?

4.10. A social choice function satisfies the **weak Pareto criterion** if, whenever every voter places one candidate above another, say ranking candidate A over candidate B, then candidate B cannot be the unique winner. (This differs from the Pareto property owing to the words "the unique".)
(a) Explain why any method that satisfies Pareto also satisfies weak Pareto.
(b) Explain why the antiplurality method satisfies weak Pareto.
(c) Explain why the agenda method violates weak Pareto.

4.11. We say that a social choice function satisfies the **top criterion** if every winner must receive at least one first-place vote.
(a) Does plurality satisfy the top criterion?
(b) Does antiplurality satisfy the top criterion?
(c) Does Borda count satisfy the top criterion?
(d) Does the Hare method satisfy the top criterion?
(e) Does the Coombs method satisfy the top criterion?
(f) Does the Copeland method satisfy the top criterion?
(g) Does a dictatorship satisfy the top criterion?

4.12. A social choice function satisfies the **bottom criterion** if, whenever a candidate is not at the bottom of the preference list of any voter, that candidate is a winner.

(a) Does the plurality method satisfy the bottom criterion?
(b) Does the antiplurality method satisfy the bottom criterion?
(c) Explain why it is impossible for a social choice function with three or more candidates to satisfy both the bottom criterion and the Pareto criterion.

4.13. Show that the Coombs method fails to be monotone, as Table 4.1 asserts.

4.14. Show that the Coombs method fails to satisfy independence, as Table 4.1 asserts.

4.15. Verify that the bloc voting method satisfies neutrality, unanimity, monotonicity, and Pareto, but not anonymity, decisiveness, majority, Condorcet, anti-Condorcet, or independence, as Table 4.1 asserts. To be definite, assume there are 25 voters who agree to pool their votes in blocs of size 9, 7, 5, 3, and 1, respectively, and assume there are 3 candidates.

4.16. Verify that the agenda method satisfies anonymity, unanimity, majority, Condorcet, anti-Condorcet, and monotonicity, but not neutrality, decisiveness, Pareto, or independence, as Table 4.1 asserts. You may assume that the number of voters is odd, in order that no ties occur as the process of head-to-head competitions evolves.

5

Arrow's Theorem

"Dictatorship naturally arises out of democracy, and the most aggravated form of tyranny and slavery out of the most extreme liberty."
— Plato

5.0 Scenario

In a certain election with 100 voters, you serve as campaign manager for one of the 5 candidates. Part of your job is to persuade voters to vote for your candidate. Your staff keeps a list that tracks voters who have made a commitment to support your candidate (by ranking them first). As soon as you recognize that this list is sufficient to guarantee that your candidate will be the unique winner of the election — no matter how the other voters feel and no matter how your group of voters ranks the remaining candidates — you treat your staff to a champagne party.

Assuming that the election uses the plurality method, how long must your list be before you can break open the champagne? What if the election uses the Hare method? What if the election uses the Borda count method? What if the election uses the antiplurality method? What if the election uses a dictatorship, but it is a secret to you which voter has the dictatorial power? What if the election uses a dictatorship and the identity of the dictator is known to all?

5.1 The Condorcet Paradox

The essential problem in the theory of social choice is to take individual preferences and distill from them a societal preference. We have not been able to find a way to do this that passes every test of appropriateness. What seems to be the underlying difficulty?

In some sense, the crucial obstacle to producing a perfect social choice function is the specter of a profile like this one.

A	C	B
B	A	C
C	B	A

Profile 5.1

It is a very simple profile, with just three voters and three candidates, but it completely captures the essential difficulty. In Profile 5.1, A is preferred by a majority of voters to B, B is preferred by a majority of voters to C, and yet C is preferred by a majority of voters to A. Hence even while we insist on rationality of voters — which means that the voters themselves must have transitive preferences — we can end up with an electorate that collectively fails to have transitive preferences. Candidates A, B, and C form a strange cycle in which the electorate prefers each candidate to the next candidate around the cycle.

The existence of these strange cycles is a phenomenon known as the **Condorcet paradox**, and it is at the heart of the theory of voting. Anytime that a profile fails to have a Condorcet candidate or an anti-Condorcet candidate, there is within that profile one of these strange cycles: a choice of three candidates A, B, and C with A preferred to B, B preferred to C, and C preferred to A. This means that even though every voter must rank the candidates individually, there is no way to rank the candidates on behalf of the collective electorate that respects majority rule. Put another way, even if we insist on rational voters, we may have an electorate that is collectively irrational.

One can think of Profile 5.1 in a different way that calls into question our underlying assumption of voter rationality. Instead of thinking of the columns as three separate voters, think of them as a ranking of three candidates by a single person rating three different aspects of the candidates. For example, imagine that the first column represents one's personal ranking of three candidates for president from the point of view of foreign policy, the second column represents one's ranking from the point of view of domestic policy, and the third column represents one's ranking from the point of view of fiscal policy. There is nothing irrational about having different views about the same candidates depending on what issues you are considering. Neither does it seem irrational to hold the view that if one candidate seems favorable to you when compared to another from at least two of these three points of view, then you prefer that candidate to the other. It then cannot be regarded as irrational for a single voter to favor A over B, B over C, and C over A simultaneously. This is the Condorcet paradox in full view. Irrationality of individual voters — intransitivity of preferences — is forced upon us by Profile 5.1.

Imagine that the US House of Representatives wishes to implement an important policy objective. Three closely related bills are introduced; call them A, B, and C. The members of the House all agree that any of the three bills will accomplish the objective, but there remains some disagreement about which does so most effectively. The House meets to decide which bill to pass. A straw poll shows that the three measures are preferred by the members according to the following tabulated profile:

145	145	145
A	C	B
B	A	C
C	B	A

Measure C is debated first, but there is an outcry for measure B, preferred to C by an overwhelming majority (by 2/3 of the members). Measure B is then substituted, but there is an outcry for measure A, preferred by a 2/3 supermajority to measure B. Finally measure A is brought to the House floor, and what occurs? An overwhelming majority demands that measure C replace measure A. The Condorcet paradox makes it difficult for the House to proceed in this situation. (By the way, it is not essential that in this example the House is divided into exact thirds. The paradox would be present even if the three appearances of the number 145 were changed to, say, 200, 150, and 85. All that is necessary is that each of these numbers represents less than half of the House.)

Even more curious is the following variant of the previous example:

145	145	145
D	C	B
E	D	C
F	E	D
G	F	E
A	G	F
B	A	G
C	B	A

Imagine that D represents a current law, C the law with an amendment, B the law with the same amendment and an additional amendment, and A the law with both preceding amendments and a further additional amendment. (E, F, and G represent other options.) The current law is rather well liked, but when the amendment is offered, it passes by a 2/3 majority, yielding alternative C. Next, the additional amendment is considered, and it passes again by a 2/3 majority, yielding alternative B. Finally, the further additional amendment is considered and passed, yielding alternative A. Each of the votes required just a simple majority to pass, and none of the votes were close. In the end, the chamber has amended the law three times and ends up with a law that is regarded poorly by every representative. It is the unanimous view of the chamber that, after amending (improving?) the law three times, the outcome is worse than the original law. This counterintuitive phenomenon is another appearance of the Condorcet paradox.

We now turn to our central result on social choice functions, the impossibility theorem of Kenneth Arrow. It turns out that every proof of every version of this theorem involves some sort of encounter with the Condorcet paradox.

5.2 Statement of the Result

In 1951, economist Kenneth Arrow (1921–) presented a doctoral dissertation that included what is now regarded as one of the central theorems in the theory of voting. Arrow was awarded the Nobel Prize in economics in 1971 for this and other seminal work. To explain what the theorem says, we need to examine carefully the independence criterion. Recall that a social choice function satisfies the **independence criterion** (or is **independent**) if the following holds: Suppose that there are two profiles between which no voter changes his mind about whether candidate A is preferred to candidate B. (In other words, if a voter puts A ahead of B in one profile, then he puts A ahead of B in the other, and vice versa.) Suppose also that when facing the first profile, the method chooses candidate A but not candidate B as a winner. Then the method must not choose candidate B as a winner when facing the second profile.

To understand independence, it may be helpful to imagine that candidates A and B are the principal candidates in the race while the other candidates are minor, insignificant, or less popular. Independence requires that the decision about whether to elect A or B must depend only on whether voters prefer A over B or vice versa. Imagine that A wins and B loses in the first profile. If the social choice function is independent, then no changes in opinions of the voters about other candidates should influence this. If news about another candidate influences voter opinion, but in a way that does not change the relative standing of A and B on preference lists, then it would be inconsistent for the electorate to switch their decision and favor B over A. The decision between A and B should be independent of the electorate's sentiments about C. It is certainly possible that favorable news about candidate C could cause candidate A to become a loser (with C winning). But independence demands that no news about candidate C should cause candidate B to become a winner.

We are now ready to state a version of the theorem of Arrow.

Theorem 5.1. **(*Arrow*)** *If a social choice function with at least three candidates satisfies Pareto and independence, then it must be a dictatorship.*

At first blush, the theorem seems to be identifying dictatorships as the optimal social choice functions, because it is unique among all possible methods in meeting certain criteria for desirability and reasonableness. But of course dictatorship is not a fair way to run democratic elections. After all, dictatorships violate the anonymity criterion that seems to be essential in any democratic voting system. Instead, the following interpretation may be more realistic: Arrow's Theorem says that if you adopt Pareto and independence as essential properties for a social choice function, then you have painted yourself into an uncomfortable corner, where only dictatorships are available. Therefore it is necessary to discard at least one of these two properties. Along these lines, one can rephrase Arrow's theorem to be a result expressing impossibility.

Corollary 5.2. *It is impossible for a social choice function with at least three candidates to be Pareto, independent, and nondictatorial.* □

The adjective "nondictatorial" here means nothing except "not a dictatorship". Because dictatorships violate anonymity, any social choice function that is anonymous is nondictatorial. Hence Corollary 5.2 implies that Pareto, independence, and anonymity are mutually incompatible. Those who search for an ideal anonymous social choice function must abandon the hope of finding one that is both Pareto and independent.

When we say that it is "impossible" to find a method satisfying certain properties, it does not mean merely that none of our favorite methods have the required properties, or that no method has yet been discovered that has the required properties. It means that no method will ever be discovered, that no person, however clever, will ever be able to devise such a method. It means that the search for a method satisfying these three criteria is over.

We devote the remainder of this chapter to a proof of the theorem. We first imagine a social choice function that satisfies Pareto and independence. Then we deduce various facts about such a function until eventually we conclude that it must be a dictatorship. The results that we deduce as we work our way along seem strange or, some would say, hard to believe. Why is that? It is because we are conducting a certain thought experiment when we imagine some method that satisfies Pareto and independence. We imagine that there is such an example, one that is not merely a dictatorship, but in the end we learn that there are no such examples! So it should not alarm us along the way when we learn that the example we are conjuring up has some strange properties. The world we create when we imagine the existence of an example that doesn't in fact exist is likely to be strange. Eventually our imagined world is shown to contain a contradiction, and it is then that we learn that our example doesn't exist at all.

5.3 Decisiveness

We begin by showing that a social choice function satisfying the criteria in the hypothesis of Arrow's Theorem cannot ever output a tie.

Lemma 5.3. ***The decisiveness lemma.*** *A method with at least three candidates that satisfies Pareto and independence must be decisive.*

Proof. Suppose that candidates A and B both win when facing profile P. Our goal is to obtain a contradiction from this supposition. In particular, we will infer from this supposition that there is another profile with no winner. But this contradicts the assumption of universal domain.

Let $\boldsymbol{X}$ be the set of voters who place A ahead of B in profile P, and let $\boldsymbol{Y}$ be the complementary set of voters who place B ahead of A. We illustrate this loosely with a graphical representation of P, where the entries in the first row there indicate the set of voters whose preference order is listed below that entry.

$\boldsymbol{X}$	$\boldsymbol{Y}$
⋮	⋮
A	B
⋮	⋮
B	A
⋮	⋮

Profile P

Note that neither $\boldsymbol{X}$ nor $\boldsymbol{Y}$ can be the empty set, because if all voters prefer candidate A to B, then the Pareto property does not permit B to be a winner, and if all voters prefer B to A, then the Pareto property does not permit A to be a winner.

Let C be some candidate different from A and B. That such a candidate C exists is due to the assumption that we have at least three candidates on the slate. We now fashion a new profile that we call Q. In that profile, every voter in $\boldsymbol{X}$ places A in first place, C in second place, and B in third place. All remaining candidates, if any, are placed below these three. (It makes no difference in what order they are placed.) Every voter in $\boldsymbol{Y}$ places C in first place, B in second place, A in third place, and the remaining candidates below in some order. We illustrate this schematically in the following chart, labeled Profile Q.

$\boldsymbol{X}$	$\boldsymbol{Y}$
A	C
C	B
B	A
$\vdots$	$\vdots$

Profile Q

Because every voter rates C higher than B and because the method satisfies the Pareto criterion, B must lose when facing Profile Q. Since every voter in Profile Q places A relative to B exactly as they do in Profile P, and because we are assuming that the method satisfies independence, A must lose also when facing Profile Q. After all, if A won and B lost when facing Profile Q, then independence would require that B lose also when facing Profile P. But B wins when facing Profile P by assumption.

So who wins in Profile Q? We claim that candidate C is the unique winner. We have seen that neither A nor B wins. Could some candidate other than A or B, call her D, win here? No, because every voter prefers C to D, and so the Pareto criterion implies that D must lose. Hence C is in the end the only candidate who can possibly win in Profile Q.

We now contemplate a third profile that we call R, which is just like Profile Q except that candidates B and C are interchanged on the preference list of every voter.

X	*Y*
A	B
B	C
C	A
⋮	⋮

Profile R

Applying the independence property to candidates A and C in Profiles Q and R, we see that A cannot win in Profile R. Comparing Profiles P and R, since A wins in P and loses in R, B must also lose in Profile R. The Pareto criterion implies that no candidate other than A, B, or C can win in Profile R. Hence C is the unique winner in Profile R. But every voter in Profile R prefers candidate B to C, so by the Pareto criterion, C cannot win. This is a contradiction. It follows that our original assumption of a tie between A and B cannot occur. □

Already we see that the world of methods that satisfy Pareto and independence is quite strange. After all, none of the familiar or natural methods is decisive. Also, according to Proposition 3.12, a method that is decisive must violate anonymity or neutrality. It is important to recognize that in general in this chapter we do not assume anonymity and neutrality. If we did, Lemma 5.3 would be the end of the story, because Lemma 5.3 has the following corollary.

Corollary 5.4. *It is impossible for a method to satisfy Pareto, independence, anonymity, and neutrality.*

Proof. The first two criteria listed force decisiveness, while the last two criteria listed are incompatible with decisiveness. □

5.4 Proving the Theorem

We now use Lemma 5.3 to provide a proof of Arrow's Theorem. The decisiveness lemma comes into play by providing us with a stronger version of independence. In our original formulation of the independence criterion, we used the phrase "the method chooses candidate A but not candidate B as a winner." In the presence of the decisiveness lemma, this becomes a redundancy. The phrase "the method chooses candidate A as a winner" means the same thing. Put another way, Lemma 5.3 has the following consequence.

Lemma 5.5. *Suppose a social choice function with at least three candidates satisfies Pareto and independence. Suppose also that there are two profiles between which no voter changes his mind about whether candidate A is preferred to candidate B. If A wins in the first profile, then B cannot win in the second profile.* □

The idea of the proof of Arrow's Theorem is that, for every pair of candidates, some single voter can force one of these candidates to lose by ranking the other candidate higher. We say that a voter has **dictatorial control** for A over B if, whenever that voter prefers A to B, no matter what other voters prefer, B will lose.

Proof of Theorem 5.2. We first show that, for every pair of candidates A and B, there is a voter who has dictatorial control for A over B.

To this end, fix candidates A and B and let C be any other candidate. Imagine a profile in which C, A, and B are the first, second, and third choices, respectively, of every voter, with the remaining candidates, if any, placed further down the preference lists. The Pareto criterion implies that C must be the unique winner in such a profile.

Now imagine that the voters, one by one, decide to lower C to third place, raising A and B to first and second places, respectively. After all the voters have changed their votes in this way, A will certainly be the winner, again by the Pareto criterion. Also, during the process, B will never be the winner, since in each of these profiles every voter prefers A to B. Moreover, no candidate other than A, B, or C could be a winner in any of these profiles. So at some critical moment, one voter makes this change and the winner changes from C to A. Call that voter v. Let $\boldsymbol{X}$ be the set of voters other than v who have not yet changed the position of candidate C, and let $\boldsymbol{Y}$ be the set of voters other than v who have already changed the position of C. Here are schematic profiles illustrating the situation before and after this moment, with the critical voter v represented by the middle column.

$\boldsymbol{X}$	v	$\boldsymbol{Y}$
C	C	A
A	A	B
B	B	C
⋮	⋮	⋮

"before"

$\boldsymbol{X}$	v	$\boldsymbol{Y}$
C	A	A
A	B	B
B	C	C
⋮	⋮	⋮

"after"

We claim that this voter has dictatorial power for A over B. To see this, consider *any* Profile P in which voter v places A over B (we do not show Profile P here). We want to show that B loses. To this end, define a new profile, called Profile Q as follows: Every voter in $\boldsymbol{X}$ places C in first place and A and B in the next two places, ordering A and B exactly as they ranked them in Profile P. Every voter in $\boldsymbol{Y}$ places C in third place and A and B in the top two places, ordering A and B exactly as they ranked them in Profile P. And voter v ranks A, C, and B in that order as top choices. Any other candidates are ranked below the top three positions by all voters. This is pictured here.

$\boldsymbol{X}$		v	$\boldsymbol{Y}$	
C	C	A	A	B
A	B	C	B	A
B	A	B	C	C
⋮	⋮	⋮	⋮	⋮

Profile Q

We now appeal to the independence criterion repeatedly, using Lemma 5.5. Since candidate C wins in the "before" profile, and all voters rank B and C in Profile Q just as they do in the "before" profile, B must lose in Profile Q. Since A wins in the "after" profile, and all voters rank A and C in Profile Q just as they do in the "after" profile, C must lose in Profile Q. Since no candidate besides A, B, and C can win in any of these profiles by the Pareto criterion, candidate A must win in Profile Q. But all voters rank A and B in Profile Q just as they do in profile P. Therefore B loses in Profile P. We have thus shown that our special voter v has dictatorial power for A over B.

We now show that there must be a single dictator whose first choice is always selected by the method. For each pair of candidates, we have shown that there is a voter with dictatorial power for the first candidate over the second, but we may worry that such a voter for one pair of candidates and such a voter for another pair of candidates may be different voters. In fact, this cannot happen. First note that a voter who has dictatorial power for A over B must also have dictatorial power for B over A. Otherwise, no candidate at all could win in a profile with the first voter ranking A first and B second, the second voter ranking B first and A second, and every other voter ranking A and B first and second (in either order). So we may prefer to use language that suggests this symmetry between the two candidates. From here on, we say that a voter has dictatorial power "between A and B" in place of "for A over B".

Finally, if one voter v has dictatorial power between A and B and another voter w has dictatorial power between B and C, then consider the following profile:

v	w	all others
C	B	C
A	C	B
B	A	A
⋮	⋮	⋮

Profile P

Voter v prevents B from winning, while voter w prevents C from winning. The Pareto criterion applied to A and C prevents A from winning. And the Pareto criterion applied to A and any other candidate prevents any other candidate from winning. This contradicts our assumption that every profile must yield at least one winner, so we have shown that we cannot have one voter with dictatorial power between A and B and another between B and C. Hence the voter with dictatorial power between A and B is the same as the voter with dictatorial power between B and C. Continuing in this way, that same voter has dictatorial power between C and any other candidate D.

The conclusion is that one voter has dictatorial power between any two candidates. This candidate is clearly a dictator, since any candidate not ranked first by this voter must lose. Hence our social choice function is a dictatorship. □

The Pareto criterion is somewhat more natural and easier to test than the independence criterion. In some cases, one may be able to recognize that a social choice function satisfies the Pareto criterion, but one has difficulty checking whether it is independent. In these cases, Arrow's Theorem comes to the rescue. Unless the method under consideration is a dictatorship, once it is determined that the method is Pareto, it cannot also be independent.

Many people would argue that the Pareto criterion is an essential criterion for a reasonable social choice function but that dictatorship is not a reasonable social choice function. According to Arrow's Theorem, such people must then be resigned to discard the independence criterion. No method that meets their notion of reasonableness can satisfy the independence criterion.

Giving up on the independence criterion is akin to tolerating third-party or minor candidate interference in elections. It is a feature of plurality elections that minor candidates can affect the outcomes of elections by stealing votes from other candidates. Often, methods like Hare and Borda count are suggested precisely to avoid this kind of interference. But Arrow's Theorem tells us that no reasonable social choice function can completely solve the difficulty. Neither the Hare method nor the

Borda count method satisfy independence, despite the best intentions of those advocating for them. But neither does any other Pareto method, except dictatorship.

In the next chapter, we investigate a number of other models of collective voting and decision-making, some involving only slight alterations of the assumptions we have made so far while others involve completely different hypotheses.

5.5 Exercises and Problems

5.1. Give examples to show that Corollary 5.2 fails if any of the three criteria mentioned there is omitted.

5.2. Consider the method that declares candidates to be winners if and only if they have at least one first-place vote. (This method, which we dubbed the "nomination method" in Problem **2.17**, often leads to ties.)
(a) Does this method satisfy the Pareto criterion?
(b) Does this method satisfy the independence criterion?

5.3. Consider the social choice function that works as follows: If one candidate is at the top of every voter's preference list, that candidate is the unique winner. Otherwise, the method declares a tie among all candidates.
(a) Is this method anonymous?
(b) Does this method satisfy the independence criterion?
(c) Does this method satisfy the Pareto criterion?

5.4. Consider a social choice function that operates as follows: Three special voters are identified. The first is the **near-dictator** and the other two are close advisors. The method selects as the winner the candidate at the top of the preference list of the near-dictator unless there is another candidate who is at the top of the preference lists of both close advisors, in which case the other candidate is the unique winner.
(a) Is this method Pareto?
(b) Is this method independent?

5.5. Imagine a country in which the ruling family has two sons, Prince A and Prince B. The ailing king decides that it is time for him to give up the throne, so he announces that elections will take place. To give the appearance of democracy, all citizens are invited to place their names on the slate of candidates, along with his two sons. What the king does not announce is that his ministers will implement a social choice function known as the **best-of-two method**, which works as follows. If a majority of voters prefer Prince A to Prince B, regardless of how they rank any other candidates, then Prince A is the winner. If a majority of voters prefer Prince B to Prince A, regardless of how they rank any other candidates, then Prince B is the winner. If exactly half of the voters prefer Prince A to Prince B, then the election is a tie between Prince A and Prince B.

(a) Show that the best-of-two method is independent.
(b) Does the best-of-two method satisfy Pareto?

5.6. At Springfield High School, there is a student election for homecoming queen. There is a large slate of candidates, and preference ballots are used. After the ballots have been collected but before they have been revealed, the principal rules all but three of the candidates ineligible on academic grounds, leaving only Angela, Bertha, and Cassandra as permissible candidates. Rather than going to the trouble of calling a new election, the school president decides to implement the **best-of-three method** in this situation: If any one of Amy, Bertha, or Cassandra is preferred to the other two by a majority of the student body, then that candidate becomes homecoming queen. Otherwise, the election is declared to be a tie among these three girls. Does the best-of-three method satisfy the independence property? Compare with Exercise **5.5**.

5.7. A **coalition** is a set of voters. Given a social choice function, let us call a coalition **influential** if whenever every voter in the coalition places a candidate in first place, that candidate becomes the unique winner.

(a) What are the influential coalitions for the plurality method?
(b) What are the influential coalitions for a dictatorship?
(c) Assuming that there are 4 candidates and 10 voters, what are the influential coalitions for the Borda count method?

5.8. Consider an election with 4 candidates and 10 voters. Given a social choice function, let us call a coalition **controlling** if, whenever every voter in the coalition places one candidate ahead of another, the second candidate must lose.

(a) If the election uses the plurality method, how large does a coalition have to be in order to be controlling?
(b) If the election uses the Borda count method, how large does a coalition have to be in order to be controlling?
(c) If the election uses the agenda method, how large does a coalition have to be in order to be controlling?

5.9. Call a social choice function **nearly Pareto** if, for any pair of candidates A and B, whenever 90% or more of the electorate favors A over B, candidate B must lose. Explain why no social choice function is nearly Pareto.

5.10. Corollary 5.4 has four hypotheses. Which of them turns out to be unnecessary (extraneous)?

5.11. Show that every profile with an odd number of voters that does not have a Condorcet candidate displays the Condorcet paradox in the following sense: There exist candidates A, B, and C such that a majority of the electorate prefers A to B, a majority of the electorate prefers B to C, and a majority of the electorate prefers C to A.

5.12. Show that only one voter can have dictatorial power between A and B.

6

Variations on the Theme

"The ideal form of government is democracy tempered with assassination."
— Voltaire

6.0 Scenario

A **social ranking function** is a function that inputs a voter profile and outputs a complete ranking (permitting ties) of the candidates. Where a social choice function chooses winners, a social ranking function ranks all candidates. A social ranking function can be thought of as a method for taking the preference lists of all the voters and producing a composite preference list representing the entire electorate. But note: Ties are forbidden on voter preference lists, even though ties are expressly permitted on the social preference list. For example, if the set of candidates is $\{A, B, C, D, E, F, G\}$, a typical output might be described by a notation like $C > D = G > A = E = F > B$ to describe the electorate's view that candidate C is best, candidates D and G are tied for second best, candidates A, E, and F come next and are equally regarded, and candidate B is worst.

Your job is to come up with a number of different social ranking functions and a number of different criteria of reasonableness for such functions. Think of as many functions (methods) and criteria (rules) as you can, and give each one a catchy name. To get you started, we suggest one method and one rule:

Definition 6.1. In the **rank-by-first-place-votes method**, we rank the candidates by the number of voters who put them at the top of their preference list.

Definition 6.2. A social ranking function satisfies the **majority-rules rule** if, whenever a majority of the voters prefer candidate A to candidate B, B may not end up ahead of A in the social preference order.

6.1 Inputs and Outputs

In our definition of a social choice function, we decided that an input to such a function should be the collection of preference lists submitted by each voter and that the output should be a nonempty subset of the slate of candidates. These are neither the only options that make sense nor the only ones worthy of study. In this chapter, we contemplate variations on the model but still with a focus on functions that in some way attempt to aggregate the views of voters. With several of these variations, there is a complete theory parallel to the one we have developed in Chapters 1 through 5. But here we content ourselves with just an introduction.

First, we consider other possibilities for the domain of the function, for the input to the function. In particular, we consider the possibility of ballots that allow voters to express equal preferences between pairs of candidates, that don't force voters to list every candidate, or that invite voters to indicate their approval (or degree of approval) for candidates. Generally, there is a spectrum of possibilities ranging from ballots that are less expressive, that allow voters to contribute a narrow or limited set of responses, to ballots that are more expressive, that allow voters to contribute a complex array of responses representing their views. (The preference ballot that we have adopted up to now might be deemed intermediate on this scale.)

At first it may seem that the more expressive a ballot is, the better it is for voters. There are two reasons why this might not be the case. The first reason is a practical one: It is difficult for voters to express views that they don't have. Even the preference ballot is not easy for a voter to fill out if there are many candidates. For example, when sportswriters vote on the NCAA Division I basketball rankings each week, they are not forced to give a complete ranking of the 180 teams. Instead, they merely have to rank the top 25. That is lucky for the sportswriters, since most of the 180 teams probably would not have come to their attention, and they would not be comfortable having to decide, for example, which team to rank 143rd and which team to rank 144th. The second reason is a theoretical one: With the most expressive ballots, there are even more striking impossibility theorems and opportunities for voters to express themselves insincerely in order to manipulate the outcome. With a more permissive ballot comes a more chaotic theory.

Next we consider various possibilities for the range of the function, for the output of the function. One possibility is to ask that the function not merely choose one or more candidates but instead rank all the candidates. We call such functions **social ranking functions**, since they produce not merely a social choice but a social ranking or preference list. There is a pleasant symmetry to a social ranking function, since it takes a set of rankings as input and delivers a ranking as output. One may want to permit social rankings to have ties, in order that social indifference can be reported. In particular, this seems especially appropriate when combined with a ballot that allows voters to express indifference between candidates. Moreover, as with social choice functions, forbidding ties conflicts with other compelling criteria and severely restricts the available methods.

Another variation considers functions whose outputs yield a probability for each candidate, with these probabilities summing to 1. One thinks of the probability associated with each candidate as the likelihood that he becomes the winner. This model permits at the extremes an output that assigns probability 1 to one single candidate, when the electorate is deemed to have expressed a clear choice, but it also permits outputs that assign, for example, probability 1/2 to each of two candidates. This can be thought of as an appropriate output when the electorate clearly favors two candidates but there is an even split between them. One can imagine resolving such an election by flipping a fair coin at the end. An objection to this model is the dependence on randomness, which is arguably undemocratic. An advantage to this model is the treatment of ties, which require no special provisos or special treatment.

6.2 Vote-for-One Ballots

An example of a type of ballot that substantially restricts what the voter may express is the familiar **vote-for-one ballot**, on which each voter is permitted to report just one choice. Because we are all accustomed to this sort of ballot, we don't often complain about the limitations it imposes on the voter. But those limitations are not trivial. Suppose that among three candidates in a close election, you find one to be much worse than the other two. How do you decide how to vote? Anticipating the plurality method, you may want to vote for whichever of the two tolerable candidates can defeat the intolerable one. But you may not have the means to guess how the remainder of the electorate is planning to vote. Polling data is notoriously inaccurate. The vote-for-one ballot does not allow you to express your view that you support either of the two candidates but feel strongly that the third is an inferior choice.

On the other hand, the advantage of the vote-for-one ballot is that, with the limited information that the ballots yield about the electorate, the spectrum of social choice functions is much narrowed. Any social choice function that uses only information about the first choices of voters is a social choice function that we can use with vote-for-one ballots. Of the social choice functions considered in Chapter 2, the only ones that depend only on first-choice data are

- the plurality method;
- dictatorships;
- the all-ties methods; and
- monarchies.

The vote-for-one ballot does not have enough information on it to work out head-to-head match-ups, transferable votes after elimination, or positional voting tallies. Hence the methods of Copeland, Hare, and Borda, among others, are not available in this setting. If you collect only the first choices of the voters, there are not so many reasonable ways to compute a winner.

But there are some other social choice functions that work in this context that we haven't considered so far for elections with three or more candidates.

Definition 6.3. The **majority method** is the social choice function in which any candidate who receives a majority of the votes is the unique winner. If no candidate receives a majority of the votes, then all the candidates are declared winners.

Here we use the option of declaring all candidates to be winners as a way of throwing up our hands in resignation. Setting 50% as a threshold for declaring one candidate to be the unique winner assures us that our winner is a Condorcet candidate. But one can alter this threshold to obtain other methods, encompassing the idea of the supermajority method from Chapter 1.

Definition 6.4. Choose a number q. The **quota method** with quota q is the social choice function in which any candidate who receives at least q votes is a winner. If no candidate receives q votes, then all the candidates are declared winners.

If the quota q is larger than half the number of voters, then at most one candidate can receive q votes. On the other hand, one can also consider q to be smaller than half the number of voters. In particular, the all ties method can be thought of as the quota method with $q = 0$, while the quota method with $q = 1$ names every candidate who receives any votes to be a winner.

The majority method and the quota method both suffer from the possibility of excessive ties. Declaring all candidates to be winners is an escape clause that can be imposed whenever one fears that making a tough decision will raise objections. We shouldn't be pleased with methods that lead to the imposition of this sort of escape clause very often. To address this concern, we may wish to impose the following version of a criterion from Chapter 1 that is meant to limit ties.

Definition 6.5. A method is **nearly decisive** if whenever two candidates are both named winners, they have received the same number of votes.

Once this criterion is imposed, we have a theorem, reminiscent of May's Theorem, that identifies the plurality method as the optimal method with vote-for-one ballots.

Theorem 6.6. *Among the social choice functions that use only information about the first choices of voters, the only one that is anonymous, neutral, monotone, and nearly decisive is the plurality method.*

Proof. Suppose we have a method that is anonymous, neutral, monotone, and nearly decisive. Suppose also that candidate A receives a votes, candidate B receives b votes, and $a > b$. We argue that B cannot be a winner.

To see why, imagine that B is a winner. Then alter the profile in two steps. First, have all those who voted for A or B switch their votes from one to the other. Second, have $a-b$ voters in this new profile change their votes from candidate B to A. After the first change, A receives b votes and B receives a votes. Owing to the assumption of neutrality, since B was a winner before the first change, A must be a winner after the first change. After the second change, A receives $b + (a - b) = a$ votes and B receives $a - (a - b) = b$ votes. But the second change involves voters switching their preference to A, and so by the monotonicity assumption, A remains a winner in this new profile. Now notice that all candidates in this new profile receive exactly as many votes as they did in the original profile. By the assumption of anonymity, a winner of the election with this new profile is a winner in the original profile. Hence A was a winner in the original profile. We have shown that if B is a winner in the original profile, then so is A. But we assumed that $a > b$, so this contradicts our assumption that the method is nearly decisive.

We have shown that candidate B cannot win this election if candidate A has more votes. This applies to any two candidates A and B, so the only possible winner of this election is a candidate with as many votes as any other candidate. Thus the only candidate who can be the unique winner of such an election is one who has a plurality of the votes. Moreover, neutrality and anonymity require that two candidates who both receive a plurality (an equal number) of the votes must both be winners. Thus the method is indeed plurality. □

6.3 Approval Ballots

An approval ballot is one on which a voter votes yes-or-no for each candidate. This type of ballot is neither stronger nor weaker than a preference ballot. Unlike the vote-for-one ballot, we cannot infer a voter's response on an approval ballot from the information contained on a preference ballot. Knowing the order in which a voter ranks the candidates does not tell us which candidates the voter deems acceptable.

Conversely, knowing which candidates a voter deems acceptable does not tell us how that voter will rank the candidates. The social choice functions we have considered in the previous chapters are therefore not usable with approval ballots. We don't even have enough information from approval balloting to compute the plurality winner of the election, because we don't know which candidate each voter ranks in first place.

So we need an entirely new theory for approval balloting. What methods come to mind that can use data from approval ballots? What criteria should such methods satisfy? Is there an impossibility theorem? Is there a theorem that identifies an ideal among the possible methods? We provide a brief synopsis of this theory in this section.

An **approval profile** is a list of all the approval ballots of all the voters. With three candidates and ten voters, it would look something like this:

A	Y	Y	Y	Y	Y	Y	N	N	N	N
B	N	N	N	N	Y	Y	Y	Y	Y	N
C	N	Y	Y	Y	N	Y	Y	Y	N	Y

where the rows are indexed by candidates, the columns are indexed by voters, and a "Y" in a cell indicates that the voter associated with that column approves of the candidate associated with that row. We may even wish to abbreviate this with a **tabulated approval profile**, as follows.

	1	3	1	1	2	1	1
A	Y	Y	Y	Y	N	N	N
B	N	N	Y	Y	Y	Y	N
C	N	Y	N	Y	Y	N	Y

An **approval social choice function** is a function that takes as input any approval profile and gives as output some subset of the candidates, regarded as the set of election winners.

The most obvious example of an approval social choice function is what we still call the plurality method.

Definition 6.7. The **plurality method** for approval voting is the approval social choice function that selects as a winner whichever candidate or candidates get the most "yes" votes.

This method is the one that is traditionally called "approval voting". It is used to select the secretary general of the United Nations as well as the presidents of a number of scientific societies (including the Mathematical Association of America and the American Mathematical Society).

Still one can envision other approval social choice functions.

Definition 6.8. The **majority method** for approval voting is the approval social choice function in which any candidate receiving a majority of "yes" votes is a winner.

Notice that the majority method can yield no winners at all. While such an outcome was expressly forbidden in Chapters 1 through 5, it makes sense to permit the empty set as a set of winners in the context of approval voting. There is the possibility with approval voting that all voters mark "no" for all candidates. In such a circumstance, it seems appropriate to declare every candidate to be a loser. Approval voting with the majority method seems especially reasonable in a context when there is no restriction on the number of winners. An example of such a situation is if the candidates are associates at a law firm, the voters are the partners, and those candidates who are approved become partners themselves. There is no single executive office for the candidates to occupy as in an election for governor, so there is in a sense no competition between the candidates. All candidates are advanced along on their own merits if they acquire enough "yes" votes. If no associates are deemed acceptable, then none needs to be promoted. If all of the associates are deemed acceptable, then that is okay as well.

The majority method is a special case of a more general idea.

Definition 6.9. Let q be any positive integer. The **quota method** for approval voting with quota q is the approval social choice function in which any candidate receiving at least q "yes" votes is a winner.

We provide one last example of an approval social choice function, this one quite different from anything we have seen before.

Definition 6.10. The **split-vote method** is the approval social choice function in which a voter who approves of exactly $k > 0$ candidates contributes $1/k$ points to each of those k candidates. (If a voter approves of $k = 0$ candidates, then no points are contributed by that voter.) Whichever candidate tallies the most points wins.

The split vote method provides an incentive to voters to approve of fewer candidates. An example illustrates the point. Consider the following tabulated approval profile.

	2	3
A	Y	N
B	N	Y
C	N	Y

With either the plurality or majority methods (for approval voting), candidates B and C tie with 3 votes, while candidate A earns only 2 and therefore loses. However, in the split vote method, candidates B and C earn only 3/2 points, because the three voters who approved of them can be regarded as having split their votes between the two, while candidate A earns 2 points, because the two voters who approved of candidate A gave their full vote to candidate A. So candidate A wins the split-vote method outright.

One feature of approval voting is that the Condorcet paradox that haunted Chapters 1 through 5 is absent here. In the setting of approval voting, candidate A is said to be **socially preferred** to candidate B if the number of voters approving A but not B exceeds the number of voters approving B but not A. The following proposition asserts that this relation is transitive.

Proposition 6.11. *In the setting of approval voting, if candidate* A *is socially preferred to candidate B and candidate B is socially preferred to candidate* C, *then candidate* A *is socially preferred to candidate* C.

Proof. To say that candidate A is socially preferred to candidate B is simply to say that A has more approval votes than B has. Hence the hypothesis in the proposition tells us that A has more approval votes than B, who in turn has more approval votes than candidate C. By the transitivity of the "greater than" relation, this implies that A has more votes than C. Hence A is socially preferred to C. □

Contrast Proposition 6.11 to Profile 5.1 in the context of preference ballots.

We now develop criteria for judging approval social choice functions using the same terms as we used earlier, with analogous definitions.

Definition 6.12. An approval social choice function is **anonymous** if the outcome is unchanged whenever any two voters exchange approval ballots.

Definition 6.13. An approval social choice function is **neutral** if the following condition holds: Imagine an approval profile, and suppose that the method names a certain candidate a winner. Call that candidate A. Let B be any other candidate, and suppose that all the voters who approved of A but not B change their approval ballots so that they approve of B but not A, while all the voters who approved of B but not A change their approval ballots so that they approve of A but not B. (All other voters leave their ballots unchanged, and all voters leave unchanged their views about all candidates other than A or B.) Then the method must name candidate B a winner when facing the new approval profile.

Definition 6.14. An approval social choice function is **monotone** if the following condition holds: Suppose that candidate A is a winner when facing an approval profile P. Let Q be another approval profile that is identical to P except for one particular voter, who changes a "no" vote for A in P to a "yes" vote for A in Q. Then the method must name candidate A a winner when facing approval profile Q as well.

Definition 6.15. An approval social choice function is **independent** if it has the following property: Suppose there are two candidates A and B and two approval profiles that are identical with regard to A and B (meaning that every voter who approves of A or B in one profile also approves of that candidate in the other). If A wins and B loses in the first profile, then B loses in the second profile.

Owing to Proposition 6.11, independence is not as elusive a criterion in the context of approval ballots as it was in the context of preference ballots. We are not plagued here by the omnipresent looming of the Condorcet paradox. The majority method, quota methods, and the plurality method are all anonymous, neutral, monotone, and independent. One way to narrow down the field is to ask that the approval social choice function not be too permissive.

Definition 6.16. An approval social choice function is **nearly decisive** if whenever it names two winners, they have exactly the same number of "yes" votes.

We are now in a position to identify the plurality method as the unique method that satisfies all of our criteria.

Theorem 6.17. *The only approval social choice function that is anonymous, neutral, monotone, independent, and nearly decisive and that always selects at least one winner is the plurality method.*

Proof. Consider an approval social choice function that meets all six criteria in the hypothesis. Let P be any approval profile. Suppose that candidate A receives at least as many approval votes as does candidate B. We want to establish that B cannot be a winner when facing P unless A is a winner as well. We partition the set of voters into 5 blocks: Let $\boldsymbol{Y}$ be the set of voters who approve of both A and B, let $\boldsymbol{N}$ be the set of voters who approve of neither A nor B, let $\boldsymbol{K}$ be the set of voters who approve of B but not A, let $\boldsymbol{L}$ be a set of the same size as $\boldsymbol{K}$ of voters who approve of A but not B, and let $\boldsymbol{M}$ be the (possibly empty) set of remaining voters. The figure below displays Profile P graphically.

	$\boldsymbol{Y}$	$\boldsymbol{N}$	$\boldsymbol{K}$	$\boldsymbol{L}$	$\boldsymbol{M}$
A	Y	N	N	Y	Y
B	Y	N	Y	N	N
⋮	⋮	⋮	⋮	⋮	⋮

Profile P

Suppose that B wins but A loses when facing Profile P. Let Q be the approval profile obtained from Profile P by interchanging all the ballots of voters in $\boldsymbol{K}$ with voters in $\boldsymbol{L}$.

	$\boldsymbol{Y}$	$\boldsymbol{N}$	$\boldsymbol{K}$	$\boldsymbol{L}$	$\boldsymbol{M}$
A	Y	N	Y	N	Y
B	Y	N	N	Y	N
⋮	⋮	⋮	⋮	⋮	⋮

Profile Q

By anonymity, B must win and A must lose when facing Profile Q. Let R be the approval profile obtained from Q by switching every voter's responses for candidate A and candidate B.

	$\boldsymbol{Y}$	$\boldsymbol{N}$	$\boldsymbol{K}$	$\boldsymbol{L}$	$\boldsymbol{M}$
A	Y	N	N	Y	N
B	Y	N	Y	N	Y
⋮	⋮	⋮	⋮	⋮	⋮

Profile R

By neutrality, A must win and B must lose when facing Approval Profile R. Finally, let S be the approval profile obtained from R by having the voters in M change their votes both for candidate A and for candidate B.

	Y	***N***	***K***	***L***	***M***
A	Y	N	N	Y	Y
B	Y	N	Y	N	N
⋮	⋮	⋮	⋮	⋮	⋮

Profile S

By monotonicity, A must win and B must lose when facing Profile S. But all voters in Profile S express the same view with regard to A and B as they did in Profile P. Therefore, by independence, B must lose in profile P as well. This contradicts our assumption.

We have shown that if any candidate wins, then any other candidate with at least as many approval votes also wins. By assumption, we are guaranteed to have at least one winner. The near decisiveness criterion goes on to say that a candidate with fewer approval votes than a winner cannot also be a winner. We conclude that the method in question is in fact the plurality method. □

Approval voting has a strong following. Proponents argue that approval voting is the optimal method for deciding elections for a single public office. They contend that approval voting is simple, practicable, and fair, that it encourages sincere voting, that it discourages negative campaigning, and that the results of approval voting elections will be widely accepted.

Another argument in favor of approval voting is that the approval ballot is expressive enough to allow voters to escape the typical predicaments that arise with vote-for-one ballots but not so expressive that Arrow's impossibility theorem interferes. The voter who may select only one candidate often faces a choice of sincerely expressing her first choice, even if it is for a candidate who is unlikely to win, or voting instead for a compromise candidate who she finds tolerable. (The former choice is derided as "throwing away your vote".) Approval voting allows such a voter to express approval for both candidates. On the other hand, approval voting constrains the expressiveness of voters in a way that some voters may find annoying. Still, the world of approval methods is free of the Condorcet paradox, and this is one advantage of forbidding the voters to submit a preference ranking.

6.4 Mixed Approval/Preference Ballots

On a vote-for-one ballot, voters may feel constrained by an inability to express themselves fully. On various other types of ballots that permit more expressiveness, voters may feel manipulated into expressing views that they don't actually have. It is difficult to allow each voter to be just as expressive as they wish to be. The ultimate effort in this direction is to allow voters to contribute their own personal essay on the candidates (perhaps with a page limit?). The only difficulty with this sort of "essay ballot" is: What do you do with the information once it has been collected?

One kind of ballot that is slightly more expressive than a preference ballot and that works well in certain situations is a mixed approval/preference ballot. On such a ballot, voters are required to give a full preference ballot and a full approval ballot for the slate. Moreover, they are required to make their approval ballot compatible with their preference ballot, in the sense that they may not disapprove of a candidate whom they rank higher than another candidate of whom they approve. When we speak of the rationality of voters in the context of this sort of ballot, we require this kind of compatibility.

A graphical representation of this kind of mixed approval/preference ballot is as follows, where the column represents the preference order of one voter, as usual, from highest to lowest ranking, and where the double line indicates where along the preference order the voter distinguishes between approval and disapproval.

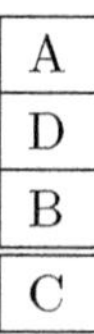

In this example, the voter ranks candidates A, D, and B in that order and regards them as all acceptable, while candidate C is ranked last and is regarded as unacceptable.

The difficulty with methods for this sort of ballot is that the considerations of the preference information may conflict with considerations of the approval information. For example, in the mixed preference/approval profile:

A	B	A	A	B
C	D	E	B	D
B	C	D	E	E
E	E	C	C	C
D	A	B	D	A

Candidate C is the Condorcet loser and has no first-place votes, yet is the only candidate approved by all 5 voters. On the other hand, in the mixed preference/approval profile:

A	A	B	D	E
D	B	C	C	C
C	C	A	E	D
E	E	E	B	B
B	D	D	A	A

Candidate C is the Condorcet winner and the Borda count winner, yet no voters approve of his election. These examples illustrate the principle that more information about voter preferences leads to more difficulty in processing that information and in more opportunities for voters to vote insincerely or strategically. It is entirely unclear what method to use in rendering a decision when facing this kind of mixed preference/approval ballots.

Another kind of mixed preference/approval ballot is one in which a voter ranks every candidate of whom he approves, but does not rank those of whom he disapproves. Voters may appreciate not having to compare two candidates both of whom they regard as unacceptable. Moreover, when a slate is large or not clearly determined, many voters do not want to rank every conceivable candidate. After all, they may have no information or opinion about some of the candidates. It is entertaining to ponder what methods one might use facing a profile consisting of this kind of ballot. The difficulty is in deciding whether an unranked candidate is regarded as disliked or unknown. Perhaps a further refinement could be considered in which approved candidates are ranked and the remaining candidates are classified as either unacceptable (disliked) or unknown. One can continue to conjure up bells and whistles to attach to a ballot to allow all kinds of subtle expression from voters, but we avoid this morass and instead move on to consider a different sort of expressive ballot.

6.5 Cumulative Voting

With a **fractional ballot**, voters are each given one vote, but they are permitted to divide it into parts as they wish and distribute it among the candidates. One way to represent a typical ballot is as follows:

A	1/2
B	1/6
C	0
D	1/3

where in this example the voter casts 1/2 of his vote for candidate A, 1/6 of his vote for candidate B, and 1/3 of his vote for candidate D. The fractions must add to at most 1, and we assume that they always add to exactly 1, since voters in elections usually want to exert as much influence as is permitted by the rules.

The fractional ballot allows more expressiveness than an approval ballot, since one can indicate degree of approval even among those candidates deemed approvable. It allows more expressiveness than a preference ballot as well, since one can indicate the degree to which one candidate is favored over another and one can also express indifference by offering equal fractions to two candidates.

Although one can implement a number of social choice functions when facing a profile using this versatile kind of ballot, the most natural method is to tally the fractional votes and declare the candidate(s) who receives the largest sum to be the winner. This is what is ordinarily referred to as **cumulative voting**.

A common variation is to give each voter as many votes as there are seats to be elected. If five representatives are to be elected, then the voters are each given five votes, which they are then entitled to spread out as they wish, all to one candidate, all to different candidates, or split in any manner at all among the candidates. This amounts to giving each voter one full vote, but inviting them to split it in five equal parts that they then distribute as they wish. We refer to this variation as **discrete cumulative voting**. This imposes a small constraint compared to the more permissive fractional ballot, in that only fractions with denominator 5 are possible in this system. There is no important reason that the fractions into which votes may be divided should relate to the number of desired winners in the election. In particular, there is no extraordinary level of extra freedom offered to the voter by permitting any fractions at all to be used, as in the fractional ballot.

Cumulative voting, usually in the discrete form, is used by many corporations in the election of officers and trustees. It is sometimes argued that it provides more rights to minorities. Because the fractional ballot is highly expressive, it is easy to argue that it gives all voters the special privilege to express approval, preference, and indifference in a more subtle way than other voting systems. One problem with cumulative voting is that voters are forced to play a certain tactical game. Suppose for example that in an election with three candidates, you favor candidate A to candidate B to candidate C. How should you fill out a fractional ballot in a cumulative voting system? It depends on what you believe about how other voters will vote. If you know that candidate C has little support, you might contribute this ballot,

A	1
B	0
C	0

throwing all your support behind your favorite candidate. If you know that your favorite candidate A has little support, you might contribute this strategic response,

A	0
B	1
C	0

in effect avoiding throwing your vote away on candidate A, in an effort to avoid a victory by candidate C. To express the view that your main fear is the election of candidate C, you might contribute this,

A	1/2
B	1/2
C	0

though you won't be pleased if candidate C ekes out a victory by 1/3 of a point. Perhaps if your interest is to express most strongly your preference ranking, you will contribute this.

A	2/3
B	1/3
C	0

But how is the voter to know whether 2/3 is precisely the right fraction to assign to candidate A in this situation? How can the voter examine her own intentions and discover that 2/3 is a more accurate reflection of her views than, say, 7/11 or 9/13?

While cumulative voting has a strong following, the down side of cumulative voting is the omnipresence of these tactical decisions it requires of the voters. It is undesirable to have a system in which voters have an incentive to express themselves dishonestly. Moreover, it is undesirable to give the voters so much freedom that their decision process becomes fraught with complexity.

6.6 Condorcet Methods

In a **Condorcet ballot**, every pair of candidates is presented to each voter, and the voter is asked to choose one of the two from each pair. Indifference is not allowed. If there are n candidates, then there are $n(n-1)/2$ pairs of alternatives on the Condorcet ballot. We call any method that uses this kind of ballot a **Condorcet social choice function** or simply a **Condorcet method**, because this type of voting was advocated by the Marquis de Condorcet.

After an election using a Condorcet ballot, we count the number of voters who preferred each candidate in each pair to obtain a **tabulated Condorcet profile**. Such a profile is shown below for an election with 4 candidates A, B, C, D and with 11 voters. With 4 candidates, there are 6 pairs of candidates for each voter to consider.

A	B
5	6

A	C
5	6

A	D
9	2

B	C
9	2

B	D
9	2

C	D
11	0

Profile 6.1

We tend to think of each pair of candidates in a tabulated Condorcet profile as having engaged in a head-to-head match, and we think of the votes each candidate gets as the score in that match.

As one might expect, a Condorcet method is called anonymous if it treats all voters equally. Clearly, any Condorcet method that depends only on a tabulated Condorcet profile satisfies the anonymity criterion. One obvious example of a Condorcet method is Copeland's method.

To implement Copeland's method in this context, we simply count how many matches each candidate wins, and the candidate who wins the most matches is the winner. For Profile 6.1, A wins one match, B wins three, C wins two, and D wins none, so B is the winner under Copeland's method. The agenda method, for any choice of an agenda, is also a Condorcet method. Another example of a Condorcet method is the method from Problem 18 in Chapter 2.

Any time we have a tabulated preference profile (the kind of profile studied in Chapters 1 through 5), we can use it to construct a tabulated Condorcet profile. For example, the tabulated preference Profile 6.2 gives the tabulated Condorcet Profile 6.1.

5	4	2
A	B	C
B	C	D
C	A	B
D	D	A

Profile 6.2

Thus, any Condorcet method is a social choice function in the sense of Chapters 1 through 5. However, the converse is false. For example, one cannot always tell from a Condorcet profile how many first-place votes a candidate will get. As Condorcet himself was well aware, voters who are queried about their opinions, one pair of alternatives at a time, may not always give "rational" responses. That is, their response may not come from a transitive preference ranking. Consequently, not every Condorcet profile comes from a preference profile. For example, the three-voter three-candidate Profile 6.3 cannot be obtained from any preference profile, since the data for A versus B and for A versus C indicate that all three voters prefer C to A to B, but this contradicts the data for B versus C.

A	B
3	0

B	C
2	1

C	A
3	0

Profile 6.3

Given any tabulated Condorcet profile, we can determine the Condorcet and anti-Condorcet candidates, if they exist. For example, in the case of Profile 6.1, the Condorcet candidate is B, while the anti-Condorcet candidate is D. However, in Profile 6.3, which is an extreme case of the Condorcet paradox, there is neither a Condorcet candidate nor an anti-Condorcet candidate. Because the notions of Condorcet and anti-Condorcet candidates make sense in Condorcet voting, we can ask whether or not a Condorcet method satisfies the Condorcet criterion or the anti-Condorcet criterion. For Copeland's method, the agenda methods, and the method of Problem **2.18**, the answer is yes for both questions. However, we will see below that this is not always the case.

Notice in Profile 6.1 that the result of the sixth match shows that everyone prefers C to D. This suggests how to phrase the Pareto criterion in the context of Condorcet voting: any candidate who gets a score of zero in one of his or her matches should not win the election. We know that Copeland's method and the method of Problem **2.18** satisfy the Pareto criterion, but we also know that the agenda method does not.

There is another way to present the data in a tabulated Condorcet profile that will be useful in explaining Condorcet methods. The **Condorcet matrix** is an n-by-n array, whose rows and columns are indexed by the candidates. The entry in, say, row A and column B is the number of voters who prefer A to B in the Condorcet profile, with a similar definition for any other entry. The diagonal entries are left blank since the candidates do not compete against themselves. Here is the Condorcet matrix for Profile 6.1.

	A	B	C	D
A	–	5	5	9
B	6	–	9	9
C	6	2	–	11
D	2	2	0	–

Profile 6.4

Any social choice function that uses only information from the Condorcet matrix is a Condorcet method. The following fact apparently annoyed Condorcet, who intensely disliked Borda.

Proposition 6.18. *The Borda count method is a Condorcet method.*

Proof. For each candidate, sum the entries in the row of the Condorcet matrix associated with that candidate. We leave as an exercise the verification that this sum is just the candidate's Borda count score. It follows that the Borda count winner can be determined from the Condorcet matrix. □

Now recall Black's method, which chooses the Condorcet candidate if there is one, and otherwise reverts to the Borda count.

Corollary 6.19. *Black's method is a Condorcet method.* □

Next we describe a method that, according to the mathematical political scientist H. P. Young, was the method that Condorcet had in mind when he described his theory of voting. If so, this is the method that should really be called "Condorcet's method". However, since this method was independently discovered by the mathematician John Kemeny in the 1960s, it is now generally called **Kemeny's method**.

Assume that we have the Condorcet profile of our electorate and have computed the corresponding Condorcet matrix. The first step in Kemeny's method is to calculate what is called the **Kemeny score** for each preference order on the set of candidates. To compute the Kemeny score of a preference order, count the number of times a voter chose a candidate over a lower-ranked candidate. For example, in the case of the preference order A > B > C > D, when a voter was asked to choose between candidates B and D, we score one point for this preference order if they chose B but not if they chose D, because B is higher up on the order than D is. We similarly score a point for each voter and for each pair when the voter selects the higher ranked candidate from that pair. The tally of these points is the Kemeny score for that preference order.

Here is another way to describe this. Every preference order gives rise to a winner in every pairwise comparison. For example, in the case of the preference order A > B > C > D, we find that A beats B, C, and D; B beats C and D; and C beats D. To obtain the Kemeny score for this preference order, we sum up the corresponding entries in the Condorcet matrix. For the Condorcet Matrix 6.4, we sum the entries in row A, columns B, C, and D; row B, columns C and D; and row C, column D. Thus the Kemeny score for the preference order A > B > C > D is $(5 + 5 + 9) + (9 + 9) + 11 = 48$.

Definition 6.20. **Kemeny's method** is the Condorcet method that works as follows: List every preference order among the candidates, and for each preference order, compute its Kemeny score, as described above. The **Kemeny preference order** is the preference order with the highest Kemeny score, and the winner under Kemeny's method is the first choice candidate in the Kemeny preference order. If there is a tie among the preference orders for highest Kemeny score, all the candidates who appear at the top of any highest scoring preference order are winners.

Clearly Kemeny's method is a Condorcet method. Here is the complete application of Kemeny's method to Profile 6.1.

A	A	A	A	A	A	B	B	B	B	B	B
B	B	C	C	D	D	A	A	C	C	D	D
C	D	B	D	B	C	C	D	A	D	A	C
D	C	D	B	C	B	D	C	D	A	C	A
48	37	41	34	30	23	49	38	50	43	31	32

C	C	C	C	C	C	D	D	D	D	D	D
A	A	B	B	D	D	A	A	B	B	C	C
B	D	A	D	A	B	B	C	A	C	A	B
D	B	D	A	B	A	C	B	C	A	B	A
42	35	43	36	28	29	23	16	24	25	17	18

The highest Kemeny score, 50, occurs in column 9, so B is the winner under Kemeny's method. However, Kemeny's method reports more than just a winner. It reports that the ranking B > C > A > D is the "correct" one. In other words, candidate C comes in second place and A comes in third. In fact, Kemeny's method is a social ranking function in the sense of Section 6.7. Notice also that the Kemeny's method winner B is also the Condorcet candidate. This is no accident.

Proposition 6.21. *Kemeny's method satisfies the Condorcet, anti-Condorcet, and Pareto criteria.* □

The proof is left as an exercise (**6.14**).

Here is the result of Kemeny's method applied to Profile 6.3, in which there is no Condorcet candidate.

A	A	B	B	C	C
B	C	A	C	A	B
C	B	C	A	B	A
5	4	2	5	7	4

The winner is C, which comes from the fifth preference column.

H. P. Young suggests that Condorcet viewed voting as an exercise in which all citizens strive to make the best choice for their society, but sometimes err in their judgements. According to Young, Condorcet would have interpreted this last result to mean that the two voters who voted for B over C in Profile 6.3 were in error. If these voters realized their errors and switched their votes from B to C the result would be as follows:

A	B
3	0

B	C
0	3

C	A
3	0

which is the Condorcet profile that derives from the preference profile where all three voters rank the candidates C $>$ A $>$ B.

Kemeny's method applied to Profile 6.1 illustrates its biggest problem. In order to compute it you must consider all possible rankings of candidates. Even when the number n of candidates is relatively small (like $n = 4$), the number $n!$ of preference orders is quite large.

6.7 Social Ranking Functions

Now let us consider changing not the input (the ballots) of the social function but rather the output (the decision). We return to the assumption that we are employing a preference ballot and that our function should accept as input the profile of preference ballots, one for each voter. Now, however, we ask for a function that computes for us not just a winner (or nonempty set of winners) but an entire ranking of the candidates.

As with social choice functions, requiring that the output never be a tie limits the scope of the methods that can be defined. For this reason we find it prudent to allow ties in the output ranking. A ranking that permits ties is called a **weak ranking**. An example of a weak ranking of candidates A, B, C, D, and E might be denoted A $=$ D $>$ E $>$ B $=$ C, to indicate that candidates A and D are best, candidates B and C are worst, and candidate E is intermediate.

We define a **social ranking function** to be a function that assigns to every profile a weak ranking of the candidates. Voters must express a (strong) ranking on their preference ballots. Ties are permitted only in the output, which is called the **social ranking**.

One can build an entire theory of voting in the category of social ranking functions. There is something attractive about the symmetry between preference ballots and preference outputs: If one is extracting a ranking from each voter, why not produce a ranking for the entire society? In fact, the version of voting theory based on social ranking functions rather than social choice functions is the version generally preferred by economists, including Kenneth Arrow. In economics, social ranking theory is regarded as the study of aggregating preference rankings of individuals (whether expressed on a ballot or not) into a unified preference ranking for the whole society: a *social* ranking. Whether this is accomplished by voting, by market forces, or by some other mechanism is of less interest to economists than how the social preference ranking relates to individual preferences. This is, in fact, the predominant modern view of social choice theory. Much of the more recent research on social choice theory has been conducted by economists rather than political scientists.

Here we offer nothing more than a brief introduction to this study. We begin by looking for examples of social ranking functions. Many of our examples are built from social choice functions.

Definition 6.22. Given any social choice function, the corresponding **win-lose social ranking function** ranks all the winners equally, ranks all the losers equally, and ranks all the winners above all the losers.

Definition 6.23. Given any social choice function, the corresponding **iterated social ranking function** ranks in the top position(s) the winning candidate(s). Then this candidate is (these candidates are) eliminated from the slate, and the process is repeated with the remaining candidates. The winners are ranked in second position and are then eliminated. The process continues until all candidates are ranked.

Definition 6.24. Given any social choice function that operates by tallying scores for each candidate (e.g., plurality, Borda count, Copeland method), the corresponding **rank-by-score social ranking function** ranks the candidates according to their scores.

Definition 6.25. The **most-popular-ranking social ranking function** selects whichever ranking appears on the most ballots. (If there is a tie, default to another method, or select the ranking that considers all candidates to be equal.)

Definition 6.26. Let v be one voter. The **dictatorship** social ranking function with v as dictator chooses whichever ranking appears on the ballot of voter v.

Because there are many possible social ranking functions, we seek to narrow the field in the same way that we did with social choice functions: by enunciating criteria for reasonableness (fairness, appropriateness) of such functions. We use the same words for these criteria — anonymity, neutrality, monotonicity, Pareto, independence — that we did with social choice functions. But the definitions must be modified to make sense in the context of social rankings. It is no longer just a question of winners and losers; the entire slate is (weakly) ranked.

Definition 6.27. A social ranking function is **anonymous** if the output is unchanged whenever two voters swap ballots.

Definition 6.28. A social ranking function is **neutral** if, whenever all voters interchange the positions of candidates A and B on their preference lists, the only change in the output is that the positions of candidates A and B are interchanged.

Definition 6.29. A social ranking function is **monotone** if the following holds: Suppose that there are two profiles, identical except that, in the second profile, one candidate (call her candidate A) is moved up on the preference ranking of one or more voters. Then no other candidate may appear on the social preference ranking behind candidate A in the first profile but ahead of candidate A in the second profile.

Definition 6.30. A social ranking function is **independent** if the following holds: Suppose that there are two profiles and two candidates (call them candidates A and B) with the property that a voter ranks A ahead of B in the first profile if and only if the voter ranks A ahead of B in the second profile. Then the outputs of the two profiles must agree on the relative standing of candidates A and B. In other words, either A is ahead of B in both outputs, or B is ahead of A in both outputs, or A and B are equivalent in both outputs.

Definition 6.31. A social ranking function is **Pareto** if whenever every voter ranks one candidate A over another candidate B, the social ranking similarly places A over B.

There is in fact a weaker version of the Pareto property that is nonetheless strong enough to yield a version of Arrow's Theorem in the context of social ranking functions.

Definition 6.32. A social ranking function is **unanimous** if whenever every voter contributes the same preference ranking, that order becomes the social preference ranking.

It is easy to see that the Pareto criterion implies unanimity.

We can now state the version of Arrow's Theorem that applies in the context of social ranking functions. For a proof, see [46].

Theorem 6.33. (Arrow) *With three or more candidates, the only social ranking functions that are unanimous and independent are dictatorships.* □

Because the unanimity criterion follows from the Pareto criterion, we have the following weaker result.

Corollary 6.34. *With three or more candidates, the only social ranking functions that are Pareto and independent are dictatorships.* □

We usually think of Theorem 6.33 as an impossibility theorem. Let us say that a social ranking function is **nondictatorial** if it is not a dictatorship. We then obtain the following restatement of Theorem 6.33.

Corollary 6.35. *It is impossible for a social ranking function with three or more candidates to be unanimous, independent, and nondictatorial.*

□

6.8 Preference Ballots with Ties

A social ranking function produces as output a weak ranking, in other words a preference list permitting ties. It may seem natural then to permit the inputs for social ranking functions to be weak rankings as well. In fact, we may want to consider permitting voters in social choice functions to contribute weak preference lists as inputs. This allows voters a small extra measure of expressiveness, allowing them to express their indifference between two candidates.

How would one design a ballot to allow voters to express their weak preference rankings? An ordinary preference ballot lists the names of each of the n candidates and requires the voters to place the whole numbers from 1 to n after the names, using each number exactly once. Let us adopt the convention that low numbers represent greater preference (a more favored choice). A **weak preference ballot** could use the same instrument, and the voter is again required to place the whole numbers from 1 to n after the names. This time, however, we dispense with the requirement that each number be used exactly once, and we replace it with the following rule: A voter is permitted to rank a candidate with the number k only if there are exactly $k - 1$ other candidates ranked with a smaller number than k. For example, a voter may rank a candidate with the number 4 — in 4th place, that is — only if exactly 3 candidates are considered preferable. A typical voter facing a slate with 8 candidates might respond to such a ballot with the numbers $1, 1, 1, 4, 5, 6, 6$, and 8, to express a preference order that has three candidates tied for first place and two candidates tied for sixth place.

These sorts of rankings with ties are familiar as the results of golf tournaments, where the competitors end up ranked but with ties a frequent occurrence. It is customary in reporting the results of a tournament to use the notation "T6" to indicate that the competitors have tied for sixth place, in other words that the competitor was beaten by exactly 5 other golfers and that at least one other golfer did exactly as well.

It is possible to develop an entire voting theory based on a weak preference ballot. In fact, this is the approach that Kenneth Arrow himself took in his groundbreaking work of 1951. One can examine voting methods, criteria by which those methods can be assessed, and impossibility theorems that ensue. The theory proceeds in parallel to the theory with (strong) preference ballots, but there are differences.

For example, how might one extend the notion of a plurality election to the context of weak preference ballots? It isn't clear how one should deal with voters who have listed more than one candidate in first place. What is the appropriate way to conduct a Borda count election? If two candidates are ranked tied for last place, do they each receive 0 Borda count points or 1 Borda count point? Or perhaps they should each receive 1/2 of a point? The Copeland method translates easily over to the context of weak preference ballots, although for each pair of candidates, the head-to-head match-up between them may involve a number of voters who are indifferent.

Similar issues arise in attempting to find the most natural analogues for the familiar criteria that we considered in Chapter 3. In the context of weak preference ballots, what should the Pareto criterion require? What about monotonicity?

One can also address the context of social ranking functions with weak preference ballots. What are the appropriate analogues for the kinds of social ranking functions we considered earlier, and what are the appropriate criteria for judging them?

We leave to the reader the project of answering such questions, of developing the methods and criteria appropriate for each of these contexts. For us, it suffices to point out that voters have the prerogative on a weak preference ballot to refrain from expressing any indifference at all by submitting a ballot with no ties. Any social choice function or social ranking function that works on all profiles comprised of weak preference ballots must yield an output even if all voters contribute (strong) preference rankings. If one assumes that the Pareto criterion and the independence criterion for weak preference ballots are defined in such a way that they agree with the Pareto and independence criteria for (strong) preference ballots when all voters contribute (strong) preference rankings, then it is a corollary of the impossibility theorems we have already proved that it is impossible to find a voting method in the context of weak preference ballots that satisfies Pareto and independence and that isn't a dictatorship. After all, it was impossible to find such a thing in the earlier context, and all we have done in allowing weak preference ballots is to make the problem harder. It is therefore still impossible.

6.9 Exercises and Problems

6.1. A **one-two ballot** is one in which a voter is allowed to pick a first choice and a second choice candidate but that is all. With a one-two ballot one can still execute the plurality method. One can also execute the vote-for-two method. One cannot execute the Borda count method, the Hare method, or the Copeland method, however, since one does not have sufficient information from the voters.

(a) What variations on the Borda count method are possible with a one-two ballot?
(b) What variations on the Hare method are possible with a one-two ballot?
(c) What variations on the Copeland method are possible with a one-two ballot?
(d) For elections with exactly three candidates, how informative is a one-two ballot by comparison with a standard preference ballot?

6.2. There are two natural social ranking functions that are related to the Borda count social choice function. The first is the one that ranks the candidates according to their Borda count score. (This is what we would call the rank-by-Borda-count social ranking function.) The second is the one that places the Borda count winner or winners in first place and then repeats the Borda count social choice function on the slate of candidates obtained by removing the winner or winners. Rank the winner or winners of this restricted contest second highest, remove this winner or these winners from the slate, and repeat again, continuing until all candidates are ranked. (This is what we would call the iterated-Borda-count social ranking function.) Show by an example that these two social ranking functions can give different outputs.

6.3. Suppose that voters submit a mixed preference/approval ballot. Show that a candidate with the majority of first-place votes can fail to win the approval voting election.

6.4. In the context of weak preference rankings, rationality of voters requires both that strict preference be transitive and that indifference be transitive. In other words, if a voter prefers A to B and B to C, the voter must also prefer A to C, while if a voter ranks A and B equally and ranks B and C equally, the voter must also rank A and C equally. Which of the following represent the views of a rational voter about a slate of four candidates A, B, C, and D?

(a) Peter is indifferent about every pair of candidates except that he prefers A to D.
(b) Paul is indifferent about every pair of candidates except that he prefers A to B and that he prefers C to D.
(c) Mary is indifferent about A and D, but regarding any other pair she prefers the one that comes first alphabetically.

6.5. Show that a social choice function satisfies the Pareto criterion (for social choice functions) if and only if the iterated social ranking function derived from it satisfies the version of the Pareto criterion appropriate for social ranking functions.

6.6. Show that a social choice function satisfies the independence criterion (for social choice functions) if and only if the iterated social ranking function derived from it satisfies the independence criterion (for social ranking functions).

6.7. A popular idea in some circles is known as range voting. A range ballot is an expressive instrument that allows each voter to score each candidate on a continuous scale from 0 (complete disapproval) to 1 (complete approval). This is a generalization of an approval ballot, on which a voter can assign only the extremes in this range, with 0 standing for "no" and 1 standing for "yes". In range voting, the candidate with the largest average score wins the election. Assume that voters contribute both a range ballot and an approval ballot and that their approval ballot is obtained from their range ballot by rounding their range score to whichever of 0 or 1 is closest. Give an example with three candidates showing that approval voting can rank candidates in the order A, B, C, while the range voting can rank the candidates in the order C, B, A.

6.8. Show that the split-vote method in the context of approval ballots is anonymous, neutral, and monotone, but not independent.

6.9. Prove that if there are n candidates, then there are $n(n-1)/2$ pairs of candidates.

6.10. Prove that if there are n candidates, then the number of preference orders, or permutations, of the candidates is $n!$.

6.11. Complete the proof of Proposition 6.18 by showing that the sum of entries in the row of a Condorcet matrix labeled by a certain candidate gives that candidate's Borda count score.

6.12. Prove Corollary 6.19, which asserts that Black's method is a Condorcet method.

6.13. Show that the Hare method is not a Condorcet method by giving an example of two profiles that yield the same Condorcet matrix but have different Hare winners.

6.14. Prove that Kemeny's method satisfies the Pareto, Condorcet, and anti-Condorcet criteria.

6.15. Formulate an appropriate concept of monotonicity for Condorcet voting and prove that Kemeny's method is monotone.

Notes on Part I

The term "democracy" originates in 6th century BC Athens, where only free male land-owners were permitted to vote. One type of election, regularly held in Athens, can be interpreted as an example of a violation of monotonicity. Voters were asked to vote for their least favorite politician, and any politician who "won" this election (by receiving 6000 or more of these negative votes) was exiled for 10 years. The modern word "ostracism" comes from the Greek word *ostraka* for the shards of broken pottery that were used as ballots.

Another early voting system was used in 13th century Venice to elect the Doge. Voting, restricted to members of a Great Council representing the noble families, consisted of an elaborate five-round system of approval voting, intended to moderate the influence of the wealthiest families. First, 30 electors were selected from the Council by drawing lots, and then this number was reduced by lot to 9. Using approval voting, these 9 electors selected another 40 members, who were subsequently reduced by lot to 12. These 12 electors chose 25 members by approval voting, who were reduced by lot to 9. These 9 elected 45, who were reduced by lot to 11. Finally, these 11 chose the 41 electors who actually elected the Doge, again by approval voting. A new Doge was introduced to the people with the statement "This is your Doge, if it please you," symbolically suggesting a popular ratification of the Council's choice. This system lasted until the 18th century.

The method that we attribute to Copeland traces its history back to the Majorcan philosopher Ramon Llull in the 13th century. [20]

Beginning with the signing of the Magna Carta in 1215, the political system in England slowly evolved toward the modern idea of a constitutional monarchy. English ideas about democracy influenced the political culture in the American colonies, which declared their independence from England on July 4, 1776. The United States Constitution, which was ratified in 1788, was a milestone in democracy, prescribing solutions to many of the practical problems of running a democratic government. Yet still, under the original Constitution, neither senators nor presidents were directly elected by the voters, and voting was restricted to white male land-owners.

A little over a decade after the American Revolution, the French revolution began in 1789. The years preceding the French revolution were a time of political conflict in France. Because King Louis XVI had nearly absolute power over civic affairs, politics tended to play out in other institutions. One of these was the French Academy of Sciences, where in 1770 a mathematician and naval engineer, Jean-Charles, chevalier de Borda proposed the voting system, now known as the Borda count, which was based on a ranked order of multiple candidates [10]. Borda's method was used by the Academy for electing its members until Napoleon imposed his own method in 1801. Mathematician Donald Saari has been a strong proponent of the Borda method in recent years [39], although the Borda method has also been criticized as being easy to manipulate.

One of Borda's political rivals in the Academy was mathematician and philosopher Marie Jean Antoine Nicolas de Caritat, marquis de Condorcet. In 1785, Condorcet published an "Essay on the Application of Analysis to the Probability of Majority Decisions" [15] which is generally regarded as the first theoretical treatment of mathematics and politics. In his essay, Condorcet hypothesized that enlightened but imperfect voters would attempt to make the right choice for the good of society, and, with a certain probability, would vote for this "superior" candidate over any inferior alternative. Provided enough voters make the correct choices when presented with pairs of alternatives, the superior candidate will beat every other candidate and become, in the terminology of Chapter 3, the Condorcet candidate. According to Condorcet, this should always guarantee a victory for that candidate. Condorcet opposed the Borda count because he knew that it did not always select the Condorcet candidate. On the other hand, Borda knew that some preference profiles yield no Condorcet candidate, and so there is no winner under the method that Condorcet was advocating. Since Condorcet also knew this — his essay also contains a description of the Condorcet paradox — there is much speculation about which method Condorcet was actually trying to describe. H. P. Young [50] makes a convincing case that Condorcet actually intended the method we call Kemeny's method in Chapter 6. In 1876, a similar method had been proposed by Charles Dodgson [18], also known as Lewis Carroll, the author of *Alice's Adventures in Wonderland.*

In 1857, English barrister Thomas Hare described a voting method called *single transferrable vote* (often abbreviated STV), intended for elections, like parliamentary elections, in which $h \geq 1$ seats need to be filled in each district. The case $h = 1$ is what we call the Hare method in Chapter 2. The STV method is supposed to enfranchise voters with minority opinions. To implement STV, suppose each of n voters submits a preference ballot. The number $q = \lfloor n/(h+1) \rfloor + 1$ is called the **drop quota**. If no candidate receives at least q first-place votes (essentially, if there is an h-way tie), then the candidate with the fewest first place votes is dropped. If a candidate receives at least q first place votes, that candidate is elected. Then the winning candidate, as well as the votes that elected the winning candidate, are dropped from the profile, resulting in a new n and a new h. The process then repeats until all seats are filled. STV is currently used throughout Britain (for local elections), Scotland, and Ireland as well as Australia and New Zealand. In the United States, it is used to elect the City Council in Cambridge, Massachusetts.

A different type of voting method, called a **proportional method**, is used in multi-seat elections when the goal is to award seats to political parties in proportion to the number of votes they receive (ballots in such systems typically name parties rather than candidates). Proportional voting methods are essentially the same as apportionment methods (the subject of Part II). Indeed two common proportional voting methods, the Sainte-Laguë method (used, for example, in New Zealand, Sweden, and Germany) and the d'Hondt method (used, for example, in Argentina, Belgium, Israel, Japan, the Netherlands, Poland, Spain, and Turkey) are essentially equivalent to Webster's and Jefferson's methods of apportionment. However, neither STV nor a proportional method is used to elect the members of parliament in the United Kingdom. Instead, each district elects one member of parliament using the plurality method, much as the US House of Representatives is chosen.

When STV is applied in elections where a single winner is desired, then $h = 1$ and $q = \lfloor n/2 \rfloor + 1$, which represents a simple majority. In other words STV becomes the Hare method, which is also widely known as *instant runoff voting* (or IRV). Recently, the Hare method has been used in some local elections in the United States, for example, in San Francisco, California, and Takoma Park, Maryland. The Hare method has many proponents, including the organization "Fair Vote America" (see http://www.fairvote.org/). However, others criticize the Hare method and STV because both violate the monotonicity criterion.

In the 1950s, economists became interested in voting methods. Economists use the word **utility** to refer to the happiness, or the good, that an individual receives from a certain outcome (e.g., an election, a social policy, or the acquisition of goods or wealth). We discuss utility theory briefly in Chapter 14 in the context of probability and game theory. One problem with utility theory is that it makes no sense to compare utilities between different individuals quantitatively. This makes it difficult to decide which social policy most benefits a society as a whole. Nevertheless, as Italian economist Vilfredo Pareto (1848–1923) observed, certain social outcomes can safely be rejected on the basis of utility. According to Pareto, an outcome should be rejected if all members of a society rank a different outcome as having a higher utility. This suggests that it might be possible to make some social choices by simply knowing the preference order that the members of a society give the set of all possible social outcomes. This is essentially the same as a preference ballot, and a method for aggregating such a listing of preferences is the same as a social ranking function.

In his 1951 work *Social Choice and Individual Values*, [2], and subsequent works, economist Kenneth Arrow considered voting methods for three or more candidates that satisfy three desirability criteria: the Pareto criterion, the independence criterion (which Arrow called "independence of irrelevant alternatives"), and non-dictatorship. In contrast to Condorcet's voters, who try to do the right thing, Arrow's voters bring their biases to the election and let the voting method sort them out. Yet instead of finding the voting methods that select the "superior" candidate (as envisioned by Condorcet), Arrow's work establishes the impossibility of actually making these ideal social choices.

A restatement of Arrow's theorem, which attracted much public attention, says that any voting method that satisfies the Pareto and independence criteria must actually be a dictatorship. This result is often quoted facetiously to say that dictatorships are the only acceptable social choice functions. However, as we saw in Chapter 4, most voting methods used in practice are actually nondictatorial, and they accomplish this by abandoning independence. Arrow was awarded the 1972 Nobel Prize in Economics for this and other seminal research.

In 1952, Kenneth May followed Arrow's work with the publication [33] that establishes Theorem 1.23 from Chapter 1. Much of the work on social choice theory since the 1950s has been concerned with extending Arrow's Theorem and also trying to explain why Arrow's Theorem isn't fatal to social choice theory.

In 2011, Balinski and Laraki [5] offered a new approach in social choice theory, proposing a voting paradigm that they called "Majority Judgment". This method employs a new kind of ballot, organized like an approval ballot except that voters express an assessment of each candidate using a word chosen from a fixed set of adjectives, such as {excellent, good, fair, poor}. Balinski and Laraki identify an optimal social choice function in this context that avoids the conundra of Condorcet and Arrow and that is resistant to strategic or insincere voting.

There are numerous sources for more detail about social choice theory in general. Among them are books by Fishburn [23], Kelly [31] and Saari (e.g., [39] or [40]). There is an excellent elementary chapter on voting in Taylor and Pacelli [46], from an earlier edition of which comes Proposition 3.13. Taylor's more recent book [45] on social choice focuses on the concept known as manipulability. An innovative book on social choice theory that follows a strictly problems-based approach is [28].

Part II

Apportionment

Introduction to Part II

The United States Constitution requires that a census be conducted every 10 years to determine the population of each state. This information is used in various ways, to determine both liabilities and assets to be accorded to the states. In particular, and most significantly, these numbers are used to determine the number of seats assigned to each state in the House of Representatives. This in turn dictates the number of electors — effectively the number of votes — assigned to each state in the presidential election process conducted via the Electoral College.

Article I, Section 2 of the Constitution says that the representatives should be apportioned to the states "according to their respective numbers". Unfortunately, the Constitution does not offer any further guidance about precisely how this should be done, and every 10 years Congress is left to implement an apportionment that meets the spirit of these words. It is not hard to determine each state's "fair share" of the House of Representatives. It is the same percentage of the entire House that the state's population is of the total population of the United States. In other words, if a state's population is 3% of the entire population of the United States, then that state's fair share should be 3% of the House of Representatives. The difficulty is that this fair share need not be a whole number, and yet the number of representatives assigned to each state must be a whole number. So some process is needed to round these fair shares to whole numbers. This is the apportionment problem. What process should we use?

At first, there does not seem to be much substance to this question. Can't we simply choose the whole number that is nearest to the state's fair share? It is only upon some reflection that one realizes that the problem is more subtle than this. Rounding the fair shares of all states to their nearest whole number may result in assigning too few or too many seats all told. What should we do? It turns out that Alexander Hamilton and Thomas Jefferson began a debate on this topic in 1792. There are a number of sensible approaches to the problem, but each has its flaws. Analogous to what we did in Part I, we will identify criteria that we would like an apportionment method to satisfy and investigate which methods satisfy which criteria. We will also prove a theorem, due to Michel Balinski and Peyton Young, that states that it is impossible, unfortunately, for any method to simultaneously satisfy all of these criteria. Like Arrow's Theorem, the theorem of Balinski and Young tells us that we must be content with an imperfect apportionment method, flawed in one way or another.

Although the method of apportionment used in the United States has remained unchanged since 1940, it remains a matter of debate to this day whether that method is in fact the optimal choice.

7

Hamilton's Method

"We are not asking the . . . federal government or anyone in Washington to bail us out. We just want our fair share." — Arnold Schwarzenegger

7.0 Scenario

Three states in a certain union have populations as follows:

- State A — 4,400,000;
- State B — 45,300,000;
- State C — 50,300,000.

The federal legislature will accord these three states, all together, 100 seats in the House of Representatives. Clearly each state must receive a whole number of seats, while a state's "fair share" of seats according to its population will not necessarily be a whole number. So a method is needed to round real numbers into whole numbers. This is the essence of the apportionment problem.

Here is the apportionment method proposed by Alexander Hamilton: Determine each state's fair share (also called its **standard quota**) by dividing its population by the total population and multiplying this by the total number of seats to be apportioned. Then round these standard quotas either up to the next largest integer (the so-called **upper quota**) or down to the next smallest integer (the so-called **lower quota**) in such a way that (1) the correct number of seats have been doled out, and (2) the benefit of the extra representative is doled out to the states in decreasing order of the fractional part of their standard quotas. (The fractional part of a number is the part after the decimal point. For example the fractional part of 3.14159 is 0.14159.)

(1) How would Alexander Hamilton's method apportion the seats?

(2) What if the legislature changes its mind and allocates 101 seats to the three states?

(3) Suppose that a new census reveals that the populations of the 3 states change to:

- State A — 4,500,000;
- State B — 45,200,000;
- State C — 49,000,000.

How does this affect the apportionment of 100 seats by Hamilton's method?

(4) Suppose we return to the original census figures, but we assume that the legislature accepts a new state (State D) with population 1,700,000 into the union, and the legislature decides to increase its size from 100 to 102 seats, accounting for the 2 seats that state D seems certain to get. How would these 102 seats be apportioned to the 4 states by Hamilton's method?

Compare your answers to the four questions above. Specifically compare your answer to question (1) to your answer to each of question (2), question (3), and question (4). Does all this seem fair and reasonable?

7.1 The Apportionment Problem

Let's look at the apportionment problem from the point of view of the state of Maryland and the 2010 census. For the purpose of apportionment, Maryland's population was determined to be 5,789,929, while the entire population of the United States was determined to be 309,183,463. Thus Maryland had a fraction 5,789,929/309,183,463, or approximately 1.873%, of the total US population. In recent years the size of the US House of Representatives has been fixed at 435 members, so Maryland's fair share of seats in the House should be 1.873% of 435, which is roughly 8.15. Of course, Maryland can't be assigned 8.15 representatives because some whole number must be assigned. It may seem most natural to choose 8, the integer nearest to 8.15. But this would leave the citizens of Maryland less well represented than they ought to be. Congressional districts in Maryland would be larger than the US average, and each Maryland representative would have an above average number of constituents. On the other hand, if 9 representatives were assigned to Maryland, the citizens of Maryland would have more representation than they have a right to expect.

In fact, it isn't sufficient to think of the apportionment problem one state at a time. The problem is to make a simultaneous assignment of whole numbers to states in approximate proportion to the populations of the states. But in the end there are only 435 seats to be distributed, so although assigning 9 seats to Maryland may seem the generous thing to do, one cannot be similarly generous to all states at the same time. To round the fair shares of all the states up to the next whole would require an extra supply of House seats. In particular, using data from the 2010 census, this "generous algorithm" would assign a total of 462 seats to the states. The difficulty with this is not so much that 462 is too large or that there aren't enough chairs in the House chamber. From a mathematical viewpoint, the difficulty is that once it is decided to dole out 462 seats, every state obtains a new fair share of this expanded House. Computing those fair shares and rounding all of them upwards leads to further escalation of the House size in a never-ending spiral of generosity.

Let us generalize the setting somewhat. An apportionment problem involves some discrete collection of identical elements that must be shared among various stakeholders. The stakeholders are assumed to have claims to those elements of varying strength. The goal is to apportion the identical elements among the stakeholders as nearly as possible in direct proportion to the strength of the stakeholder's claim. In the case of the apportionment of the US House of Representatives, the stakeholders are the 50 states and the identical elements are the 435 House seats. The strength of the claim that each state has on House seats is given by the population of that state.

Doling out representatives in a federal system of government is only one application of a more general apportionment problem. It is important to recognize that the problem has broader applicability. When a school system doles out teachers to schools based on their enrollment figures, it is solving an apportionment problem. When a transit system doles out buses to bus routes based on ridership, it is solving an apportionment problem. When a group of children doles out candies from a box, based on each child's initial investment, it is solving an apportionment problem.

All of these applications have two common features. First, the elements to be apportioned are indivisible. When money is to be divided amongst parties, it is more or less possible to give each party their exact fair share, because money is divisible (at least down to the penny). So we do not consider problems of dividing money or liquid or anything else that can be separated into parts along a continuous scale. The apportionment problem is a problem of allocation of discrete elements. Second, the elements are identical. When heirs come to divide up an estate, the items to be divided may include furniture, automobiles, artwork, and other possessions. There is no reason to expect that these items would be of equal value, nor, for that matter, that any individual item would be equally valued by the various beneficiaries. So our study of apportionment does not apply directly to the more complex problems of division of estates. We confine our attention to problems of allocating identical items.

As we proceed to develop this theory, we will use the language of the political apportionment problem, because it is the main motivating example for our study. So we will speak of apportioning the seats in the House, even as we are aware that we may be apportioning teachers or buses or candies. We will speak of states, even as we are aware that our stakeholders may be schools or bus routes or purchasers. And we will speak of populations, even though the strength of a claim on the elements may be based on money or age or even good deeds.

7.2 Some Basic Notions

Let us now attach some notation to the various parameters in this problem in order to be precise about what we mean when we speak of an "apportionment method". We will consistently use the same letters to stand for the same quantities in the problems. We begin with a positive integer $n > 1$, which stands for the number of states (or stakeholders) in the problem. The value of n appropriate for the political apportionment problem in the United States over the past half century has been 50. We also have a whole number $h > 0$, which stands for the number of seats in the House of Representatives (the number of identical elements to be shared). The value of h for the current political apportionment problem in the United States is 435, the number of seats in the House of Representatives. We imagine that the n states are listed in a certain order (alphabetical, say, or in decreasing order of population). Let p_k be the population of the kth state, so that the population of the first state is denoted p_1, the population of, say, the 17th state is denoted p_{17}, and the population of the last state is denoted p_n. We use the letter p, without a subscript, to denote the total population of all the states. Hence

$$p = p_1 + p_2 + \cdots + p_n.$$

We refer to the numbers h, n, and $p_1, p_2, p_3, \ldots, p_n$ collectively as a **census**. The census is the input to an apportionment problem. The output is to be a list of n whole numbers that we denote by $a_1, a_2, a_3, \ldots, a_n$. We think of the number a_k as the number of seats accorded to the kth state. In general, we permit the possibility that a_k might equal 0 for some k. The letters we use here were chosen from the alphabet in accordance with our motivating example: h is for house size, n is for number of states, p is for population, and a is for apportionment.

Definition 7.1. An **apportionment method** is a function whose domain is the set of all possible values of h, n, and $p_1, p_2, p_3, \ldots, p_n$, where h and n are positive integers and the numbers $p_1, p_2, p_3, \ldots, p_n$ are positive real numbers, and whose output is a sequence of n nonnegative integers $a_1, a_2, a_3, \ldots, a_n$ satisfying

$$a_1 + a_2 + \cdots + a_n = h.$$

This equation represents the requirement that the total number of seats accorded to the states must equal the prescribed house size, exactly.

We will consider a variety of apportionment methods. All involve numerical calculations followed by comparing the sizes of various quantities. We will often uncritically instruct the reader to choose the larger of two quantities without pausing to ask what happens if the two quantities are precisely equal. One approach to this difficulty is to ignore this possibility, defending this position by arguing that an exact tie is unlikely. Another approach is to make the choice both ways and to deliver two different outputs to the function. While the latter approach is more like the way we handled ties in the theory of voting, the former approach is the one we will take in our treatment of the theory of apportionment.

If we did want to include ties, we could allow an apportionment method to report two (or more) outputs as protection against the possibility that the input forces a decision where there is no difference. In that case, we can simply agree to do it both ways and report two (or more) answers instead of one. The user of this information (Congress, perhaps) can decide for itself how to break ties.

Let us practice using our notation. Assume that the 50 United States are listed in alphabetical order, so that Maryland is the 19th state. The ratio p_k/p represents the fraction of the total population that resides within the kth state. For example, the fraction of the United States population that resides in Maryland is $p_{19}/p = 5{,}789{,}929/309{,}183{,}463 \approx 1.873\%$, using census data from the year 2010. The ratio a_k/h represents the fraction of the total number of representatives that have been apportioned to the kth state. Since Maryland is assigned 8 representatives, this ratio for Maryland is $a_{19}/h = 8/435 \approx 1.839\%$. Our aim is to make these two percentages as close to one another as possible. Equality would be perfect, but it isn't possible, because the numbers p, p_{19}, and h are given to us, and the number a_{19}, that we get to choose, must be a whole number. The 8 seats assigned to Maryland leave the state slightly short of the representation it deserves. But we can't simply attain perfection by setting a_k equal to $h(p_k/p)$, because there is no reason to suppose that this quantity will be a whole number. This is the essence of the difficulty with apportionment.

Here is another way to look at it. The ratio p/h represents the average size of a congressional district across the entire United States. Using 2010 census data, this number is 309,183,463/435 $\approx$ 710,767. If it were possible to carve up the country into h districts of exactly equal population, each district would have population p/h and could have one representative. But the United States is a federation of states, and our laws require that congressional districts do not cross state lines. The ratio p_k/a_k represents the average size of a congressional district in the kth state. For Maryland, this is $p_{19}/a_{19} = 5{,}789{,}929/8 \approx 723{,}741$. Our aim is to make this fraction as close to p/h as possible. But again we cannot achieve perfection, because the number a_{19} must be a whole number. Note that the size of an average congressional district in Maryland is somewhat larger than would be ideal, which is another way of seeing that Maryland residents are somewhat underrepresented in Congress.

Taking the reciprocals of these fractions leads to an alternative interpretation. The fraction a_k/p_k can be thought of as a measure of the strength of the voice in Congress of a single citizen in the kth state, assuming that the state is divided into a_k districts of equal size. That size would be p_k/a_k, and each resident of that district may be thought of as having a fraction $1/(p_k/a_k) = a_k/p_k$ of a representative in Congress. In Maryland, this fraction comes out to be $a_{19}/p_{19} = 8/5{,}789{,}929 \approx 1/723{,}741 \approx .00000138$. We would like to keep this ratio independent of k, so that each citizen has the same voice in Congress, regardless of which state they reside in. If these could all be precisely equal, they would all be $h/p = 435/309{,}183{,}463 \approx .00000141$, which represents the voice in Congress for each US resident if the congressional districts could cross state lines and be exactly equal in size. Once again, the obstacle to perfection is that the apportionment number a_{19} must be an integer.

The real number $q_k = h(p_k/p)$ is called the fair share or **standard quota** of the kth state. The standard quota for the state of Maryland is 435(5,789,929/309,183,463) $\approx$ 8.15. Sometimes it is convenient to represent the standard quota in the form $q_k = p_k/(p/h)$. The denominator in this last expression is called the **standard divisor**, and we designate it by the letter $s = p/h$. Thus, we obtain the standard quota of a state by dividing its population by the standard divisor:

$$q_k = p_k/s.$$

The word "quota" here is related to the word "quotient"; our quotas are indeed the results of dividing.

The standard divisor represents the population of an ideal district. The standard quota for a state represents the ideal number of seats to assign to that state. If we could set a_k exactly equal to the standard quota q_k of the kth state, then we would automatically have $a_1+a_2+\cdots+a_n = h$ because

$$\begin{aligned} &q_1 + q_2 + q_3 + \cdots + q_n \\ &= h(p_1/p) + h(p_2/p) + h(p_3/p) + \cdots + h(p_n/p) \\ &= (h/p)(p_1 + p_2 + p_3 + \cdots + p_n) = (h/p)p = h. \end{aligned}$$

But we can't do that, because we are restricted to choose a_k to be a whole number. Two natural choices for a_k are the **lower quota**, which is the standard quota rounded down, or the **upper quota**, which is the standard quota rounded up. Maryland's standard quota from the 2010 census is approximately 8.15, so Maryland's lower quota is 8 and Maryland's upper quota is 9.

It is important to recognize the high stakes involved in the political apportionment problem. As an example, the 2010 census affords Montana a standard quota of approximately 1.40. Whether this is rounded up to 2 or down to 1 is critically important to the welfare of Montanans. If Montana is allocated just one seat in the House, the lone representative from this state will represent 994,416 people, making Montana the largest congressional district in the nation (by far), much larger than the average size of a district nationwide, the standard divisor, which in the United States today is approximately 710,000. On the other hand, if Montana is allocated two seats, each congressional district in Montana will contain approximately 497,000 residents, far fewer than the national average. It seems that the only choices for Montana are extremes, making the residents either the most represented or the least represented Americans. Moreover, the apportionment decision based on 2010 census data has an impact on Montana's Electoral College representation in presidential elections of 2012, 2016, and 2020. (In fact, the apportionment method used in 2010 assigned just one seat to Montana.)

7.3 A Sensible Approach

The apportionment method proposed by Alexander Hamilton in 1792 assigns to each state either its lower quota or its upper quota. Any method that does this is called a **quota method**. Of course, every state would prefer to get its upper quota, but not every state can. Since the standard quotas sum to h, if we rounded them all up, the sum would exceed h, which is forbidden. So some states will necessarily get fewer representatives than their standard quota. There will be some losers in this process.

The **integer part** of a real number is the greatest integer less than or equal to it. The **fractional part** of a real number is the difference between that number and its integer part. For example, the integer part of 3.14159 is 3 and the fractional part of 3.14159 is .14159. Hamilton argued that a state should be awarded a number of representatives that is at least the integer part of its standard quota (i.e., its lower quota), and then it should be considered for one additional seat based on the fractional part of the standard quota. For example, if the standard quota of a state is 5.9, Hamilton felt that we should definitely award at least 5 representatives to this state and that we should be inclined to give that state its upper quota of 6 representatives. This is because the fractional part of the standard quota 5.9 is 0.9, which is large (almost 1). If, on the other hand, another state has standard quota equal to 14.1, then that state's claim to its upper quota is smaller, represented by fractional part of just 0.1. So under Hamilton's method we should round 5.9 up to 6 before we round 14.1 up to 15. This makes a lot of sense.

Definition 7.2. Hamilton's method: As a provisional apportionment, assign every state its lower quota. Then assign the seats that remain to the states, at most one per state, in decreasing order of the size of the fractional parts of their standard quotas.

We illustrate this method with an example.

Example 7.3. Apportion $h = 10$ seats to $n = 3$ states with populations $p_1 = 264$, $p_2 = 361$, and $p_3 = 375$.

The total population p is $264 + 361 + 375 = 1000$, so the standard divisor is $s = p/h = 1000/10 = 100$. This indicates that there is one representative for every 100 people across the three states, so an ideal district would have precisely 100 residents. When the standard divisor is a round number like this, the computations are particularly simple. The standard quotas are obtained by dividing the state populations by the standard divisor s. We obtain $q_1 = 2.64$, $q_2 = 3.61$, and $q_3 = 3.75$ for the three states. The lower quotas are 2, 3, and 3, respectively. According to Hamilton's method, we provisionally apportion 8 of the 10 seats in the House according to these lower quotas. Two seats remain unassigned, and we give them to the two states the fractional part of whose standard quota is largest. These fractional parts are 0.64, 0.61, and 0.75, so it is the first and the third states that get extra seats. These states are both assigned their upper quota instead of their lower quota. The final Hamilton apportionment is therefore $a_1 = 3$, $a_2 = 3$, and $a_3 = 4$. The process is summarized in the Table 7.1.

Table 7.1 Working out Hamilton's method.

k	p_k	Standard Quota	Lower Quota	Upper Quota	Fractional Part of Standard Quota	Hamilton Apportion-ment
1	264	2.64	2	3	0.64	3
2	361	3.61	3	4	0.61	3
3	375	3.75	3	4	0.75	4

$p = 1000$ $h = 10$

Hamilton's method is the first method one can imagine, and it might be regarded as the obvious one. In fact, at first it is difficult to imagine any other way to solve an apportionment problem. You simply round every state's fair share up or down, rounding up the states that are closer to their upper quotas, naturally. What could be wrong with that? Plenty, it turns out.

7.4 The Paradoxes

One curious feature of Hamilton's method is that it can behave strangely in the face of changes to the house size h, the number n of states, or the populations $p_1, p_2, p_3, \ldots, p_n$ of the states. In fact, this strange behavior is sufficiently disturbing that Hamilton's method has nowadays been all but eliminated from consideration as the method of choice in the United States. But it was not always so. Hamilton's method was in use to some degree in the United States during the last part of the 19th century and into the 20th, during which time these strange behaviors were actually encountered occasionally in the course of discussions about apportionment in Congress. To understand the weaknesses of Hamilton's method, we investigate an example.

Example 7.4. Let us suppose that we have a union of $n = 3$ states, with $h = 10$ seats to be apportioned. Imagine that the population data is given in Table 7.2.

The populations have been chosen to sum to exactly $p = 10{,}000{,}000$ for ease of computation. This means that the standard divisor $s = p/h = 1{,}000{,}000$, which we interpret as the desired size of every congressional district. Thus the standard quotas are obtained by dividing the standard divisor 1,000,000 into the populations of the states. (This is easy to do

Table 7.2 Computing the Hamilton apportionment with $h = 10$.

State	Population	Standard Quota	Lower Quota	Hamilton Apportionment
1	$p_1 = 1{,}450{,}000$	1.45	1	2
2	$p_2 = 3{,}400{,}000$	3.40	3	3
3	$p_3 = 5{,}150{,}000$	5.15	5	5
	$p = 10{,}000{,}000$			$h = 10$

without a calculator!) Assigning the lower quota to each state leaves 1 seat unassigned. According to Hamilton's method, the tenth seat should be assigned to state 1, since the fractional part of its standard quota is 0.45, larger than that of either of the other states. Thus the Hamilton apportionment assigns 2, 3, and 5 seats to the three states, respectively, as shown in Table 7.2. So far, so good.

Now imagine that the Congress authorizes an increase in the size of the House from $h = 10$ to $h = 11$. One imagines that such an addition would be welcomed by these states, although one might also imagine that a battle would ensue over the 11th seat. Which state will gain from this addition? Which state will be granted the extra seat?

It is not correct to look at Table 7.2 and select state 2 on the basis that the fractional part of its standard quota is second largest. In fact, the new value of h results in a new standard quota for each state. Each state's fair share of 11 seats is naturally larger than its fair share of 10 seats. The new standard divisor is $s = p/h = 10{,}000{,}000/11 \approx 909{,}091$. The new quotas and apportionments are shown in Table 7.3.

Table 7.3 Computing the Hamilton apportionment with $h = 11$.

State	Population	Standard Quota	Lower Quota	Hamilton Apportionment
1	$p_1 = 1{,}450{,}000$	1.595	1	1
2	$p_2 = 3{,}400{,}000$	3.740	3	4
3	$p_3 = 5{,}150{,}000$	5.665	5	6
	$p = 10{,}000{,}000$			$h = 11$

The lower quotas remain unchanged from Table 7.2, but this time there are two seats that remain to be allocated after the lower quota is granted to each state. At the same time, each state's standard quota has increased (by exactly 10%), so each state's justification for an extra representative is stronger than it was before. The surprise is that the winners this time are states 2 and 3, and the Hamilton apportionment assigns 1, 4, and 6 seats to the three states, respectively, as shown in Table 7.3.

There is a curious anomaly in the data from Tables 7.2 and 7.3. When the House size increased from 10 to 11, the Hamilton apportionment allotted to state 1 decreased from 2 to 1. Why should one state lose a seat when there are more seats to go around? Since state 1 was already entitled to 2 seats when $h = 10$, how could it lose this entitlement when one additional seat became available for sharing? If we imagine doling out the 11 seats to the three states, one seat at a time, it is as if we would have to retract our decision about the 10th seat when deciding who has the greatest claim for the 11th. So, in asking which state would get the 11th seat, we found rather that, when $h = 10$ was increased to $h = 11$, states 2 and 3 both got an extra seat, while state 1 was forced to relinquish one.

This phenomenon, where a state loses a seat when the House size is increased, has come to be called the **Alabama paradox**, after the state that was to be harmed by an increase in the value of h in the 1880 reapportionment. The loss of a seat in the face of an increase in h seems unreasonable for a fair apportionment method. We therefore articulate a criterion for apportionment methods that forbids this kind of unreasonableness. An apportionment method is called **house monotone** (or satisfies the **house monotonicity criterion**) if an increase in h, all other parameters remaining fixed, can never result in a decrease in any a_k. Thus an instance of an Alabama paradox is an example showing that Hamilton's method violates the house monotonicity criterion.

Further study of Tables 7.2 and 7.3 demonstrates the source of the difficulty. The standard quota for state 1 increases from 1.45 to 1.595 as h increases from 10 to 11. Simultaneously, the standard quota for state 3 increases from 5.15 to 5.665. So while both of these represent a 10% increase, the latter increase is much larger in absolute terms than the former increase. Notice how the fractional part of the standard quota of state 3 is smaller than that of state 1 when $h = 10$ but not when $h = 11$. The 10% increase has helped state 3 much more than it helped state 1, and only after h is increased from 10 to 11 does state 3 overtake state 1 in the ranking for extra seats. State 2 similarly overtakes state 1, which is knocked from first place into last place in its claim for an extra seat. Hence the Alabama paradox.

A related and equally disturbing paradox involves changes in population. Return to the data from Table 7.2 and compare it to the data in Table 7.4. We might imagine that the data in Table 7.4 is the result of a new census. One observes that the population of state 1 increased slightly from Table 7.2 to 7.4, while the population of states 2 and 3 decreased, in state 2 slightly but in state 3 substantially. Now what is the result of these changes on the Hamilton apportionment? State 1, whose population grew, loses a seat to state 2, whose population fell. What sense does this make, when the two states are merely trying to find the right way to share 5 seats together? How can it be that the stronger claim of state 1 to the extra seat, coupled with a weaker claim of state 2 to the extra seat, result in state 1 being forced to relinquish a representative to state 2?

Table 7.4 Computing the Hamilton apportionment after a population shift.

State	Population	Standard Quota	Lower Quota	Hamilton Apportionment
1	$p_1 = 1{,}470{,}000$	1.55	1	1
2	$p_2 = 3{,}380{,}000$	3.56	3	4
3	$p_3 = 4{,}650{,}000$	4.89	4	5
	$p = 9{,}500{,}000$			$h = 10$

This phenomenon is called the **population paradox**. When one state gains in population while another loses, and yet in the transition the first state loses a seat while the second gains a seat, we have an instance of the population paradox. The example of Tables 7.2 and 7.4 shows that the Hamilton method is susceptible to the population paradox.

We would hope for an apportionment method that avoids the population paradox. Such a method is said to satisfy the **population monotonicity** criterion. Hamilton's method is susceptible to both the Alabama paradox and the population paradox. Hamilton's method is neither house monotone nor population monotone.

Let us return to Table 7.2 and ask how this union of three states might greet the addition of a new state. Suppose a new state is acquired with a population of 2,600,000. Since every 1,000,000 citizens entitle a state to one seat in the House, we can predict that this new state will be assigned 3 seats (or perhaps just 2). One way to handle this additional state would be to increase the House size h by 3 to a total of 13 seats, so that the new state could get the 3 seats it seems likely to earn, and so that the original three states could retain their original 10 seats. Table 7.5 shows a calculation of the Hamilton apportionment for this new union with $h = 13$.

The first thing to notice is that state 4 did indeed get the 3 seats that we predicted, leaving the original three states with their original

Table 7.5 Computing the Hamilton apportionment after the addition of a new state.

State	Population	Standard Quota	Lower Quota	Hamilton Apportionment
1	$p_1 = 1{,}450{,}000$	1.50	1	1
2	$p_2 = 3{,}400{,}000$	3.51	3	4
3	$p_3 = 5{,}150{,}000$	5.31	5	5
4	$p_4 = 2{,}600{,}000$	2.68	2	3
	$p = 12{,}600{,}000$			$h = 13$

10 seats. But something strange has gone on. A comparison of the data in Tables 7.2 and 7.5 shows that in the process of adding state 4, state 2 acquired a seat from state 1, this despite the fact that neither their populations nor their combined number of seats changed. In Table 7.2, state 1 got 2 seats and state 2 got 3, while in Table 7.5, state 1 got 1 seat and state 2 got 4. On what basis can state 2 defend its acquisition of a seat from state 1?

This phenomenon is called the **new states paradox**. It is also known as the **Oklahoma paradox**, owing to the impact of the addition of Oklahoma to the union. The apportionment for the first decade of the 20th century was computed using Hamilton's method. When Oklahoma entered the union in 1907, it was awarded 5 representatives in the House. Had it been added in the manner of state 4 in Table 7.5, however, New York would have been ordered to concede a seat to Maine, this despite the fact that the calculation would use the population data for New York and Maine from the 1900 census without alteration. This illustrates a serious flaw with Hamilton's method.

The new state's paradox can be seen as an instance of the population paradox, since it involves two states, call them state i and state j, whose populations p_i and p_j remain unchanged and yet a_i decreases and a_j increases. So the new state's paradox is seen to be a special case of the population paradox.

The point of this discussion is not so much to denigrate and discard Hamilton's method as to cast doubt about it as the obvious method of choice. It remains an appealing alternative, owing in part to its simplicity. A fair question is: Is there a clearly better method? Is it possible to find a method that satisfies these monotonicity criteria, that avoids the Alabama and population paradoxes? The answer to this last question is yes. In the next chapter, we introduce a class of apportionment methods that are both house monotone and population monotone. Unfortunately, these methods turn out to have difficulties of their own.

7.5 Exercises and Problems

7.1. If a nation has 17 equally populous states and a house size of 60, then what are the standard quota, the lower quota, and the upper quota for each state?

7.2. If a state has 17% of the population of a nation with a house size of 60, then what are the standard quota, the lower quota, and the upper quota for this state?

7.3. Which of the following are apportionment methods?
(a) Every state is assigned its lower quota.
(b) The largest state is assigned all of the representatives.
(c) The smallest state is assigned all of the representatives.
(d) Any state whose standard quota has fractional part at least 0.5 gets its upper quota. All other states get their lower quota.

7.4. Alice, Barbara, and Carolyn purchase a bag of pearls for $10,000. Alice contributed $1250, Barbara contributed $3650, and Carolyn contributed $5100. They take the bag home and pour 20 pearls onto their kitchen table.
(a) How should they apportion the pearls, using Hamilton's method?
(b) While Alice and Barbara are out of the room, Carolyn notices that the bag contains one more pearl. How would the apportionment be affected if a 21st pearl were introduced?

7.5. A small country that uses Hamilton's method of apportionment has three states of population 540, 2430, and 7030, respectively.
(a) Compute the apportionment with house size 10 and with house size 11. Which state got the extra representative?
(b) Suppose the country stays with a house size of 11 but the population of the smallest state decreases by 10 while the population of the other two states increases by 70 and by 940, respectively, yielding new populations of 530, 2500, and 7970. How is the apportionment affected?

7.6. A certain country has states with populations 6000, 11,000, 16,000, 21,000, and 26,000. Work out the Hamilton apportionment for $h = 15$, $h = 16$, and $h = 17$.

7.7. A certain county wants to establish a local bus service consisting of

6 routes (denoted A through F) using 130 buses. Research shows that the average daily ridership for each route is:

A	B	C	D	E	F
45,300	31,070	20,490	14,160	10,260	8,720

Explain the interpretation of all of the quantities: p, h, the standard quota, the upper quota, the lower quota, and especially the standard divisor s in the context of this bus route apportionment problem. How would the bus schedule be apportioned by Hamilton's method?

7.8. In the country of Begonia, the national assembly must decide how to apportion its membership across the various states.

(a) A certain Begonian state has population 5,800,000. The total population of Begonia is 310,000,000. How large does the house size h have to be for the state to have a lower quota of at least 8?

(b) A state has population 5,800,000. The house size in Begonia is 435. How small does the total population p of Begonia have to be for the state to have a lower quota of at least 8?

(c) The total population of Begonia is 310,000,000. The house size is 435. How large does the population of a state have to be for the state to have a lower quota of at least 8?

7.9. Three children pool their change to buy a box of candy that costs \$1.10. Bobby contributes a dime, and Jessie and Sandy each contribute two quarters. When they get the box home, they pour out the 27 pieces of candy. It is decided that they should apportion the pieces according to their monetary contributions. Not having any background in making such decisions, they adopt Hamilton's method to distribute the candies.

(a) How many candies does each child receive?

(b) It is suddenly discovered that the box contains a 28th piece of candy. The children are delighted and eagerly compute the Hamilton apportionment for 28 candies to see which child is entitled to the 28th piece. Who gets it?

7.10. In a particular country, there are three states with populations 241,000, 339,000, and 420,000.

(a) Hamilton's method is to be used to apportion $h = 10$ seats. What is the result?

(b) The largest state secedes from the union, leaving just two states with populations 241,000 and 339,000. How would Hamilton's method apportion $h = 6$ seats to this new two-state union? What anomaly do you notice?

7.11. Assume that Hamilton's method is adopted as the apportionment method. The Alabama paradox shows that a state can lose a seat when

h is increased. Show that a state can never lose *more* than one seat when h is increased.

7.12. Show that the Hamilton apportionment for a two-state union assigns to each state the whole number that is closest to the state's standard quota.

7.13. Show by example that a state the fractional part of whose standard quota is 0.1 can obtain its upper quota via Hamilton's method.

7.14. Show by example that a state the fractional part of whose standard quota is 0.9 can obtain its lower quota via Hamilton's method.

7.15. The **Gross method** of apportionment works as follows: As a provisional apportionment, assign every state its lower quota. If k seats remain to be assigned, then assign them to the k largest states. Give an example to show that this method does not coincide with Hamilton's method.

7.16. The apportionment method of Lowndes is a quota method that works as follows. As a provisional apportionment, give each state its lower quota. Then rank the states in decreasing order of the ratio of the fractional part of their standard quota to their lower quota. In other words, if a state has standard quota 3.1, this ratio is $0.1/3 = 1/30$. Give whatever seats remain to the states, at most one per state, in the order of this ranking.

(a) Compute the Lowndes apportionment for $n = 2$, $h = 10$, $p_1 = 1200$, and $p_2 = 8800$.

(b) Which states are likely to favor Lowndes's method over Hamilton's method?

7.17. Create an example of a census involving three states with $h = 8$ in which Hamilton's method gives an exact tie between an apportionment of 1, 2, 5 and an apportionment of 1, 3, 4.

7.18. Obtain from the web the data for apportionment populations of the United States from the year 2010 census. Construct a spreadsheet that computes the apportionment numbers using Hamilton's method. Compare these numbers with the actual apportionment using Hill's method (i.e., the current method used in the United States).

8

Divisor Methods

"Friendship makes prosperity more shining and lessens adversity by dividing and sharing it." — Cicero

8.0 Scenario

Imagine a country consisting of 5 states with populations 1.3 million, 2.7 million, 4.7 million, 8.1 million, and 16.9 million, respectively. Suppose that the constitution of this country directs that there shall be a legislative body of representatives from the states but forbids any representative from representing fewer than 700,000 people.

The parliament decides to allot to each state the maximum number of representatives permitted by the "700,000 clause" in the constitution. This means, for example, that the state with 4.7 million residents will be assigned 6 representatives, since each of these representatives could represent 1/6 of 4.7 million, or roughly 783,000 residents. Seven representatives are forbidden, because 1/7 of 4.7 million is roughly 671,000, and there is no possibility of having at least 700,000 people in each of 7 districts in this state.

(1) How many representatives does the parliament assign to each of the states?

(2) How many representatives does the parliament assign all told? Call the answer h.

(3) How would Hamilton's method have assigned h seats to these five states? In light of the Hamilton apportionment, which state is likely to complain about the parliament's procedure?

(4) How does the largest state fare under the parliament's procedure? Compare the apportionment for this state with the standard quota for this state.

After the new parliament is seated, a national capitol is built with a parliamentary hall seating precisely h representatives. After that, a new census is obtained, the populations of the 5 states are seen to have grown to 2 million, 3 million, 5 million, 10 million, and 17 million. The parliament recognizes the need to recompute the apportionment numbers, but the constitutional rule would allocate more representatives all told, and their parliamentary hall cannot accommodate any more representatives. So they decide to amend their constitution to increase the number 700,000 in such a way that the prior rule can still be used and yet the number of representatives that they apportion comes to equal exactly the prior value of h.

(5) What number should replace 700,000 in the constitution?

(6) What apportionment results from the constitutional amendment with the new census?

8.1 Jefferson's Method

In our framework for studying apportionment, the house size h is regarded as part of the input of an apportionment function. This is to say that we assume that h is fixed and given to us in advance, and our job is to distribute exactly that many seats. This was not always the viewpoint in the United States. For most of our history, the size of the House of Representatives was not fixed in advance, but was determined by congressional apportionment legislation. In 1792, the first apportionment of the House was based on a census that allotted $h = 105$ members. The number of members rose gradually over the ensuing decades, reaching its current value of $h = 435$ in 1930.

If h is not predetermined, then another approach to apportionment is suggested. Why don't we just round the share of representation in every state according to some sort of systematic rule and allow the value of h to fall where it may? According to this viewpoint, the house size h is part of the output of the apportionment process rather than part of the input.

This was the idea of Thomas Jefferson. Instead of beginning with a fixed value of h, Jefferson proposed to begin with a fixed idea of how large a congressional district should be, at least. When we have a fixed value for h, we use the standard divisor $s = p/h$ to represent this quantity. If we do not have a fixed value for h, then we choose an arbitrary number d that we regard as an appropriate size for a congressional district. Such a number d is called a **modified divisor**. (The word "modified" refers to the fact that the divisor is not the standard one.) Just as we did with the standard divisor s, we divide the modified divisor into the population of each state to obtain what we call a **modified quota**. In particular, if p_k is the population of state k, then the modified quota for state k is p_k/d.

This quota represents the appropriate number of seats to assign to state k, if each district is to have population d. But unfortunately this number is not necessarily an integer. So Jefferson proposed that we round these numbers down to the next whole number. This uniform rule about rounding is in contrast to Hamilton's method, with its winners and losers. We are able to round the modified quotas down because we are not forced in advance to settle on any particular number h of house seats.

In Jefferson's original conception, the number of house seats h was determined only at the end, by adding the seats allotted to each state. That is to say, in the end we get a house size $h = a_1 + a_2 + a_3 + \cdots + a_n$. But Jefferson's idea can also be used for an apportionment for which h is specified in advance. This is accomplished by selecting the divisor d strategically. This leads to the following apportionment method.

Definition 8.1. Jefferson's method: Choose a modified divisor d. Compute the modified quotas p_k/d. Round each of these numbers down to obtain a_k. If $a_1 + a_2 + a_3 + \cdots + a_n = h$, then we have the Jefferson apportionment. Otherwise, modify the divisor d and try again.

Our description of Jefferson's method makes it sound like the divisor d is chosen at random, by trial and error. In practice, choosing d is not so difficult, because if it is too large, then the total number of seats, $a_1 + a_2 + a_3 + \cdots + a_n$ will be smaller than h, while if it is too small, the total number of seats will be larger than h. It is therefore easy to home in on a correct value of d rather quickly. In fact, since the standard divisor $s = p/h$ always results in an apportionment of the lower quota to each state, a correct value of d is always slightly less than the standard divisor s.

The problem of finding a modified divisor d for Jefferson's method is not a problem with a unique answer. There is a range of values for d that will work. This begs the question of whether the method as described is well defined, that is, whether it always produces the same values of the numbers a_k. The answer is yes.

Proposition 8.2. *Suppose that h, n, and $p_1, p_2, \ldots, p_n$ are given. If d and d' are two different divisors yielding Jefferson apportionments $a_1, a_2, \ldots, a_n$ and $a'_1, a'_2, \ldots, a'_n$, respectively, then $a_k = a'_k$ for all states k.*

Proof. Suppose without loss of generality that $d \le d'$. Then, for every

k, the modified quota p_k/d for state k with the first divisor d can be no smaller than the modified quota p_k/d' of state k with the second divisor d'. When we round these numbers down, respectively, we see that $a_k \geq a'_k$. Yet we have both $a_1 + a_2 + a_3 + \cdots + a_n = h$ and $a'_1 + a'_2 + a'_3 + \cdots + a'_n = h$. Were we to have $a_k > a'_k$ for even a single value of k, we would have

$$h = a_1 + a_2 + a_3 + \cdots + a_n > a'_1 + a'_2 + a'_3 + \cdots + a'_n = h,$$

which is a contradiction. Therefore $a_k = a'_k$ for all k. □

Jefferson found support for his method in the words of the Constitution. Article I, Section 2 says that the "number of Representatives shall not exceed one for every thirty thousand", which is to say that the divisor d must be at least 30,000. It even suggests 30,000 as the anticipated value. Nowadays a typical congressional district has a population of roughly 730,000, so this clause in the Constitution is now largely extraneous. If Jefferson's method with $d = 30{,}000$ were used today, we would have a House of Representatives with $h = 9356$ members. But in Jefferson's time, a congressional district of roughly 30,000 may have seemed reasonable. Using this divisor, a state with population 30,000 should be allocated 1 seat in the House. On the other hand, a state with population 60,000 should be allocated 2 seats in the House, and so forth. But what should be done with a state that has a population of, say, 57,000? Dividing by the modified divisor of 30,000, we obtain a modified quota of 1.9. This seems closer to 2 and rather far from 1, but Jefferson argued that it must be rounded down to 1. The Constitution seems to say that it may not be assigned 2 seats, because, if it were, then it would "exceed one for every thirty thousand". So Jefferson concluded that rounding down is mandated by the Constitution itself.

Let us work out an example of Jefferson's method in action. We keep the numbers small so that we can follow clearly the choice of divisors.

Example 8.3. Suppose that $n = 3$, $h = 10$, $p_1 = 1{,}500{,}000$, $p_2 = 3{,}200{,}000$, and $p_3 = 5{,}300{,}000$. Let's try various divisors d and hunt for one that gives the Jefferson apportionment. Table 8.1 tracks our computations.

We first try the standard divisor $s = 1{,}000{,}000 = p/h$. With the standard divisor, the quotas are called standard quotas and are 1.5, 3.2, and 5.3, respectively. Rounded down, these give 1, 3, and 5, whose sum is 9. This is smaller than the predetermined $h = 10$, so we need to choose a smaller divisor d. (A smaller divisor results in a larger quotient, which is to say a larger quota.) The standard divisor is always too large a guess for the modified divisor of Jefferson's method, unless by coincidence all the standard quotas are whole numbers. We next try a smaller divisor

Table 8.1 Hunting for the Jefferson divisor.

	Divisor	s = 1,000,000		d = 800,000		d = 850,000	
State	Population	Quota	Round Down	Quota	Round Down	Quota	Round Down
1	p_1 = 1,500,000	1.50	1	1.87	1	1.76	1
2	p_2 = 3,200,000	3.20	3	4.00	4	3.76	3
3	p_3 = 5,300,000	5.30	5	6.62	6	6.24	6
	p = 10,000,000		sum=9		sum=11		sum=10

d = 800,000. The modified quotas are now 25% larger than they were before, since dividing by 800,000 yields a number exactly 25% larger than dividing by 1,000,000. Now when we round the modified quotas down, we obtain 1, 4, and 6, whose sum is 11. This time the divisor was chosen too small, so the quotas (quotients) became too large. We try a third time and guess $d = 850,000$. The modified quotas are now 1.76, 3.76, and 6.24, which round down to 1, 3, and 6, whose sum is $h = 10$. This is the Jefferson apportionment for this problem: $a_1 = 1$, $a_2 = 3$, and $a_3 = 6$. Note that this is different from the Hamilton apportionment, which can be inferred from the third and fourth columns of Table 8.1. The lower quotas are 1, 3, and 5, respectively, and state 1 is the winner of the last seat up for grabs according to Hamilton's method. So the Hamilton appointment for this problem would be $a_1 = 2$, $a_2 = 3$, and $a_3 = 5$. Note also that 850,000 is not the only number that can serve as modified divisor for Jefferson's method. Any number sufficiently close to 850,000 will do.

8.2 Critical Divisors

There is another way of thinking about calculating the Jefferson apportionment that doesn't involve the trial and error associated with guessing the modified divisor. Instead of hunting for an appropriate modified divisor, one can employ instead the following algorithmic approach.

Let us ask what makes the divisor d work in the example of Table 8.1. When the modified divisor d = 850,000 is selected there, state 3 gets assigned 6 seats. That is because p_3/d = 5,300,000/850,000 = 6.24, which rounds down to 6. Another way to say this is that 5,300,000/6 $\geq$ 850,000 but 5,300,000/7 $<$ 850,000. We note that there are exactly 6 positive integers m such that $p_3/m \geq 850{,}000$, namely 1, 2, 3, 4, 5, and 6.

Let us call a number of the form p_k/m, where $1 \leq k \leq n$ and m is a positive integer, a **Jefferson critical divisor** for state k or, when it is clear that we are dealing with Jefferson's method, simply a **critical divisor** for state k. We now have a new way of thinking about the hunt for an appropriate modified divisor. When a modified divisor d is chosen, each state will be awarded a number of seats equal to the number of critical divisors for that state that are greater than or equal to d. Therefore, a modified divisor that is suitable for Jefferson's method is a number d such that there are exactly h Jefferson critical divisors (over all states) that are greater than or equal to d.

This gives a more systematic way of computing the Jefferson divisor. List all the Jefferson critical divisors for all states — you can limit m to relatively small positive integers, numbers from 1 to h, say — and place them in decreasing order. Pick d so that there are exactly h Jefferson critical divisors larger than d. Such a number d will work as the Jefferson modified divisor.

In fact, if we choose any divisor d at all and set a_k equal to the number of critical divisors for state k that are at least as large as d, then we have the Jefferson apportionment for the house size $h = a_1 + a_2 + \cdots + a_n$. Only if we regard h as prescribed in advance do we need to hunt for an appropriate size for d.

Let's try this alternative approach with the example in Table 8.1. The populations of the states are $p_1 = 1{,}500{,}000$, $p_2 = 3{,}200{,}000$, and $p_3 = 5{,}300{,}000$. The first few critical divisors for state 3 are as follows.

$$\begin{aligned}
5{,}300{,}000/1 &= 5{,}300{,}000\\
5{,}300{,}000/2 &= 2{,}650{,}000\\
5{,}300{,}000/3 &= 1{,}766{,}667\\
5{,}300{,}000/4 &= 1{,}325{,}000\\
5{,}300{,}000/5 &= 1{,}060{,}000\\
5{,}300{,}000/6 &= 883{,}333\\
5{,}300{,}000/7 &= 757{,}143
\end{aligned}$$

This gives us complete information about how many seats state 3 will be awarded for various values of d. In particular, if $d = 1{,}000{,}000$, then $a_3 = 5$, because there are exactly 5 critical divisors for state 3 that are larger than $d = 1{,}000{,}000$. In Table 8.2, we list the critical divisors for all three states.

We can now use Table 8.2 to determine the Jefferson apportionment for any house size at all. We simply list all the critical divisors over all states in decreasing order and we count off the top h numbers on the list. If we choose d smaller than the top h critical divisors but larger than the $h + 1$st critical divisor, then d is an appropriate choice for a Jefferson modified divisor, and the Jefferson apportionment will have a_k equal to the number of critical divisors for state k among the top h critical divisors on the list. An advantage of the critical divisor approach is that one can easily determine the Jefferson apportionment for varying values of h, and one can determine the precise range of values of d that correspond to any particular choice of h.

Table 8.2 All the critical divisors.

1,500,000/1 = 1,500,000	3,200,000/1 = 3,200,000	5,300,000/1 = 5,300,000
1,500,000/2 = 750,000	3,200,000/2 = 1,600,000	5,300,000/2 = 2,650,000
1,500,000/3 = 500,000	3,200,000/3 = 1,066,667	5,300,000/3 = 1,766,667
1,500,000/4 = 375,000	3,200,000/4 = 800,000	5,300,000/4 = 1,325,000
1,500,000/5 = 300,000	3,200,000/5 = 640,000	5,300,000/5 = 1,060,000
1,500,000/6 = 250,000	3,200,000/6 = 533,333	5,300,000/6 = 883,333
1,500,000/7 = 214,857	3,200,000/7 = 457,143	5,300,000/7 = 757,143
⋮	⋮	⋮

The 10th largest critical divisor in Table 8.2 is 883,333, and the 11th largest critical divisor is 800,000. Choosing d to be any number between these two critical divisors will lead to the Jefferson apportionment for $h = 10$. One such number is 850,000, which is the modified divisor that we chanced upon in Table 8.1. We now see that any number greater than 800,000 but not greater than 883,333 would also work.

Not only can we determine an appropriate Jefferson modified divisor d via Table 8.2, we can also read off the final apportionment numbers. In that table, 1 number in the first column, 3 numbers in the second column, and 6 numbers in the third column are greater than d. These represent the apportionment numbers $a_1 = 1$, $a_2 = 3$, and $a_3 = 6$.

The advantage of the analysis with critical divisors is that the process of finding an appropriate modified divisor is no longer a trial and error search. Instead one can imagine beginning with a very large value of d and slowly decreasing it. Each time d encounters a critical divisor, one more seat is given away. Continue until you have given away exactly h seats. Actually, one need not start with a very large value of d to execute Jefferson's method, because the Jefferson modified divisor is known to be smaller than the standard divisor. So one can begin with the d equal to the standard divisor s, and then reduce d until h seats are apportioned.

This suggests a somewhat more efficient way to organize the information from Table 8.2. We demonstrate it on the example in Table 8.1. We begin with the provisional apportionment obtained using the standard divisor. This results in an apportionment of 9 seats, one short of the desired $h = 10$. Then for each state k, we calculate the Jefferson critical divisor for the next seat for state k, seat $a_k + 1$. This will be $p_k/(a_k + 1)$. Table 8.3 tracks the calculation.

Table 8.3 Computing the Jefferson apportionment.

	Divisor	$s = 1{,}000{,}000$			$d = 850{,}000$	
State	Population	Standard Quota	Lower Suota	$\dfrac{p_k}{a_k + 1}$	Modified Quota	Jefferson
1	1,500,000	1.50	$a_1 = 1$	750,000	1.76	$a_1 = 1$
2	3,200,000	3.20	$a_2 = 3$	800,000	3.76	$a_2 = 3$
3	5,300,000	5.30	$a_3 = 5$	883,333	6.24	$a_3 = 6$

The next critical divisor below 1,000,000, which is to say the largest number in the fifth column of Table 3, is 883,333, is associated to state 3, so state 3 gets the next seat. For a corresponding modified divisor, we select a number slightly less than 883,333, making certain that we do not select a number less than the second largest critical divisor 800,000. We select 850,000. Notice that all the numbers in this column appear in Table 8.2, and also notice where they appear.

If we wanted to allot one more seat, the 11th seat, we could start this process over. We would then make a new column of values of d and select the largest. But our work can be reduced if we notice that the only value of d that changes is the one for the state that got a seat on the previous step, state 3 in this case. Its next value is

$$d = p_3/(a_3 + 1) = 5{,}300{,}000/(6 + 1) = 757{,}143.$$

This is smaller than the critical divisor 800,000 for state 2, so the next seat would go to state 2.

8.3 Assessing Jefferson's Method

One objection to Jefferson's method is that it seems to favor large states, at least by comparison with Hamilton's method. A small example illustrates the point. Suppose that $n = 2$, $h = 10$, $p_1 = 1{,}800{,}000$, and $p_2 = 8{,}200{,}000$. Table 8.4 summarizes the calculations.
The standard divisor is 1,000,000 here and the modified divisor used to compute the modified quotas is 910,000. Many people would argue that

Table 8.4 Comparing Jefferson to Hamilton.

State	Population	Standard Quota	Hamilton Apportion-ment	Modified Quota	Jefferson Apportion-ment
1	1,800,000	1.8	2	1.98	1
2	8,200,000	8.2	8	9.02	9

$d = 910{,}000$

Jefferson's method is giving state 2 an extra seat that it doesn't really deserve. Hamilton's method gives the extra seat to state 1, and it isn't a close call, since 0.8 is a much larger fractional part than 0.2.

An even more striking example is given in Table 8.5, with $h = 10$ again. Here the standard divisor is 1,000,000. This time the modified

Table 8.5 Jefferson's method violates quota.

State	Population	Standard Quota	Hamilton Apportion-ment	Modified Quota	Jefferson Apportion-ment
1	1,500,000	1.5	2	1.88	1
2	1,400,000	1.4	1	1.75	1
3	1,300,000	1.3	1	1.62	1
4	5,800,000	5.8	6	7.25	7

$d = 800{,}000$

divisor is $d = 800{,}000$, and Jefferson's method again favors the large state. But this time we have a striking result for state 4. Its standard quota is 5.8. Yet the Jefferson apportionment for that state is 7, which is neither the lower quota of 5 nor the upper quota of 6. Hamilton's method had already rounded state 4 up to 6 seats. On behalf of what constituency should they receive a 7th representative? Where are the residents to justify such an apportionment for this state? Jefferson's method favors this large state to such a great degree that it assigns it a number of seats larger than its upper quota. We call this a **quota violation**. In particular, this is a violation of what seems like a reasonable criterion that we should expect an apportionment method to satisfy.

Definition 8.4. An apportionment method satisfies the **quota crite-**

rion or **quota rule** if it assigns every state either its lower quota or its upper quota.

Sometimes it is useful to think of this criterion as having two separate aspects.

Definition 8.5. An apportionment method satisfies the **upper quota criterion** or **upper quota rule** if it never assigns any state an apportionment that exceeds the upper quota for that state.

Definition 8.6. An apportionment method satisfies the **lower quota criterion** or **lower quota rule** if it never assigns any state an apportionment that is less than the lower quota for that state.

A violation of the first of these two criteria is naturally called an **upper quota violation** and a violation of the second is called **lower quota violation**. We often use the word "quota" in place of the phrase "the quota criterion", and we similarly refer to the lower quota criterion and the upper quota criterion simply as "lower quota" and "upper quota". The example in Table 8.5 shows that Jefferson's method can violate upper quota (and therefore quota). That's the bad news. The good news, however, is that Jefferson's method cannot lead to a lower quota violation.

Proposition 8.7. *Jefferson's method satisfies the lower quota rule.*

Proof. Start by giving every state a provisional apportionment equal to its lower quota. This is the Jefferson apportionment using the standard quota for h as a modified divisor, but it apportions fewer than h seats. Since we want to apportion h seats, we must lower the modified divisor. This results in additional seats for certain states, but it does not reduce the apportionment for any states. Thus each state gets at least its lower quota. □

An important advantage of Jefferson's method is that it does not succumb to the paradoxes of Hamilton's method.

Proposition 8.8. *Jefferson's method satisfies the house monotonicity criterion.*

We offer two proofs, depending on whether Jefferson's method is understood via a modified divisor process or via a critical divisor process.

First Proof. An increase in h requires a smaller modified divisor d. Decreasing d increases the modified quota p_k/d for every k. Rounding down a larger number never results in a smaller answer, so no state can lose a seat by the prospect of an increase in h. □

Second Proof. Imagine listing the critical divisors in decreasing order. The critical divisor p_k/m is naturally associated with state k. If you choose the first h critical divisors in the list, then the seats are doled out by Jefferson's method according to the states that are associated to these critical divisors. If h is increased to $h+1$, then you must instead choose the first $h+1$ critical divisors in the list. That obviously includes the first h critical divisors in the list, and then one more, and the new seat is then doled out by Jefferson's method to the state associated with that new critical divisor. No state loses a seat in this process. □

We will see in the next chapter that Jefferson's method also avoids the population paradox and the new states paradox. So Jefferson's method manages to avoid the three major pitfalls of Hamilton's method. But unfortunately it violates the quota rule, which Hamilton's method clearly does satisfy. So while we may take sides between these two methods, it is important to recognize that a decision between them involves a trade-off. Both methods have flaws.

8.4 Other Divisor Methods

Jefferson's method is one of a class of methods that all work in the same general way. These are known as **divisor methods**. The populations of the states are divided by a modified divisor to obtain modified quotas. Those quotas are then rounded to whole numbers to obtain a provisional apportionment. Let t be the number of seats allotted all told in this provisional apportionment. If $t = h$, we have our apportionment. Otherwise, we modify the divisor d, increasing it if $t > h$, and decreasing it if $t < h$.

The difference between divisor methods is in the interpretation of the phrase "rounded to whole numbers". Jefferson's method does this by rounding down to the nearest whole number, which is to say by truncating the decimal representation of the modified quota at the decimal point. We describe several other ways to round numbers, and each way gives rise to an apportionment method.

Instead of rounding down, it seems natural to ask what happens if we round all numbers up to the nearest whole number. This leads to the apportionment method suggested by John Quincy Adams.

Definition 8.9. Adams's method: Choose a modified divisor d. Compute the modified quotas p_k/d. Round each of these numbers up to obtain a_k. If $a_1+a_2+a_3+\cdots+a_n = h$, then we have the Adams apportionment. Otherwise, modify the divisor d and try again.

Adams's method lies at the opposite extreme from Jefferson's method on the spectrum of divisor methods. Just as Jefferson's method seems to have a bias in favor of large states, Adams's method seems to have a bias in favor of small states. Table 8.6 shows an example like the one in Table 8.3 but illustrating the opposite bias.

Table 8.6 Comparing Adams to Hamilton.

State	Population	Standard Quota	Hamilton Apportionment	Modified Quota	Adams Apportionment
1	1,200,000	1.2	1	1.04	2
2	8,800,000	8.8	9	7.65	8

$d = 1{,}150{,}000$

In this example, the Hamilton apportionment is 1 and 9 (and it is not a close decision) while a modified divisor of $d = 1{,}150{,}000$ yields the Adams apportionment of 2 and 8. (It is instructive to hunt for an appropriate divisor in this example, and the reader is invited to experiment to see how this divisor was found and to discover what range of values of d is suitable.) Some would regard these Adams apportionment numbers as an inappropriate favoritism toward state 1, which has only 12% of the population but gets 20% of the representatives. It is also possible to create an example like the one in Table 8.5, illustrating the possibility that Adams's method can violate the lower quota rule and hence the quota rule. The reader is left to provide this example for herself.

Adams's method, unlike Jefferson's method, meets the Constitutional requirement to assign every state, no matter how small, at least one representative. A state with positive population will have a positive modified quota no matter how large a divisor is chosen, and rounding up a positive number leads to a positive integer. This can be seen as an advantage of Adams's approach, but it can also be seen as a disadvantage. The problem is that Adams's method cannot be implemented unless there are enough seats to go around, at least one per state. Thus, we require $h \geq n$ in order to be able to find an appropriate Adams modified divisor. This could be seen as a minor violation of our universal domain requirement, but it is of the same nature as the difficulty one encounters if two states have identical populations. We do generally have h much larger than n, and we don't expect to see two states having precisely equal populations.

If Jefferson's method has a bias in favor of large states, while Adams's method has a bias in favor of small states, then there is a natural choice for a method that is precisely halfway between these two methods. Instead of making a decision to round down always or to round up always, why not round in the way that real numbers are usually rounded? Numbers whose fractional part is 0.5 or greater are rounded up to the next whole number, while numbers whose fractional part is less than 0.5 are rounded down to the preceding whole number. Put simply, this is just rounding to the nearest integer. Most people would regard this as the ordinary way to round numbers. This kind of rounding gives rise to the method of Daniel Webster.

Definition 8.10. Webster's method: Choose a modified divisor d. Compute the modified quotas p_k/d. Round the modified quota p_k/d to the nearest whole number a_k. (Round up if there is a tie.) If $a_1 + a_2 + a_3 + \cdots + a_n = h$, then we have the Webster apportionment. Otherwise, change the divisor d and try again.

Webster's method can be thought of as an effort to improve on Hamilton's method. With Hamilton's method, we round up standard quotas when the fractional parts are large and we round down standard quotas when the fractional parts are small. Unfortunately, this is not always ordinary rounding, since a state with fractional part 0.48 may be rounded up. Webster's method fixes this flaw, at the cost of changing from standard to modified quotas. When Hamilton's method is achieved by chance using ordinary rounding, then Webster's method agrees with Hamilton's. Otherwise, Hamilton's modified quotas can be regarded as an adjustment of the standard quotas that allows ordinary rounding to be used.

8.5 Rounding Functions

When we speak about a procedure for "rounding" numbers, we are speaking about a function f whose domain is the set of real numbers, whose output is always an integer, and that satisfies the following two conditions:

(1) If x is an integer, then $f(x) = x$.

(2) If $x > y$, then $f(x) \geq f(y)$.

Think of $f(x)$ as the real number x rounded to a neighboring integer.

We call such a function a **rounding function**. The first of these conditions requires that an integer must be rounded to itself. The second condition, a kind of monotonicity requirement, forces there to be a cutoff between the integers m and $m+1$ above which we round up to $m+1$ and below which we round down to m.

Three examples of rounding functions f are the ones that are used in the apportionment methods of Jefferson, Adams, and Webster:

Example 8.11. $f(x) = \lfloor x \rfloor$, the greatest integer less than or equal to x.

Example 8.12. $f(x) = \lceil x \rceil$, the least integer greater than or equal to x.

Example 8.13. $f(x)$ is the integer nearest to x.
To resolve ties in this last example, it is customary to take $f(m+0.5) = m+1$ for any integer m. This is the same convention we use for all rounding functions that involve a cutoff that is not a whole number: We round up when the cutoff is hit exactly.

We refer to the function in Example 8.13 as **arithmetic rounding**, since the dividing line between the integers m and $m+1$ is their arithmetic mean or average $m+1/2$. (The word "arithmetic" in this context is an adjective and is pronounced with the accent on the third syllable.) Rounding down, rounding up, and arithmetic rounding are three obvious rounding functions, but they are not the only possibilities. We can easily conjure up an infinite number of other rounding functions, depending on the cutoffs that we set between m and $m+1$. We now introduce two other rounding functions that give rise to particularly significant apportionment methods.

Example 8.14. $f(x) = m$ where m is the integer satisfying $\sqrt{m(m-1)} \leq x < \sqrt{m(m+1)}$.

The idea here is to use the number $\sqrt{m(m+1)}$ as the cutoff between m and $m+1$. We call this **geometric rounding**, because $\sqrt{m(m+1)}$ is the geometric mean of m and $m+1$. The apportionment method that comes from this rounding function is named after Joseph Hill, who was the chief statistician of the Census Bureau from 1909 to 1921. In 1941, Congress passed a bill that made Hill's method the permanent apportionment method in the United States. The law is still in force, and Hill's method has been used every decade since then.

Definition 8.15. Hill's method: Choose a modified divisor d. Compute the modified quotas p_k/d. Geometrically round the modified quota p_k/d to obtain a_k. If $a_1 + a_2 + a_3 + \cdots + a_n = h$, then we have the Hill apportionment. Otherwise, change the divisor d and try again.

We also consider the following curious rounding function.

Example 8.16. $f(x) = m$ where m is the integer satisfying

$$2m(m-1)/(2m-1) \leq x < 2m(m+1)/(2m+1).$$

This rounding function uses the cutoff $2m(m+1)/(2m+1)$ between the whole numbers m and $m+1$. Where does the formula $2m(m+1)/(2m+1)$ come from? It turns out that this is the reciprocal of the average of the reciprocal of the two neighboring integers m and $m+1$. To see this, note that

$$\frac{1}{\frac{\frac{1}{m}+\frac{1}{m+1}}{2}} = \frac{2}{\frac{m+1}{m(m+1)}+\frac{m}{m(m+1)}} = \frac{2m(m+1)}{2m+1},$$

The reciprocal of the average of the reciprocals of two numbers is called the harmonic mean of those numbers. We refer to this function f as **harmonic rounding**, because it uses the harmonic mean of m and $m+1$ as a cutoff.

The harmonic mean is the kind of average of two quantities that is relevant when it is the reciprocals of the quantities that are regarded as important. The apportionment method that comes from this rounding function is named after James Dean, who was a professor of mathematics at the University of Vermont in the early 19th century.

Definition 8.17. Dean's method is the apportionment method that works as follows. Choose a modified divisor d. Compute the modified quotas p_k/d. Harmonically round the modified quota p_k/d to obtain a_k. If $a_1 + a_2 + a_3 + \cdots + a_n = h$, then we have the Dean apportionment. Otherwise, change the divisor d and try again.

Let us compare the various methods of rounding to get a feeling for how they work. We know that the whole number m must round to itself and that the whole number $m+1$ must also round to itself. What we don't know is at what point between m and $m+1$ we make the transition from rounding down to m to rounding up to $m+1$. Table 8.7 provides a numerical list of these cutoffs for the five divisor methods we have encountered. For example, the number 5.455 indicates that numbers between 5 and 5.455 are harmonically rounded down to 5 while numbers between 5.455 and 6 are harmonically rounded up to 6. Hence, when computing Dean's method, if a modified quota of 5.47 is obtained for a given state, then that state will be entitled to 6 representatives. The numbers in the third and fourth columns of Table 8.7 are approximations to three decimal places.

One sees from Table 8.7 that harmonic rounding is more likely to round up than arithmetic rounding is — the cutoffs are smaller — and this tendency is more pronounced for small numbers. Geometric rounding tends to round up more than arithmetic rounding but less than harmonic rounding. Note that the number 1.42 is rounded arithmetically

Table 8.7 Cutoffs for rounding between small whole numbers.

	Rounding Function and Method				
	Rounding Up	Harmonic Rounding	Geometric Rounding	Arithmetic Rounding	Rounding Down
	Adams	Dean	Hill	Webster	Jefferson
0–1	0	0	0	0.5	1
1–2	1	1.333	1.414	1.5	2
2–3	2	2.400	2.449	2.5	3
3–4	3	3.429	3.464	3.5	4
4–5	4	4.444	4.472	4.5	5
5–6	5	5.455	5.477	5.5	6
6–7	6	6.462	6.481	6.5	7
7–8	7	7.467	7.484	7.5	8

down to 1 but geometrically and harmonically up to 2. The number 2.42 is rounded arithmetically and geometrically down to 2 but harmonically up to 3. The number 3.42 is rounded down to 3 arithmetically, geometrically, and harmonically.

In Chapter 11, we will contrast these five divisor methods, identify what is special about each of them, and attempt to discern which of them is the superior method. For now, let us see what these five methods have in common. When we speak of a divisor method, we mean any method that operates like the methods of Jefferson, Adams, Webster, Dean, or Hill. These five methods are known as the five historic divisor methods. The only difference between one divisor method and another is the manner in which numbers are rounded.

The divisor methods have been described as if they are based on trial and error guessing of the appropriate modified divisor. Actually, just like Jefferson's method, all of the divisor methods can be made algorithmic by adopting a critical divisor approach. We illustrate how this works for Hill's method. With Hill's method, every state with a positive population gets at least one seat, because no positive modified quota is ever rounded down to 0. (The same is true with the methods of Adams and Dean.) Imagine that the census data $p_1, p_2, p_3, \ldots, p_n$ has been collected but that the house size h is not yet determined. Begin with an extremely large modified divisor d and slowly decrease d while keeping track of the corresponding value of h and the Hill apportionment numbers at each stage. At what point does state k become entitled to a second seat? The answer is: Just when the modified quota p_k/d reaches $\sqrt{2}$, which is the cutoff between being rounded down to 1 and being rounded up to 2. To put it another way, state k earns its second seat when the modified divisor d reaches $p_k/\sqrt{2}$. At what point does state k obtain its third seat? When the modified divisor p_k/d reaches $\sqrt{6}$, that is, when $d = p_k/\sqrt{6}$. In general, state k will get its $(m+1)$st seat when d is reduced to

$$p_k/\sqrt{m(m+1)}.$$

Numbers of the form $p_k/\sqrt{m(m+1)}$ where $1 \leq k \leq n$ and m is a nonnegative integer, are called **Hill critical divisors**, because they are the values of d at which a state acquires an additional seat. (Think of division by 0 as giving an infinitely large number.) Associate the Hill critical divisor $p_k/\sqrt{m(m+1)}$ to state k. Write all of the Hill critical divisors in descending order, including the n infinitely large ones — one for each state — at the beginning. An appropriate value for the Hill modified divisor is any number between the element in position h in this list and the element in position $h+1$ in this list. Among the first h elements in this list, the number of them associated to state k is the number of seats apportioned to state k by Hill's method.

Let us study an example of a computation via Hill's method, comparing the outcome with the methods of Hamilton and Webster. Imagine that there are $h = 10$ seats in the house and that there are three states, with populations $p_1 = 1{,}385{,}000$, $p_2 = 2{,}390{,}000$, and $p_3 = 6{,}225{,}000$, respectively, competing to share those seats. Table 8.8 shows the computation of the methods of Hamilton, Webster, and Hill.

Table 8.8 Comparing Webster, Hill, and Hamilton.

				$d = 957{,}000$		$d = 977{,}000$	
k	p_k	q_k	Hamilton Method	Webster Quota	Webster Method	Hill Quota	Hill Method
1	1,385,000	1.385	1	1.447	1	1.418	2
2	2,390,000	2.390	3	2.497	2	2.446	2
3	6,225,000	6.225	6	6.505	7	6.372	6

The example is striking because no two of these three methods offer the same outcome. The standard divisor in this problem is precisely 1,000,000, the quotient of the total population $p = 10{,}000{,}000$ and the house size $h = 10$. Thus the standard quotas are obtained simply by dividing the state populations by 1,000,000. The lower quotas are thus 1, 2, and 6, respectively, and Hamilton gives the extra seat to state 2, the fractional part of whose standard quota is largest (beating state 1 by just a hair). The standard divisor will not work for Webster's method, because all three fractional parts are less than 0.5. Hence a modified divisor is needed for Webster's method. The divisor 957,000 gives modified quotas of approximately 1.447, 2.497, and 6.505, which round arithmetically to 1, 2, and 7, respectively. Now it is state 3 that benefits. Finally, note that the Webster divisor is inappropriate for Hill's method, because the three numbers 1.447, 2.497, and 6.505 all round harmonically up, to 2, 3, and 7, respectively. (See the cutoffs in Table 8.7.) We need smaller quotas for harmonic rounding, so we need a larger modified divisor. The number 977,000 does the job, yielding modified quotas of 1.418, 2.446, and 6.372, respectively. These quotas round harmonically to 2, 2, and 6, respectively, which are the Hill apportionment numbers for this example. Note that 1.418 is rounded up to 2 while 2.446 is rounded down to 2. The priority for the extra seat goes to the first state, even though the fractional part of 1.418 is smaller than 2.446.

We now rework the computation of Hill's method in this example through an analysis of critical divisors. Here are the Hill critical divisors for this example:

$$
\begin{array}{lll}
\frac{1{,}385{,}000}{0} = \infty & \frac{2{,}390{,}000}{0} = \infty & \frac{6{,}225{,}000}{0} = \infty \\
\frac{1{,}385{,}000}{\sqrt{2}} = 979{,}343 & \frac{2{,}390{,}000}{\sqrt{2}} = 1{,}689{,}985 & \frac{6{,}225{,}000}{\sqrt{2}} = 4{,}401{,}740 \\
\frac{1{,}385{,}000}{\sqrt{6}} = 565{,}424 & \frac{2{,}390{,}000}{\sqrt{6}} = 975{,}713 & \frac{6{,}225{,}000}{/\sqrt{6}} = 2{,}541{,}346 \\
\frac{1{,}385{,}000}{\sqrt{12}} = 399{,}815 & \frac{2{,}390{,}000}{\sqrt{12}} = 689{,}934 & \frac{6{,}225{,}000}{\sqrt{12}} = 1{,}797{,}003 \\
\frac{1{,}385{,}000}{\sqrt{20}} = 309{,}695 & \frac{2{,}390{,}000}{\sqrt{20}} = 534{,}420 & \frac{6{,}225{,}000}{\sqrt{20}} = 1{,}391{,}952 \\
\frac{1{,}385{,}000}{\sqrt{30}} = 252{,}865 & \frac{2{,}390{,}000}{\sqrt{30}} = 436{,}352 & \frac{6{,}225{,}000}{\sqrt{30}} = 1{,}136{,}524 \\
\frac{1{,}385{,}000}{\sqrt{42}} = 213{,}710 & \frac{2{,}390{,}000}{\sqrt{42}} = 368{,}785 & \frac{6{,}225{,}000}{\sqrt{42}} = 960{,}538 \\
\vdots & \vdots & \vdots
\end{array}
$$

One is looking for a divisor d with the property that exactly 10 numbers in the table exceed d. The number 977,000 is seen to meet the requirement. In fact, any number between 975,713 and 979,343 can serve the role of Hill modified divisor for this example.

Table 8.9 summarizes how the five divisor methods work using the critical divisor approach. Imagine that we have already apportioned a certain number of seats and we are contemplating which state deserves to be assigned the next seat. The critical divisor for state k indicates the state's priority for receiving that next seat. This depends on the state's population p_k and on the number a_k, which is the number of seats state k has already been assigned. The state with the highest priority wins the next seat.

Table 8.9 Critical divisors for the five historic divisor methods.

Method	Critical Divisor for State k
Adams	p_k/a_k
Dean	$p_k(2a_k+1)/2a_k(a_k+1)$
Hill	$p_k/\sqrt{a_k(a_k+1)}$
Webster	$p_k/(a_k+1/2)$
Jefferson	$p_k/(a_k+1)$

We have focused our attention on five principal divisor methods, but there are infinitely many others. It is not difficult to conjure up other rounding functions that lead to new divisor methods of apportionment. Marquis de Condorcet (the same Condorcet noted for his contributions to voting theory) suggested the method that comes from rounding numbers up if and only if their fractional parts are at least 0.4. Another method aimed at supporting smaller states would be to round numbers up if they are less than 3, say, and down otherwise. (Such a method would clearly favor smaller states.) There are two reasons that we pay special attention to the methods of Adams, Dean, Hill, Webster, and Jefferson. The first reason is that they are the answers to five different interpretations of the question "Which divisor method is best?" We take up this question in Chapter 11. The second reason is that they are the divisor methods that have been used or considered in attempts to settle the United States political apportionment problem.

8.6 Exercises and Problems

8.1. Suppose that $n=2$, $h=5$, $p_1=13$, and $p_2=27$.
(a) What is the standard divisor for this apportionment problem?
(b) What is the standard quota for state 1?
(c) What is the Hamilton apportionment?
(d) What modified divisor works for Jefferson's method?
(e) What is the Jefferson apportionment?
(f) What is the smallest possible number that can serve as a modified divisor for Adams's method for this problem?
(g) What is the Adams apportionment?

8.2. Compute the Hamilton, Adams, and Webster apportionments for $h=10$ and the census $p_1=119{,}000$, $p_2=241{,}000$, and $p_3=640{,}000$.

8.3. Consider a country with four states whose populations are given by:

A	B	C	D
3,310,000	2,670,000	1,330,000	690,000

(a) What is the total population p?
(b) If $h = 160$, what is the standard divisor?
(c) Find the standard quota, upper quota, and lower quota for each state.
(d) Compute the apportionment for this data, using the methods of Hamilton, Jefferson, Webster, and Adams.

8.4. Determine the Dean and Hill apportionments for a federation whose three states have populations 1,340,000, 2,390,000, and 6,260,000 and whose house has size $h = 10$.

8.5. Assume that $h = 10$ and that $p_1 = 120$, $p_2 = 140$, $p_3 = 150$, and $p_4 = 590$.
(a) What is the standard divisor for this problem?
(b) What is the upper quota for state 4?
(c) What is the Hamilton apportionment?
(d) What modified divisor works for Jefferson's method?
(e) What is the Jefferson apportionment?
(f) Does the Jefferson apportionment for this problem exhibit a quota violation?
(g) What is the largest number that can be a Jefferson divisor for this problem? (Give the exact answer.)
(h) What modified divisor works for Adams's method?
(i) What is the Adams apportionment?

8.6. Here is a census for an apportionment problem in a hypothetical country comprised of four states.

- State of Ambivalence: 8,000;
- State of Boredom: 9,000;
- State of Confusion: 24,000;
- State of Depression: 59,000.

Assume that the house has $h = 10$ seats to apportion to these four states. What apportionment is determined by the method of:
(a) Hamilton?
(b) Jefferson?
(c) Adams?
(d) Webster?

8.7. A small college has one mathematics professor who can teach a total of five sections. The subjects that she teaches are calculus, set theory, and topology, and the number of students who wish to enroll in these three subjects are 52, 33, and 15, respectively. Using the apportionment methods of Hamilton and Jefferson, determine how many sections of each subject she should teach.

8.8. Round the following percentages to whole numbers using the methods of Hamilton, Jefferson, Adams, and Webster:

$$92.15\% + 1.59\% + 1.58\% + 1.57\% + 1.56\% + 1.55\% = 100\%$$

The rounded numbers must be whole numbers and must add to exactly 100. (It may help to think about apportioning 100 seats to 6 states.)

8.9. Alice, Barbara, and Carolyn purchase a bag of pearls for $10,000. Alice contributed $1250, Barbara contributed $3650, and Carolyn contributed $5100. They take the bag home and pour 20 pearls onto their kitchen table.

(a) How should they apportion the pearls, using the methods of Jefferson and Adams?

(b) While Alice and Barbara are out of the room, Carolyn notices that the bag contains one more pearl. How would the apportionment be affected if a 21st pearl were introduced?

Compare your answers to the answers to Problem **7.4** in Chapter 7.

8.10. A certain union of three states has populations $p_1 = 1{,}350{,}000$, $p_2 = 2{,}380{,}000$, and $p_3 = 6{,}270{,}000$.

(a) What is the Hamilton apportionment with $h = 10$?

(b) It turns out that the standard divisor in this problem is an appropriate modified divisor for Dean's method. What is the Dean apportionment with $h = 10$?

(c) What is the Webster apportionment with $h = 10$?

8.11. Suppose that there are just two states in a nation with total population $p = 10{,}000{,}000$, and suppose that the house size is 10 seats. Let us ask at what point the apportionment changes from 3-and-7 to 4-and-6. For the apportionment methods of Jefferson, Webster, Hill, Dean, and Adams, find the value of x that has the following property: If the population p_1 of state 1 is less than x, then state 1 gets 3 or fewer seats, while if the population p_1 of state 1 is greater than x, then state 1 gets 4 or more seats.

8.12. Give the populations of two states such that the Hill apportionment with $h = 5$ gives a precise tie between $a_1 = 1$, $a_2 = 4$ and $a_1 = 2$, $a_2 = 3$.

8.13. Give an example of a census involving two states that, for $h = 10$, gives Hamilton apportionment 3-and-7 and gives Adams apportionment 4-and-6. Identify your Adams modified divisor.

8.14. Give the populations of two states such that the Dean apportionment for $h = 5$ is $a_1 = 2$, $a_2 = 3$ but the Hill apportionment is $a_1 = 1$, $a_2 = 4$.

8.15. Give an example of a census for which Adams's method yields a quota violation.

8.16. Compute the Jefferson, Adams, Webster, and Dean apportionments for the US using $h = 435$ and data from the most recent census.

8.17. The **identric mean** of two numbers a and b is defined to be

$$\frac{1}{e}\left(\frac{b^b}{a^a}\right)^{\frac{1}{b-a}},$$

where e is Euler's constant, approximately 2.71828. A method of apportionment suggested by Robert Agnew is to use the rounding function that rounds down to the integer m those numbers between m and $m+1$ that are less than the identric mean of m and $m+1$, that is, less than

$$\frac{1}{e}\left(\frac{(m+1)^{m+1}}{m^m}\right).$$

(a) Compute the apportionment using this method for a census with $p_1 = 148$, $p_2 = 248$, $p_3 = 604$ and $h = 10$.
(b) Give an example of a census for which Agnew's method differs from Webster's method.

8.18. The **logarithmic mean** of two numbers a and b is defined to be $(b - a)/(\ln b - \ln a)$, where $\ln x$ is the natural logarithm of x. In the same paper in which the apportionment method of Problem **8.17** was introduced, Robert Agnew suggested the rounding function that rounds down to the integer m those numbers between m and $m+1$ that are less than the logarithmic mean of m and $m+1$, that is, less than

$$\frac{1}{\ln((m+1)/m)}.$$

(a) Compute the apportionment using this method for a census with $p_1 = 148$, $p_2 = 248$, $p_3 = 604$ and $h = 10$.
(b) Give an example of a census for which Agnew's logarithmic method differs from Agnew's identric method of Problem **8.17**.

8.19. Explain why no divisor method can give a quota violation with just two states.

9

Criteria and Impossibility

"We are an impossibility in an impossible universe." — Ray Bradbury

9.0 Scenario

Four states in a certain union have populations as given in the following chart:

State A:	42,251,600
State B:	33,751,400
State C:	22,753,000
State D:	1,244,000

The federal legislature will accord these four states in total 200 seats in the House of Representatives. How should these seats in the House be apportioned between the four states? Use the methods of Hamilton, Jefferson, Adams, Webster, Dean, and Hill.

(1) What modified divisors are used in each case? Are these smaller or larger than the standard divisor?

(2) How do the apportionments in each case compare to the upper and lower quotas?

(3) What biases do you notice? Which method seems the fairest to you, and why?

9.1 Basic Criteria

In this chapter, we address the dilemma of sorting through the various apportionment methods we have encountered and determining which of them is ideal. We will articulate various competing criteria that spell out different notions of fairness or reasonableness. But in the end, we will prove that it is impossible to find a method that satisfies all of these criteria.

We begin by enunciating a number of compelling but rather weak criteria. These are criteria that we can feel comfortable imposing on our search for a good method, but they are not particularly discriminating, in that all of the methods we have encountered so far satisfy them.

Definition 9.1. An apportionment method satisfies the **neutrality criterion** (or is **neutral**) if permuting the input populations results in permuting the output apportionment numbers in the corresponding way.

Suppose that a neutral method assigns 6, 5, 3, and 2 seats, respectively, to four states with populations, say, 153, 126, 73, and 41. Then that method must assign 3, 6, 2, and 5 seats, respectively, to four states with populations 73, 153, 41, and 126.

Neutrality says, in effect, that the apportionment treats all states the same. In particular, an apportionment must depend only on the populations of states and not on any other factors. One needs to be careful with this idea. For example, neutrality does not prevent bias in favor of large states or of small states. But a neutral method must not have a bias in favor of, say, western states or progressive states. Neutrality in apportionment is clearly analogous to neutrality (and to anonymity) in voting theory. It is a compelling criterion, and one that is certainly implicitly required by the Constitutional requirement that apportionment assign representatives to states "according to their respective numbers". Neutrality is the requirement that only the number of people in the state may be a basis for the state's final assignment. It is easy to see that Hamilton's method, as well as all divisor methods, are neutral.

Although the neutrality criterion is compelling in many settings, one can imagine situations where a lack of neutrality might be appropriate. For example, in apportioning teachers to schools, one may wish to impose certain special requirements on schools that serve an underrepresented student body. One could understand using an apportionment process in this situation that imposes a clear bias in favor of such schools.

It may be argued that a criterion called proportionality is also implicitly mandated by the Constitution. This criterion says that the end result of an apportionment can depend only on the proportion of the total population that each state has. Suppose that the population in state k is denoted by p_k, as usual, and let $p = p_1 + p_2 + \cdots + p_n$. The **population distribution** is the list of numbers $p_1/p, p_2/p, \ldots, p_n/p$.

The kth entry in the population distribution gives the fraction of the total population that resides in state k.

Definition 9.2. An apportionment method satisfies the **proportionality criterion** (or is **proportional**) if when the method is applied to any two censuses that have the same population distributions and the same house size, the outputs of the method are the same.

Suppose that between two censuses the population of every state grows by a certain fixed percentage, say 10%. Then the total population of the nation also grows by 10%, and hence the fraction p_k/p does not change, since both its numerator and denominator increase by that same 10%. Therefore the two censuses have the same population distribution. Proportionality requires that the apportionment remain the same when facing this kind of uniform population growth.

We claim that Hamilton's method and all the divisor methods are proportional. To see this, note that the standard quota for state k is given by $q_k = p_k/s$. Since $s = p/h$, it follows that $q_k = h(p_k/p)$. In other words, the standard quotas are equal to the numbers in the population distribution times the house size. Thus, two censuses will have the same population distribution if and only if they have the same standard quotas, assuming that the house size h is constant. It follows that an apportionment method that depends only on the standard quotas must be proportional. This is clearly the case for Hamilton's method. A similar argument shows that any method that depends only on a modified quota must be proportional, for the list of modified quotas that can be obtained from one census can also be obtained from any other census with the same population distribution. Thus all divisor methods are proportional.

The next criterion is of a different nature. It compares the apportionments for two different states.

Definition 9.3. An apportionment method satisfies the **order-preserving criterion** (or is **order-preserving**) if, whenever $a_i > a_j$, it follows that $p_i > p_j$.

This says that the only way that one state is allowed to get more seats than another is if it has a greater population. We call this criterion "order-preserving" because it requires that the apportionment method respects the order in which the states are ranked by population. In other words, the states can be ranked by input, that is, in population order, and they can also be ranked by output, that is, by apportionment numbers. The order-preserving property demands that these two rankings coincide (although the ranking by apportionment numbers will usually contain ties). Most readers of the Constitution will interpret the words on apportionment there as mandating the order-preserving criterion. As in the case of neutrality and proportionality, it is clear that Hamilton's method and all the divisor methods satisfy the order-preserving criterion. With all these methods, a larger population leads to a larger quota, which in turns rounds to at least as large an apportionment. It is easy to argue that all reasonable apportionment methods satisfy the order-preserving criterion. Still, it is not difficult to make up (unreasonable) methods that don't.

Note that the order-preserving property does not require that a larger state necessarily gets more seats than a smaller one. It merely requires that the larger gets at least as many seats as the smaller. From the 2010 census, Idaho has population 1,573,499 and Rhode Island has population 1,055,247. The order-preserving property requires that Idaho gets at least as many seats as Rhode Island. In fact, both got 2 seats using Hill's method, a method that is indeed order-preserving.

One is tempted to call the order-preserving criterion "population monotonicity", because it is a kind of monotonicity related to population. We reserve the term "population monotonicity" for a stronger notion, however, and we continue to use the term "order-preserving" (alternatively "weak population monotonicity") to describe this compelling but weak criterion.

9.2 Quota Rules and the Alabama Paradox

One criterion that we have already encountered is the following:

Definition 9.4. An apportionment method satisfies the **quota rule** if the method assigns to each state either its lower quota (its standard quota rounded down) or its upper quota (its standard quota rounded up).

A violation of the quota rule is called a **quota violation**. It occurs when a state is assigned a number of seats that is more than one unit away from its fair share. In the first United States apportionment in 1790, Virginia's fair share of 112 seats was 19.53 but it received 21 seats. The method used was Jefferson's. This example shows that Jefferson's method violates the quota rule. On the other hand, by its very design, Hamilton's method satisfies the quota rule.

We sometimes have reason to refer to two refinements of the quota rule, which are known as the upper quota rule and the lower quota rule.

Definition 9.5. An apportionment method satisfies the **upper quota**

rule if no state may be assigned by the method a number of seats greater than its upper quota.

Definition 9.6. An apportionment method satisfies the **lower quota rule** if no state may be assigned by the method a number of seats smaller than its lower quota.

A method that satisfies both the upper and lower quota rules satisfies the quota rule and is called a **quota method**.

Quota rules seem compelling at first blush, and quota violations seem to be evidence of unfairness. But the example in Table 8.5 and the example of the 1790 United States apportionment illustrate that Jefferson's method sometimes violates the quota rule. We will see below that all divisor methods violate the quota rule. Among these divisor methods are Hill's method, which is currently in use in the United States today. Hence restricting the search for the ideal apportionment method to those that satisfy the quota rule forces us to throw away a number of familiar methods, including the one in current use in the United States. We should therefore not reject too hastily all methods that violate the quota rule.

The next criterion stipulates that a method should avoid the Alabama paradox.

Definition 9.7. An apportionment method satisfies the **house monotonicity** criterion (or is **house monotone**) if, whenever h increases and all other variables remain unchanged, the method does not assign a smaller value of a_k for any k.

House monotonicity guarantees that no state may be harmed by a decision to increase the size of the House. In particular, if h is increased by one unit, then the resulting apportionment must look just like the apportionment before the increase except for the number associated to one state, which is increased by one.

The Alabama paradox of 1880 is an illustration of a violation of house monotonicity. Had h increased from 299 to 300 in 1880, Alabama's apportionment using Hamilton's method would have decreased (paradoxically) from 8 seats to 7 seats. This shows that Hamilton's method violates the house monotonicity criterion. This was one of the central reasons why Hamilton's method was abandoned in the United States.

Proposition 8.7 asserts that Jefferson's method satisfies house monotonicity. This result can be generalized to apply to all divisor methods.

Proposition 9.8. *All divisor methods satisfy house monotonicity.*

Proof. Consider a divisor method applied to the populations $p_1, p_2, \ldots, p_n$.

Consider how the method would cope with an increase in the house size h. Such an increase would require a smaller modified divisor d. Decreasing d increases the modified quota p_k/d for every k. Rounding a larger number by any means never results in a smaller answer, owing to property (2) of rounding functions. Hence no state can lose a seat when facing an increase in h. □

9.3 Population Monotonicity

Population monotonicity is a concept intended to forbid the population paradox (and the related new states or Oklahoma paradox) in the same way that house monotonicity forbids the Alabama paradox. Initially it seems that there are two different versions of the population paradox and hence two versions of the population monotonicity criterion. We call these criteria "population monotonicity" and "relative population monotonicity". It turns out that these criteria are equivalent in the presence of proportionality, and so we usually need not distinguish between them.

To discuss population monotonicity, we want to compare the output of a method applied to two different censuses. We employ a simple notational device to manage the barrage of symbols: We continue to use the symbol p_k to denote the population of state k in the first census, but we introduce the symbol p'_k to stand for the population of state k in the second census. Think of the superscript (the "prime" symbol) as denoting the second census. The corresponding apportionment outputs are given by a_k and a'_k. Here we make no assumption about the house sizes — these are denoted by h and h' — except that they are both positive.

Definition 9.9. An apportionment method satisfies the **population monotonicity** criterion (or is **population monotone**) if, whenever $a'_i < a_i$ and $a'_j > a_j$, it follows that either $p'_i < p_i$ or $p'_j > p_j$.

A violation of population monotonicity is said to be a **population paradox**. The idea is this: The only situation in which state i should be able to lose a seat in the house to state j is if the population of state i decreases or the population of state j increases (or both). A violation of this criterion for Hamilton's method is seen in the example from Tables 7.2 and 7.5, in which one state gains population while another loses population and yet the first state is forced to relinquish one of its seats to the second state. The existence of such examples proves the following result.

Proposition 9.10. *Hamilton's method does not satisfy population monotonicity.* □

On the other hand, divisor methods do not suffer from this flaw.

Proposition 9.11. *All divisor methods satisfy population monotonicity.*

Proof. Suppose that $a'_i < a_i$ and $a'_j > a_j$. By property (2) of rounding functions, this must be because the modified divisor p_i/d of state i decreased and the modified divisor p_j/d of state j increased. This is to say that $p'_i/d' < p_i/d$ and $p'_j/d' > p_j/d$ where d' and d are, respectively, the modified divisors used to apportion the two censuses. Rearranging these two inequalities yields

$$p'_i < \left(\frac{d'}{d}\right) p_i \quad \text{and} \quad p'_j > \left(\frac{d'}{d}\right) p_j.$$

Now compare the size of the divisors d and d'. If $d'/d \leq 1$, then $p'_i < p_i$. On the other hand, if $d'/d \geq 1$, then $p'_j > p_j$. Hence we have either $p'_i < p_i$ or $p'_j > p_j$. This implies that the method satisfies the population monotonicity criterion. □

Corollary 9.12. *Hamilton's method is not a divisor method.* □

The assertion of Corollary 9.12 might seem obvious. After all, Hamilton's method is described to us as a quota method. There are no modified divisors, and there is no rounding function. What Corollary 9.12 guarantees is that there is no rounding function leading to a divisor method that always computes the same results as Hamilton's method. In other words, Hamilton's method is not a divisor method in disguise.

Proposition 9.13. *Any method that satisfies population monotonicity must satisfy house monotonicity.*

Proof. Suppose that an apportionment method satisfies population monotonicity, and consider the situation in which h increases to $h' = h + 1$. Suppose also that no populations change, or in other words that

$$p'_i = p_i \text{ and } p'_j = p_j.$$

The increase in house size necessitates an increase in apportionment for at least one state, so assume this is state j and that $a'_j > a_j$. The population monotonicity criterion guarantees that we do not have $a'_i < a_i$ for any i. In other words, no state can lose a seat (to any other state). Thus, the apportionments of all states stay the same, except for the one state j, which gains the new seat. This shows that the method is house monotone. □

Proposition 9.14. *Any method that satisfies population monotonicity and neutrality must also satisfy the order-preserving criterion.*

Proof. Suppose that an apportionment method is neutral and population monotone, and suppose that we are given a census with $p_j > p_i$. We construct a second census by swapping the populations of states i and j. In other words, we define $p'_i = p_j$, $p'_j = p_i$, and $p'_k = p_k$ for all k not equal to i or j. This results in an increase in population for state i and a decrease in population for state j. Since $p'_i \geq p_i$ and $p'_j \leq p_j$, it follows from the population monotonicity assumption that it is impossible for both $a'_i < a_i$ and $a'_j > a_j$. Hence either $a'_i \geq a_i$ or $a'_j \leq a_j$.

We now appeal to the neutrality assumption. Since the new census was constructed by permuting the populations of states i and j, their apportionments must simply be interchanged. In other words $a'_j = a_i$ and $a'_i = a_j$. Therefore $a_j = a'_i \geq a_i$. We have shown that $p_j > p_i$ implies $a_j \geq a_i$, which is the order-preserving criterion. □

9.4 Relative Population Monotonicity

There are two ways to think about a population change from p_k to p'_k. The **absolute change** in population is given by $\Delta p_k = p'_k - p_k$, which is nothing but the difference between the future population and the past population. The **relative change** in population is given by $\Delta p_k / p_k$. (We assume that $p_k > 0$.) This is the usual way people talk about population change, and moreover $\Delta p_k / p_k$ is customarily expressed as a percentage. These changes can be positive, negative, or zero.

To understand the reason for measuring population change in two ways, consider the situation in which over a certain time interval two states both experience a population growth of 100,000 persons. We say that their populations both undergo an absolute change of 100,000. If the first state has an initial population of $p_1 = 1{,}000{,}000$ (so that $p'_1 = 1{,}100{,}000$) then it has experienced a relative growth of $\Delta p_1 / p_1 = 100{,}000/1{,}000{,}000 = 0.1$ or 10%. On the other hand, if the second state has an initial population of $p_2 = 10{,}000{,}000$, then its relative population growth is only $\Delta p_2 / p_2 = 100{,}000/10{,}000{,}000 = 0.01$, which is 1%. An addition of 100,000 new residents is more significant to a small state than it is to a large one. Relative population monotonicity is a criterion that requires the apportionment method to respect relative changes in populations of states.

Definition 9.15. An apportionment method satisfies the **relative pop-**

ulation monotonicity criterion (or is **relative population monotonicity**) if, whenever there are two states i and j with positive populations and with $a_i' < a_i$ and $a_j' > a_j$, it follows that$\Delta p_j/p_j > \Delta p_i/p_i$.

The phrase "with positive populations" is necessary because one cannot measure an absolute population increase from zero as a relative population increase at all, since division by 0 is forbidden. It may seem silly to consider the possibility of states with population 0, but we will in fact have cause to consider this possibility when we come to the new states paradox. Hence the restriction imposed by this phrase in the definition above is not entirely trivial.

In words, the relative population monotonicity criterion asserts that only when one state has had a larger relative population increase than another state can it increase its apportionment at the expense of that other state. The first thing to notice is that relative population monotonicity seems to be a strengthening of population monotonicity.

Proposition 9.16. *If an apportionment method satisfies relative population monotonicity, then it satisfies population monotonicity.*

Proof. Suppose that $a_i' < a_i$ and $a_j' > a_j$ and that all populations are positive. By the relative population monotonicity criterion, we have $\Delta p_j/p_j \geq \Delta p_i/p_i$. In particular, we do not have $\Delta p_j/p_j$ negative and $\Delta p_i/p_i$ positive. Hence we either have $\Delta p_i/p_i$ negative or $\Delta p_j/p_j$ positive (and possibly both). In the first case, we have $p_i' < p_i$; in the second case, we have $p_j' > p_j$. Thus the method is population monotone. □

One may guess that this stronger criterion is too strong, overly restrictive. In fact, it is a property possessed by all the divisor methods.

Proposition 9.17. *All divisor methods satisfy relative population monotonicity.*

Proof. Suppose that we are using a divisor method to compute apportionments of two censuses, and suppose that $a_i' < a_i$ and $a_j' > a_j$. By property (2) of rounding functions, the modified quota p_i/d of state i must have decreased and the modified quota p_j/d of state j must have increased. This is to say that $p_i'/d' < p_i/d$ and $p_j'/d' > p_j/d$, where d and d' are the modified divisors used to apportion the two censuses. Rearranging these inequalities (and using the assumption $p_i \neq 0$ and $p_j \neq 0$) yields $p_i'/p_i < d'/d < p_j'/p_j$. Removing the middle term d'/d here and subtracting 1 from both sides, we obtain

$$\frac{p_i'}{p_i} - \frac{p_i}{p_i} < \frac{p_j'}{p_j} - \frac{p_j}{p_j},$$

which is the same as $\Delta p_i/p_i < \Delta p_j/p_j$. □

The relative version of population monotonicity seems to be stronger than the absolute version, but this is illusory. The two criterion are essentially the same, in the sense that the converse of Proposition 9.16 holds in the presence of the proportionality criterion.

Proposition 9.18. *If an apportionment method is proportional and satisfies population monotonicity, then it satisfies relative population monotonicity.*

Proof. Suppose that a given apportionment method is proportional but not relatively population monotone. In particular, suppose that $a'_i < a_i$ and $a'_j > a_j$. while at the same time that $\Delta p_j/p_j \leq \Delta p_i/p_i$. Let $r = 1 + \Delta p_i/p_i$ and $s = 1 + \Delta p_j/p_j$. Then

$$p'_i = p_i + p'_i - p_i = p_i + \Delta p_i = \left(1 + \frac{\Delta p_i}{p_i}\right) p_i = rp_i,$$

and

$$p'_j = p_j + p'_j - p_j = p_j + \Delta p_j = \left(1 + \frac{\Delta p_j}{p_j}\right) p_j = sp_j.$$

Also $0 < s \leq r$.

Now consider a third ("double-prime") census such that the population of state k is given by $p''_k = p'_k/r$ for all k. This third census has the same population distribution as the second census, since the populations of the states are simply rescaled, uniformly, by the factor $1/r$, leaving the population distribution unchanged. Thus, since the apportionment method is proportional, the resulting apportionment satisfies $a''_k = a'_k$ for all k. We also have $p''_j = p'_j/r = (s/r)p_j \leq p_j$, since $s/r \leq 1$, and $p''_i = p'_i/r = p_i$. In other words, $a''_i < a_i$ and $a''_j > a_j$, but neither $p''_i < p_i$ nor $p''_j > p_j$. It follows that the method in question is not population monotone. □

Since proportionality is a criterion that we regard as compelling, we come to regard population monotonicity and relative population monotonicity as equivalent criteria, and we drop the adverb "relatively" from most of the further discussion.

9.5 The New States Paradox

A particularly interesting version of a population paradox can occur when some state's population increases from 0 to a positive number while the populations of all other states remain unchanged. Suppose state k is the state such that $p_k = 0$ and $p'_k > 0$. We imagine that state k is just joining the union (i.e., as a new state). Let states i and j be among the states whose populations are unchanged. We say that a **new states paradox** occurs if $a'_i < a_i$ and $a'_j > a_j$ (state i loses a seat to state j).

Such a thing could have occurred in 1907 when Oklahoma joined the union. If Hamilton's method had been rerun in 1907, after the admission of Oklahoma but using the 1900 census data for all other states, New York would have had to relinquish a seat to Maine. The surprise here is not simply that New York would have lost a seat; after all, Oklahoma entered the union and might have been seen to be new competition for seats. Neither is the surprise simply that Maine would have gained a seat; after all, the house size was increased in expectation of Oklahoma's need. The surprise is that *both* of these changes would have occurred. Maine and New York had unchanged populations and were sharing together the same number of seats, yet the presence of Oklahoma was interfering in the sharing of seats between Maine and New York. This is the anomaly that we might want to prohibit. The new states paradox is often called the **Oklahoma paradox.**

Now suppose the apportionment method satisfies population monotonicity. Since the populations of states i and j do not change, state i cannot lose a seat to state j. Thus a method that satisfies population monotonicity cannot result in an Oklahoma paradox.

It follows from Proposition 9.10 that no divisor method can result in the Oklahoma paradox. On the other hand, history tells us that Hamilton's method can result in an Oklahoma paradox. From this we can conclude, along the same lines as Corollary 9.12, that Hamilton's method is not a divisor method.

9.6 Impossibility

The method of Hamilton is not the only possible quota method. One can think of other ways of deciding which states to round up to upper quota and which states to round down to lower quota. One such method, which we introduced in Problem **7.16**, was suggested by William Lowndes, a member of the United States House of Representatives from South Carolina from 1811 to 1822.

Definition 9.19. Lowndes's method: As a provisional apportionment, assign every state its lower quota. Then assign the seats that remain to the states, at most one per state, in decreasing order of the ratio of the fractional parts of their standard quotas to their lower quotas.

This differs from Hamilton's method in a way that gives a distinct advantage to smaller states. For example, if two states have standard quotas 2.3 and 7.9, Hamilton's method tells us to give the second state a higher priority for an extra seat than the first state, because $0.9 > 0.3$. Lowndes's method, on the other hand, favors the first state over the second, because the ratio $0.3/2$ is greater than the ratio $0.9/7$. The justification for Lowndes's idea is that the people represented by the 0.3 in the first state, were that state to receive its lower quota, would have to be spread out over just two districts. Each of these two districts would have a relative population increase of $0.3/2 = 15\%$ owing to the people represented by the 0.3 fractional part. In the second state, each district would have a relative increase of $0.9/7 \approx 12.86\%$ owing to the people represented by the 0.9 fractional part. For this reason, according to Lowndes, rounding down imposes more pain on the first state than on the second.

We leave as an exercise to show that Lowndes's method violates the various monotonicity criteria, just as Hamilton's method does.

Table 9.1 summarizes which properties hold for the apportionment

Table 9.1 Which apportionment methods satisfy which criteria.

	Neutral and Proportional	Quota Rule	Lower Quota Rule	Upper Quota Rule	House Monotone	Population Monotone
Hamilton	Yes	Yes	Yes	Yes	No	No
Lowndes	Yes	Yes	Yes	Yes	No	No
Adams	Yes	No	No	Yes	Yes	Yes
Dean	Yes	No	No	No	Yes	Yes
Hill	Yes	No	No	No	Yes	Yes
Webster	Yes	No	No	No	Yes	Yes
Jefferson	Yes	No	Yes	No	Yes	Yes

methods that we have considered so far. The data here are easy to summarize. All the methods are neutral and proportional. The two quota methods — Hamilton and Lowndes — clearly satisfy the quota rules, but they do not satisfy the monotonicity criteria, owing to the Alabama and Oklahoma paradoxes for Hamilton's method and similar examples for Lowndes's. The divisor methods — of Adams, Dean, Hill, Webster, and Jefferson — satisfy the monotonicity criteria, as we proved earlier, but not the quota rule, as examples demonstrate.

While it is true that Adams's method satisfies the upper quota rule and Jefferson's method satisfies the lower quota rule, this is not often taken as evidence for the reasonableness of the methods of Adams and Jefferson. Rather, these facts simply put a positive spin on what is actually a defect of the methods, their noted bias in favor of small and large states, respectively.

Thus we have methods that satisfy quota but suffer from paradoxes, and we have methods that avoid paradoxes but violate quota. It would be desirable to find a method that satisfies all of these criteria. Unfortunately that desire cannot be fulfilled. This is the content of the following impossibility theorem.

Theorem 9.20. (Balinski and Young) *It is impossible for a neutral apportionment method to satisfy the quota rule and also satisfy population monotonicity.*

Proof. Imagine a neutral apportionment method that satisfies the quota rule and also population monotonicity. Consider how this method would apportion seats for a 4-state union with a house size of $h = 10$ and "before and after" populations as follows:

BEFORE	AFTER
State 1: $p_1 = 69{,}900$	State 1: $p'_1 = 68{,}000$
State 2: $p_2 = 5{,}200$	State 2: $p'_2 = 5{,}500$
State 3: $p_3 = 5{,}000$	State 3: $p'_3 = 5{,}600$
State 4: $p_4 = 19{,}900$	State 4: $p'_4 = 5{,}700$

First consider the "before" situation. The total population is exactly 100,000, so it is easy to read off the standard quotas.

STANDARD QUOTAS, BEFORE
State 1: $q_1 = 6.99$
State 2: $q_2 = 0.52$
State 3: $q_3 = 0.50$
State 4: $q_4 = 1.99$

State 1's apportionment must be 7 seats or less, because its standard quota is 6.99 and the quota rule guarantees at most upper quota for State 1. Similarly, State 4's apportionment must be at most 2 seats, its upper quota. Therefore, States 2 and 3 together will be apportioned at least 1 seat. Proposition 9.14 implies that our method is order-preserving. Hence State 2, with more population than State 3, must receive at least 1 seat.

Next consider the "after" situation. The standard quotas are obtained by dividing the populations by the standard divisor, which is equal to 8480. We obtain:

STANDARD QUOTAS, AFTER
State 1: $q_1' = 8.02$
State 2: $q_2' = 0.65$
State 3: $q_3' = 0.66$
State 4: $q_4' = 0.67$

State 1's apportionment must be at least 8 seats, because its standard quota is 8.02 and the quota rule guarantees at least lower quota for State 1. Therefore, States 2, 3, and 4 together will be apportioned 2 seats or fewer. By the order-preserving property again, State 2 will receive no seats at all. State 2, which was guaranteed at least one seat before, is certain to get no seats at all after. Meanwhile, State 1, which was guaranteed to get at most 7 seats before, gets at least 8 seats after. State 1 must gain seats while state 2 must lose seats.

Yet the population of state 1 has declined while the population of state 2 increased. We have constructed an example where $a_1' > a_1$ and $a_2' < a_2$ and yet where neither $p_1' \geq p_1$ nor $p_2' \leq p_2$. This demonstrates the violation of population monotonicity. □

So we are left with a choice: In the presence of neutrality, we must choose between the quota rule and population monotonicity, and we cannot have them both. The theorem does not mention house monotonicity, and it leaves open the possibility that there may exist a method that satisfies the quota rule and house monotonicity. In fact there is such a method, and we encounter it in the next chapter.

9.7 Exercises and Problems

9.1. Consider the simple apportionment method that assigns all seats to the largest state.
(a) Is this method neutral?
(b) Is this method proportional?
(c) Is this method population monotone?
(d) Does this method satisfy the quota rule?

9.2. Consider an apportionment method that we call the **odd method** that works as follows: Find a divisor d and divide it into the population of each state to obtain a modified quota p_i/d for state i. Then round these numbers to the nearest odd whole number (breaking ties, if the modified quota should happen to equal an even whole number, by rounding up).

(a) Does the odd method satisfy population monotonicity?
(b) Does the odd method satisfy the quota rule?
(c) What is the apportionment according to the odd method for the census data $h = 4$, $n = 2$, $p_1 = 123{,}456{,}789$, and $p_2 = 123{,}456{,}788$?
(d) What difficulty would you encounter if you attempted the odd method on the data from the most recent US census with $n = 50$ and $h = 435$?

9.3. Consider a union of 20 states with populations as follows:

State 1: Population 1301
State 2: Population 1302
State 3: Population 1303
⋮
State 18: Population 1318
State 19: Population 1319
State 20: Population 75,110

Assuming a house size of 100, compare the fair share of State 20 to her share by the methods of Hill, Jefferson, Adams, and Webster.

9.4. Imagine a divisor method that rounds modified quotas x in between the whole numbers m and $m+1$ down to m if and only if $x < \sqrt{(m^2 + (m+1)^2)/2}$ (the **root-mean-square** of m and $m+1$).

(a) Would this method satisfy the quota rule?
(b) Would this method be house monotone?
(c) Would this method be population monotone?

9.5. Consider a hybrid apportionment method that works as follows: Use the method of Jefferson unless a quota violation results. In that case, use the method of Hamilton instead.

(a) Does the method satisfy the quota rule?
(b) Does the method satisfy the order-preserving property?
(c) Is the method population monotone?

9.6. Consider a hybrid apportionment method that we call the **Hamster method**: Use the method of Webster unless a quota violation results. In that case, use the method of Hamilton instead.

(a) Does the method satisfy the quota rule?
(b) Does the method satisfy the order-preserving property?
(c) Is the method population monotone?

9.7. Consider an apportionment method that we call the **deal method**, named because it resembles dealing cards. Begin by dealing out one seat to each state, from largest to smallest. Then deal out a second seat to each state, again from largest to smallest. Continue in this way until the seats run out. So, for example, if $h = 10$ and there are $n = 3$ states with state 1 the largest, then the apportionment would be $a_1 = 4$, $a_2 = 3$, and $a_3 = 3$.

(a) Does this method satisfy the quota rule?
(b) Is this method house monotone?
(c) Does this method satisfy the order-preserving criterion?
(d) What is the result of applying the Deal method with $n = 50$ and with $h = 100$?

9.8. Call an apportionment method **nonzero** if it never assigns zero seats to any state. (The US Constitution mandates this criterion.) Assuming that $h \geq n$, which divisor methods that we have encountered satisfy the nonzero criterion?

9.9. Jefferson's method fails to meet the US Constitution's nonzero requirement. One way to modify it to do so is to implement a method we call **Jefferson-plus-one**. Assume that $h > n$. First assign 1 seat to each state. Then apportion the remaining $h - n$ seats according to Jefferson's method.

(a) Is the Jefferson-plus-one method proportional?
(b) Does the Jefferson-plus-one method satisfy the quota rule?
(c) Is the Jefferson-plus-one method house monotone?
(d) Is the Jefferson-plus-one method population monotone?
(e) Show that the Jefferson-plus-one method always gives the same result as Adams's method.

9.10. Another way to modify Jefferson's method so that it meets the nonzero requirement is to use the rounding function that rounds all numbers down except numbers between 0 and 1, which are rounded up.

(a) Is this method proportional?
(b) Does this method satisfy the quota rule?
(c) Is this method house monotone?
(d) Is this method population monotone?

9.11. Prove that Adams's method satisfies the upper quota rule but not the lower quota rule.

9.12. For any positive integer k, we can ask about any method whether it always gives apportionment numbers that are within k units of the standard quota. When $k = 1$, this is what we call the quota rule. In general, we call this the within-k-of-quota rule.

(a) Does the Jefferson method satisfy the within-5-of-quota rule?
(b) Does the Webster method satisfy the within-5-of-quota rule?
(c) Show that any neutral and proportional method that satisfies the within-2-of-quota rule must violate population monotonicity.

9.13. Consider the following apportionment method that attempts to compromise between the methods of Hamilton and Lowndes: Just like the methods of Hamilton and Lowndes, the lower quota is assigned provisionally to each state and then the states are ranked to establish the priority for obtaining one of the seats not yet assigned. In this method, that priority is measured by the value of $x/\sqrt{m}$, where m is the lower quota and x is the fractional part of the standard quota. Give an example showing that this method can differ from both Hamilton's method and Lowndes's method. Is this method population monotone?

9.14. Prove that Lowndes's method does not satisfy population monotonicity.

9.15. Consider the **Klein method** that first assigns each state its lower quota and then, if exactly k seats remain undistributed, gives one extra seat to each of the k smallest states.
(a) Is the small method order-preserving?
(b) Is the small method population monotone?
(c) Is the small method house monotone?
(d) Is the small method a quota rule?

10

The Method of Balinski and Young

"Life isn't fair. It's just fairer than death, that's all." — William Goldman

10.0 Scenario

Since 1940, the US House of Representatives has been apportioned using Hill's method. Balinski and Young have advocated for a switch from this method to Webster's method. Webster's method has much to recommend it. It is population monotone and seems to violate quota only rarely.

Unfortunately, Webster's method does not automatically satisfy the Constitutional requirement that each state receive at least one representative. How can Webster's method be altered to create an apportionment function that always assigns at least one seat to every state? The idea here is to adhere as much as possible to the concept of Webster's method yet patch this one potential Constitutional flaw. Assume that $h \geq n$; otherwise, no method can assign at least one representative to each state.

More generally, suppose that a Constitutional amendment is passed requiring each state to receive a certain minimum number of seats. This minimum could be set, for example, to be equal to the population of the state divided by 1,000,000, rounded up. How can Webster's method be altered to accommodate the amendment?

Although rare, quota violations can arise when using Webster's method. In a sense, nothing can be done about this, because the Impossibility Theorem asserts that all population monotone methods are subject to quota violations. But suppose that we regard lower quota violations as much more serious than upper quota violations. After all, a lower quota violation is likely to lead to litigation, since a state whose apportionment violates lower quota lacks appropriate representation, while a state whose apportionment violates upper quota has no reason to complain. How can Webster's method be altered to avoid all or most lower quota violations?

10.1 Tracking Critical Divisors

The quota methods of Hamilton and Lowndes are designed to satisfy the quota rule, but they violate house monotonicity and population monotonicity. On the other hand, the divisor methods, among them the five historical methods of Jefferson, Webster, Hill, Dean, and Adams, satisfy house monotonicity and population monotonicity, but they violate the quota rule. In 1975, Michel Balinski and Peyton Young discovered a new apportionment method that satisfies both the quota rule and house monotonicity, unlike any of the methods above. Of course, this new method does not satisfy population monotonicity; Theorem 9.20 tells us that no method that satisfies the quota rule can do so.

The basic idea of the Balinski and Young method is to make a small alteration to Jefferson's method, in its critical divisor form, in order to eliminate the upper quota violations. In order to explain the Balinski and Young procedure, we first review the critical divisor approach to Jefferson's method. Then we describe an iterative algorithm to produce the Jefferson apportionment, explaining how to produce the apportionment for house size $h + 1$ assuming that the apportionment for house size h has already been computed.

To begin, if $h = 0$, then the Jefferson apportionment is $a_j = 0$ for all j. Now suppose that the Jefferson apportionment for house size h is given by $a_1, a_2, a_3, \ldots, a_n$. The quotient $p_j/(a_j + 1)$ is a Jefferson critical divisor for state j and may be interpreted as the strength of the claim of state j for the next seat. The $(h + 1)$st seat goes to the state whose claim is strongest. The number $p_j/(a_j+1)$ represents how large an average congressional district in state j would be if state j were to earn the subsequent seat. Thus, Jefferson's method has us dole out the seats in order to the states that will put the representative to the best use, representing the most constituents. This certainly seems like a sensible approach.

Let us work a small example. Suppose that a country has three states A, B, and C with populations 7, 22, and 71, respectively, and a house size of $h = 15$. Here is how Jefferson's method works when implemented with critical divisors. The first seat is given to state C, because 71/1 is larger than 22/1 and 7/1. The second seat is given also to state C, because $71/2 = 35.5$ is still larger than 22/1 and 7/1. The third seat once again goes to state C, because $71/3 \approx 23.7$ is still larger than 22/1 and 7/1. The fourth seat goes to state B, because 22/1 is larger than both 71/4 and 7/1. The district created by this seat will serve 22 citizens of state B. Had this seat gone to state C, the district created by this seat would have served only 71/4 = 17.75 citizens. In this sense, state B makes the best use of the seat.

Table 10.1 shows a spreadsheet calculation of this process continued until $h = 15$. The three right-most columns of each row show the Jefferson apportionment for varying house sizes h. Once the Jefferson apportionment has been determined for some h, we compute the Jefferson apportionment for $h + 1$ as follows: We first compute the priorities — the Jefferson critical divisors — for each of the three states. These are shown in columns 2 through 4 of Table 10.1. The state for which this number is largest is the state that has its apportionment incremented by one.

Table 10.1 The Jefferson critical divisor method.

h	Jefferson Critical Divisors			Jefferson Apportionment		
	A	B	C	A	B	C
0				0	0	0
1	7	22	71	0	0	1
2	7	22	35.5	0	0	2
3	7	22	23.7	0	0	3
4	7	22	17.8	0	1	3
5	7	11	17.8	0	1	4
6	7	11	14.2	0	1	5
7	7	11	11.8	0	1	6
8	7	11	10.1	0	2	6
9	7	7.33	10.1	0	2	7
10	7	7.33	8.88	0	2	8
11	7	7.33	7.89	0	2	9
12	7	7.33	7.1	0	3	9
13	7	5.5	7.1	0	3	10
14	7	5.5	6.45	1	3	10
15	3.5	5.5	6.45	1	3	11

This method of computing the Jefferson apportionment is algorithmic; it does not have the trial-and-error feel of our original description of Jefferson's method, which involves guessing a modified divisor and testing to see if the resulting apportionment numbers sum to the proper house size. Yet the two ways of computing lead to the same result. One can find an appropriate modified divisor for Jefferson's method by looking at Table 10.1. In fact, one can describe precisely the range of values

d that give an appropriate modified divisor for any house size h. Here's how. The largest possible value of d is the largest critical divisor for that value of h. The smallest possible value of d is the second largest critical divisor for that value of h. For example, in Table 10.1, the critical divisors for $h = 12$ are 7, 7.33, and 7.1. Thus a modified divisor for Jefferson's method can be chosen among any of the real numbers greater than 7.1 but less than 7.33.

It is easy to see why any divisor in this range gives the Jefferson apportionment for $h = 12$. Any number between 7.1 and 7.33 is certainly greater than 7, so when such a number is divided into the population of state A, the result is less than 1. When rounded down, this yields 0 seats for state A. When such a number is divided into 22, the population of state B, it yields a number that is greater than 3 but less than 4. When this quotient is rounded down, we obtain 3 seats for state B. Similarly, when such a number is divided into 71, the population of state C, it yields a number that is greater than 9 but less than 10. When this quotient is rounded down, we obtain 9 seats for state C. Thus the resulting Jefferson apportionment will be 0, 3, and 9.

Jefferson's method is prone to upper quota violations. The apportionment of the largest state in almost all realistic examples violates upper quota. For example, the Jefferson method allots 55 seats to California using the 2010 census, although the standard quota for California is only about 52.54. Also, the 6 seats apportioned to state C in Table 10.1 with $h = 7$ violates upper quota. This is because state C has 71% of the population and 71% of 7 seats is 4.97 seats, the standard quota for state C. Thus the upper quota for state C is 5 seats. If we want to create a method that avoids upper quota violations, we can't allow state C to receive the 7th seat in the example of Table 10.1. This is precisely the idea behind the apportionment method of Balinski and Young.

Definition 10.1. The **method of Balinski and Young:** If $h = 0$, then $a_j = 0$ for all j. Suppose that the Balinski–Young apportionment for house size h is given by $a_1, a_2, a_3, \ldots, a_n$. Let the quotient $p_j/(a_j+1)$ represent the strength of the claim of state j for $(h+1)$st seat. We declare a state to be ineligible to receive a seat if receiving that seat would result in an upper quota violation. The $(h+1)$st seat goes to the eligible state with the strongest claim.

In other words, state k is **eligible** for the next seat if and only if

$$a_k + 1 \leq \lceil (h+1)p_k/p \rceil .$$

The left side of this inequality is the number of seats that state k would have if the next seat were given to it, and the right side is the upper quota of state k after the next seat is given out. A violation of this inequality is thus a violation of upper quota for state k.

The idea behind the method of Balinski and Young is both natural and simple. Forbid upper quota violations but otherwise conduct the apportionment process like the Jefferson method implemented with critical divisors. If at any time the strongest claim for the next seat is held by an ineligible state, skip to the second strongest claim. If that state is in turn ineligible, continue to the third strongest claim. Continue in this manner until an eligible state is encountered.

An astute observer will notice that the description of the method appears to be incomplete, since it does not explain what occurs if all the states turn out to be simultaneously ineligible for the subsequent seat. The next proposition asserts that, in fact, this never occurs.

Proposition 10.2. *At each stage in the execution of the method of Balinski and Young, at least one state is eligible to receive the next seat.*

Proof. Consider a census with populations $p_1, p_2, \ldots, p_n$, and suppose that, for a fixed value of h, the method of Balinski and Young assigns $a_1, a_2, \ldots, a_n$ seats to the n states. After one more seat is distributed, the standard divisor will be $s = p/(h+1)$, and so the standard quota for state k will be $p_k/s = (h+1)p_k/p$. State k is ineligible for that next seat if $a_k + 1$, the number of seats state k will have if it is assigned the next seat, is greater than $\lceil (h+1)p_k/p \rceil$, the upper quota of state k. This is to say that the whole number $a_k + 1$ is at least one unit greater than $(h+1)p_k/p$. Ineligibility for the $(h+1)$st seat is therefore expressible via the inequality

$$a_k \geq (h+1)p_k/p.$$

If we have this inequality for every state k, then we have

$$\begin{aligned} a_1 &\geq (h+1)p_1/p, \\ a_2 &\geq (h+1)p_2/p, \\ a_3 &\geq (h+1)p_3/p, \\ &\vdots \\ a_n &\geq (h+1)p_n/p. \end{aligned}$$

Adding these n inequalities yields

$$\begin{aligned} a_1 + a_2 + \cdots + a_n &\geq (h+1)p_1/p + (h+1)p_2/p + \cdots + (h+1)p_n/p \\ &= (p_1 + p_2 + \cdots + p_n)(h+1)/p \\ &= p(h+1)/p = h+1. \end{aligned}$$

This is impossible, because the left side of this inequality equals exactly

h, which is not as large as $h+1$. Hence at least one state must be eligible to receive the $(h+1)$st seat. □

Let us examine the method of Balinski and Young at work on the example of Table 10.1. We expand Table 10.1 to produce Table 10.2, which begins with the same columns but also includes data about standard quotas. Jefferson's method makes no reference to standard quotas at all, but the method of Balinski and Young requires these data in order to test whether numbers will exceed upper quota. Table 10.2 highlights the places where the Jefferson method violates upper quota. In the rightmost three columns of Table 10.2, one sees the Balinski–Young apportionment numbers.

Table 10.2 The method of Balinski and Young.

h	Critical Divisors			Jefferson			Standard Quotas			B-Y		
	A	B	C	A	B	C	A	B	C	A	B	C
0				0	0	0				0	0	0
1	7	22	71	0	0	1	0.07	0.22	0.71	0	0	1
2	7	22	35.5	0	0	2	0.14	0.44	1.42	0	0	2
3	7	22	23.7	0	0	3	0.21	0.66	2.13	0	0	3
4	7	22	17.8	0	1	3	0.28	0.88	2.84	0	1	3
5	7	11	17.8	0	1	4	0.35	1.10	3.55	0	1	4
6	7	11	14.2	0	1	5	0.42	1.32	4.26	0	1	5
7	7	11	11.8	0	1	6	0.49	1.54	4.97	0	2	5
8	7	11	10.1	0	2	6	0.56	1.76	5.68	0	2	6
9	7	7.33	10.1	0	2	7	0.63	1.98	6.39	0	2	7
10	7	7.33	8.88	0	2	8	0.70	2.20	7.10	0	2	8
11	7	7.33	7.89	0	2	9	0.77	2.42	7.81	0	3	8
12	7	7.33	7.1	0	3	9	0.84	2.64	8.52	0	3	9
13	7	5.5	7.1	0	3	10	0.91	2.86	9.23	0	3	10
14	7	5.5	6.45	1	3	10	0.98	3.08	9.94	1	3	10
15	3.5	5.5	6.45	1	3	11	1.05	3.30	10.65	1	3	11

It is instructive to compare the Jefferson apportionments in Table 10.2 to those of Balinski and Young. When $h = 10$, the two apportionment methods agree, and both give the apportionment 0, 2, 8. To increment h from 10 to 11, we compute the Jefferson critical divisors and obtain 7, 7.33, and 7.89, respectively, for A, B, and C. Thus state C has the strongest claim, and the Jefferson method for $h = 11$ is 0, 2, 9. But the computation of the standard quotas at $h = 11$ demonstrates why, according to the method of Balinski and Young, state C is ineligible for the 11th seat. The standard quota for state C at $h = 11$ is 7.81, and therefore state C's upper quota at $h = 11$ is 8. Assigning state C a 9th seat would violate upper quota. Thus, the 11th seat goes to the state with the next highest priority, which is state B. The Balinski and Young apportionment for $h = 11$ is therefore 0, 3, 8.

One apparent disadvantage of the method of Balinski and Young is that it must be computed sequentially. While the Jefferson method can be computed sequentially as well via critical divisors, one can always revert to the modified divisor approach and jump directly to the final answer. There is no parallel approach to the method of Balinski and Young. It is therefore somewhat more difficult to calculate the Balinski and Young apportionment than it is the Jefferson method for the US political apportionment problem, since it seems to require iterating the algorithm 435 times. In the modern era of computers, however, the extra time needed to execute the algorithm for the Balinski and Young apportionment can be measured in billionths of a second. So this is hardly a barrier to feasibility.

10.2 Satisfying the Quota Rule

It is clear that the method of Balinski and Young is house monotone. The seats are given away, one at a time, and once a seat is given away, it is never taken back for a subsequent value of h. In other words, increasing h can never result in a state losing a seat. It is also clear that the method of Balinski and Young satisfies the upper quota rule. This is built into the method itself. No upper quota violation can occur, because the method declares any state about to violate upper quota to be ineligible to receive the seat that would cause it to do so. What is amazing is that the method satisfies the lower quota rule as well.

Proposition 10.3. *The method of Balinski and Young satisfies the lower quota rule.*

Proof. Imagine that at house size h, state k violates lower quota. This means that the number a_k of seats assigned to state k is at least a full unit smaller than its standard quota. In other words,

$$a_k \leq \frac{hp_k}{p} - 1.$$

Our goal is to derive a contradiction from this assumption, thus demonstrating that no lower quota violations are possible.

Let I be the set of states whose apportionment a_i at house size h is greater than its standard quota hp_i/p. In symbols, i is in I if and only if $a_i > hp_i/p$. We now show that no state in I received the hth seat. To see this, note that state k was eligible to get the hth seat because $a_k + 1 \leq (h+1)p_k/p$. Moreover, state k had a stronger claim for the hth seat than any state i belonging to I, because

$$p_i/((a_i - 1) + 1) = p_i/a_i < p/h \leq p_k/(a_k + 1).$$

Now let g be the smallest house size such that all the states in I have the same apportionment that they do at house size h. Because no state in I received the hth seat, we have $g < h$. Suppose that state ℓ received the gth seat, and suppose that the apportionments for the house size g are given by the numbers $b_1, b_2, \ldots, b_n$. State ℓ is in I, since otherwise g would be smaller. Let J be the set of states that receive at least one of the seats $g+1, g+2, \ldots, h$. Now no state in J can be in I, so for every state j in J, we have

$$a_j \leq hp_j/p. \tag{10.1}$$

Let us now look at the strength of the claims for the gth seat. For every j in J, state j had fewer seats for house size g than for house size h, since $b_j < a_j$ whenever j is in J. So the strength of the claim of state j for the gth seat was at least

$$p_j/(a_j - 1 + 1) = p_j/a_j \geq p/h > p_\ell/a_\ell = p_\ell/(a_\ell - 1 + 1),$$

which is exactly the strength of the claim of state ℓ for the gth seat. Since state ℓ did, in fact, receive the seat, it must be that state j was ineligible for the seat for every state j in J. In other words, $b_j \geq gp_j/p$ or

$$-b_j \leq -gp_j/p \tag{10.2}$$

for every j in J. Adding (10.2) to (10.1) yields

$$a_j - b_j \leq hp_j/p - gp_j/p = (h-g)p_j/p \tag{10.3}$$

for every j in J. Summing $a_j - b_j$ over all j in the set J yields the total number of seats given away between the gth seat and the hth seat, in other words it adds up to $h - g$. Summing $(h-g)p_j/p$ over all j in the set J yields something smaller than summing $(h-g)p_j/p$ over all values of j from 1 to n, because k is not in J and $p_k > 0$. But summing $(h-g)p_j/p$ over all values of j from 1 to n yields exactly $h-g$. Therefore summing (10.3) over all values of j in J yields exactly $h-g$ on the left side but a quantity smaller than $h-g$ on the right side. This implies that $h - g < h - g$, which is the contradiction we sought. □

The apportionment method of Balinski and Young is designed to avoid upper quota violations, but Proposition 10.3 shows that it miraculously avoids lower quota violations as well. This shows that the method of Balinski and Young is a quota method. (Balinski and Young refer to it as "the quota method".) The method is designed also to satisfy the house monotonicity criterion, since it doles out house seats one by one without ever withdrawing seats that were previously assigned. It doesn't satisfy every criterion, however.

Corollary 10.4. *The method of Balinski and Young violates population monotonicity.*

Proof. Since the method satisfies the quota rule, Theorem 9.20 implies that it must violate population monotonicity. □

To compute the Balinski-Young apportionment for $h = 435$, one steps through the Balinski-Young apportionment first for $h = 1$, then for $h = 2$, and continuing inductively until $h = 435$ is reached. There does not seem to be any simple short-cut. Even when the Jefferson apportionment does not violate upper quota, this does not imply that the Balinski-Young apportionment agrees with the Jefferson apportionment, as the example of Problem **10.11** shows. Moreover, even though we know that the Balinski-Young apportionment will assign each state at least its lower quota, there does not seem to be any easy way to determine directly which states will get an extra seat, that is, their upper quota. As h increases, there may be a state that is ineligible for its next seat, but when the value of h is finally reached when it becomes eligible, another state becomes eligible at the same moment and the state must wait for h to increment further before it obtains the seat for which it is waiting.

No divisor method satisfies the quota rule. On the other hand, neither the method of Hamilton nor the method of Lowndes — the two quota methods we have encountered — is either population monotone or house monotone. Part of that is no surprise: The impossibility theorem of Balinski and Young asserts that no quota method can satisfy population monotonicity. But the method of Balinski and Young is a breakthrough: It is a quota method that satisfies house monotonicity.

10.3 Exercises and Problems

10.1. Compute the apportionment according to the method of Balinski and Young for a country with three states, using a house size of 15 and populations as follows:

- State A: Population 10,000,000
- State B: Population 11,000,000
- State C: Population 79,000,000

10.2. Compute the apportionment according to the method of Balinski and Young for a country with five states, using a house size of 20 and populations as follows:

- State A: Population 49,000,000
- State B: Population 15,000,000
- State C: Population 13,000,000
- State D: Population 12,000,000
- State E: Population 11,000,000

10.3. Give an example of a census with three states in which the Jefferson apportionment yields an upper quota violation even with $h = 3$.

10.4. Give an example of a census in which the Jefferson apportionment yields an upper quota violation even with $h = 2$.

10.5. The method of Balinski and Young regards a state as being ineligible to receive a seat if that would result in an apportionment for that state that exceeds its standard quota by more than one unit. Suppose that we attempted to raise the standard for eligibility by forbidding a state from exceeding its standard quota by more than 0.9 units. Show by an example that this could lead to a circumstance in which all states are simultaneously ineligible.

10.6. Prove that, in any step of the method of Balinski and Young on a three-state union, at most one state can be ineligible to receive a seat.

10.7. Show that the method of Balinski and Young is proportional.

10.8. Show that the method of Balinski and Young satisfies the order-preserving property.

10.9. Compute the Balinski and Young apportionment for the United States using census data from the year 2010 and a House size of $h = 17$. Contrast the result with the Jefferson apportionment using the same input data.

10.10. Use a spreadsheet and census data from the most recent census to compute the Balinski and Young apportionment for the United States with $h = 435$.

10.11. Consider a census in which $p_1 = 49$, $p_2 = 33$, $p_3 = 3$, and all remaining populations are smaller than 3, with a total population of $p = 100$. Compute the Balinski-Young apportionment and the Jefferson apportionment for $h = 7$.

11

Deciding among Divisor Methods

"Politics is not the art of the possible. It consists in choosing between the disastrous and the unpalatable." — John Kenneth Galbraith

11.0 Scenario

Three states in a certain union have populations as follows:

State A: 519,000
State B: 341,000
State C: 140,000

The federal legislature will accord these three states, all together, 10 seats in the House of Representatives. How should these seats in the House be apportioned between the three states? Use the methods of Hamilton, Lowndes, Jefferson, Adams, Webster, Dean, Hill, and Balinski and Young. What divisors are used? What modified quotas do these divisors yield?

11.1 Why Webster Is Best

The criteria we have articulated for measuring the reasonableness of apportionment methods do not help us to distinguish among the divisor methods. Any divisor method satisfies house monotonicity and population monotonicity. According to the theorem of Balinski and Young, therefore, no divisor method can satisfy the quota rule. This raises the question of whether there is any appropriate way to choose among them.

At the extreme ends of the spectrum live Jefferson's method and Adams's method. One argument in favor of Jefferson's method is that it adheres most closely to a literal reading of the Constitution. The words "no more than one in thirty thousand" suggest that fractional parts of quotas should be rounded down. Another argument in favor of Jefferson's method is that it is the unique divisor method that satisfies the lower quota rule. When lower quota is violated, a state earns fewer seats than its population warrants. Such a state has a strong and valid basis upon which to lodge a complaint. With Jefferson's method, no state will ever be able to object to the apportionment with this sort of claim. There is also an argument for Adams's method at the other end of the spectrum: It is the only divisor method that satisfies the upper quota rule.

But the methods of Jefferson and Adams are also the most biased. Jefferson's method is biased in favor of large states, and Adams's method is biased in favor of small ones. When these methods are implemented on realistic data, Jefferson's method will almost always violate upper quota (large states getting more seats than their upper quota), and Adams's method will almost always violate lower quota (large states getting fewer seats than their lower quota). Had Jefferson's method or Adams's method been in use in the United States since 1790, the apportionment numbers would have revealed quota violations every single decade.

The methods of Dean, Hill, and Webster, by contrast, would never have violated quota on any of the 23 US censuses since 1790. This doesn't mean that they satisfy the quota rule. Examples show that they can violate quota, sometimes to a large degree. But the examples needed to produce quota violations from these methods tend to have data that is peculiar and contrived. In natural examples, these three methods usually give numbers that fall within quota. So even though a significant criticism of these methods is that they *can* violate the quota rule, the fact that they rarely do so is a powerful mitigating factor.

We will see in this chapter that each of the methods of Dean, Hill, and Webster can be identified as the unique divisor method — not just among the five historical divisor methods but among all possible divisor methods — that satisfies a certain optimization property. Along the way, we will encounter the critical divisor implementation of each of these methods, solving the algorithmic question of how to compute these apportionments without resorting to trial and error to search for a modified divisor.

We begin by investigating Webster's method. This method is characterized by the fact that every a_i is within $1/2$ of the corresponding modified quota p_i/d. In other words,

$$p_i/d - 1/2 < a_i \leq p_i/d + 1/2 \tag{11.1}$$

for every i. The careful placement of the "$<$" and "$\leq$" in display (11.1) reflects the usual convention that a real number whose fractional part is exactly .5 is customarily rounded up. When $a_i > 0$, these inequalities can be rearranged to read

$$p_i/(a_i + 1/2) < d \leq p_i/(a_i - 1/2).$$

Since both of these inequalities hold for every i and since d is independent of i, we can also say that

$$p_j/(a_j + 1/2) < d \leq p_i/(a_i - 1/2) \tag{11.2}$$

holds for every i and every j (whether or not $i = j$). Eliminating d, this in turn rearranges to

$$(a_i - 1/2)/p_i < (a_j + 1/2)/p_j,$$

which holds, trivially, even when $a_i = 0$. Multiplying both sides by 2 yields

$$(a_i + a_i - 1)/p_i < (a_j + a_j + 1)/p_j,$$

which rearranges to

$$a_i/p_i - a_j/p_j < (a_j + 1)/p_j - (a_i - 1)/p_i. \tag{11.3}$$

The connection between inequalities (11.1) and (11.3) works both ways. If (11.1) holds, in other words if we are using Webster's method, then (11.3) follows. Conversely, if (11.3) holds for every i and j, then there exists a modified divisor d such that (11.2) holds, in which case (11.1) holds in turn. To find such a divisor, just pick any real number greater than all of the numbers of the form $p_i/(a_i + 1/2)$ and smaller than all of the numbers of the form $p_j/(a_j - 1/2)$.

The punch line to this story comes in the interpretation of inequality (11.3). The quantity a_i/p_i represents what we have called the **degree of representation** in state i. It is the fraction of a congressional seat that can be regarded as belonging to each citizen in state i. If this number is 1/700,000, roughly a typical size in the US political apportionment problem, it means that there is one representative for every 700,000 people. One can think of each resident of such a state as controlling 1/700,000 of a seat in the House. States fight to make this ratio as large as possible. Suppose that the left side of (11.3) is positive, which means that the degree of representation in state i exceeds the degree of representation in state j. This means that state i gets a better deal than state j. The left side of inequality (11.3) is a measure of the extent to which the degree of representation in state i exceeds that of state j. Now one can ask whether the apportionment would be fairer if state i would relinquish a seat to state j. Notice that the right side of inequality (11.3) can be interpreted as the extent to which the degree of representation in state j would exceed that of state i were state i to give one seat to state j. Inequality (11.3) asserts first that the right side would have to be positive also, which means that a transfer of a single seat would pass the advantage from state i to state j. Inequality (11.3) further asserts that the imbalance after the transfer would be greater than the imbalance before the transfer. That is, the advantage of state j over state i after the change would be greater than the advantage of state i over state j before.

The fact that inequality (11.3) holds for any two states i and j indicates that we have reached a stable point in the assignment of seats in which no state can relinquish any seats to any other state without increasing the imbalance between states. In particular, if we postulate in advance that our apportionment method must minimize the difference in degrees of representation between every pair of states, then inequality (11.3) will hold for every i and j. This implies that (11.1) holds for some divisor d, and so we must actually be using Webster's method.

We have proved the following result.

Theorem 11.1. *Webster's method is the unique apportionment method with the property that the difference between the degrees of representation of any two states cannot be decreased by transferring a seat from the better represented state to the worse represented state.* □

Here is an interesting consequence of Theorem 11.1. If an apportionment is done that does not give to each state the same number of seats as Webster's method, then there will be a pair of states such that a seat could be transferred from one state to the other with the result that the degrees of representation in the two states would be more nearly equal than they were before the transfer.

Let us consider a simple example with two states. Imagine that $p_1 = 10$, $p_2 = 25$, and $h = 5$. The issue is whether the apportionment should be $a_1 = 2$ and $a_2 = 3$ or it should be $a_1 = 1$ and $a_2 = 4$. In the first case, the degrees of representation are $2/10$ and $3/25$, respectively, for the two states, and $2/10$ is larger. But in the second case, the degrees of representation are $1/10$ and $4/25$, respectively, and the advantage switches to the second state, since $4/25$ is larger. Which imbalance is smaller? In the first case, the difference is $2/10 - 3/25 = 4/50$ in favor of the first state, while, in the second case, the difference is $4/25 - 1/10 = 3/50$ in favor of the second state. Webster's method selects the second apportionment, because the difference of $3/50$ is smaller.

Inequality (11.2) above also shows how to compute Webster's method via critical divisors. Suppose that we have apportioned h seats via Webster's method, obtaining the assignment $a_1, a_2, \ldots, a_n$. In that case, inequality (11.2) demonstrates the range of possible values of the modified divisor d that lead to this apportionment. Any modified divisor d that is greater than every number of the form $p_i/(a_i + 1/2)$ but less than every number of the form $p_j/(a_j - 1/2)$ will work. Suppose that the house size is increased from h to $h+1$. The modified divisor must then be decreased correspondingly until one more seat is doled out. This requires that d is moved down the number line past the largest value of $p_i/(a_i + 1/2)$. At the moment that d reaches this value, a_i can be incremented by 1 to $a_i' = a_i + 1$, and then we have inequality (11.2) holding again but with a_i' replacing a_i. In other words, state i will receive the next seat if $p_i/(a_i + 1/2)$ is larger than any other number of this form with a different subscript. The numbers $p_i/(a_i + 1/2)$ are the **Webster critical divisors**.

Because apportionment numbers must be whole numbers, it is generally not possible to keep the degrees of representation in all states precisely equal. Therefore some states must do better than others. But if you believe that the ultimate goal of an apportionment method is keeping the degrees of representation of the states close to one another, then you must advocate for the method of Webster. It is the unique method that does this the best.

11.2 Why Dean Is Best

The argument above in favor of Webster's method would be overwhelmingly persuasive if it were not for the fact that there is a similar persuasive argument for Dean's method. Dean's method is characterized by the fact that the modified quota p_i/d is rounded down to a_i whenever

$$\frac{p_i}{d} < \frac{2a_i(a_i+1)}{2a_i+1}$$

and is rounded up to a_i whenever

$$\frac{2(a_i-1)a_i}{2(a_i-1)+1} \leqslant \frac{p_i}{d}.$$

(As in the case of Webster's method, we choose to round up if the left and right sides of these inequalities are precisely equal.) This implies that

$$\frac{2(a_i-1)a_i}{2(a_i-1)+1} \leq \frac{p_i}{d} < \frac{2a_i(a_i+1)}{2a_i+1}$$

for every i. Note that no positive quota is ever rounded down to 0 by Dean's method, so we assume $a_i > 0$ for all i. When $a_i > 1$, the previous inequalities can be rearranged to read

$$\frac{2a_i + 1}{2a_i(a_i + 1)} < \frac{d}{p_i} \leq \frac{2a_i - 1}{2(a_i - 1)a_i}$$

or

$$p_i\left(\frac{2a_i + 1}{2a_i(a_i + 1)}\right) < d \leq p_i\left(\frac{2a_i - 1}{2(a_i - 1)a_i}\right).$$

Since this holds for every i, it is also true that for every i and j we have

$$p_j\left(\frac{2a_j + 1}{2a_j(a_j + 1)}\right) < d \leq p_i\left(\frac{2a_i - 1}{2(a_i - 1)a_i}\right). \tag{11.4}$$

Eliminating d and multiplying both sides by 2 yields

$$p_j\left(\frac{2a_j + 1}{a_j(a_j + 1)}\right) < p_i\left(\frac{2a_i - 1}{(a_i - 1)a_i}\right),$$

which rearranges to

$$p_j\left(\frac{1}{a_j} + \frac{1}{a_j + 1}\right) < p_i\left(\frac{1}{a_i} + \frac{1}{a_i - 1}\right),$$

or

$$\frac{p_j}{a_j} - \frac{p_i}{a_i} < \frac{p_i}{a_i - 1} - \frac{p_j}{a_j + 1}, \tag{11.5}$$

for all i and j. We may regard (11.5) as holding trivially in the case when $a_i = 1$, in which case the right side of the inequality is infinite.

Inequality (11.5) is just like inequality (11.1) except that all the fractions have been inverted. As before, we ask what the quantity p_i/a_i represents, and in this case it is clear. The quantity p_i/a_i represents the size of an average congressional district in state i. It is the reciprocal of the degree of representation in state i. States fight to make this ratio as small as possible. Suppose that the left side of (11.5) is positive, which means that the district size in state j exceeds the district size in state i. State i gets a better deal than state j in this case, and the left side of (11.5) can be interpreted as a measure of this advantage, the difference between the district size in state j and the district size in state i. We argue just as we did for Webster's method. We ask again whether it would be fairer if state i would relinquish a seat to state j. The right side of inequality (11.5) describes the extent to which the district size in state i would exceed that of state j were state i to lose one seat and state j to gain one seat. Inequality (11.5) asserts first that the right side would have to be positive also, which means again that a transfer of a single seat would pass the advantage from state i to state j. But it asserts also that the imbalance, the extent to which the consideration of district size would favor state j over state i, would be greater than the imbalance before the seat was transferred.

In other words, inequality (11.5) indicates that we have reached a stable point in the assignment of seats in which no state can relinquish any seats to any other state without increasing the imbalance between states. This time, however, our measure of imbalance is not the difference between degrees of representation but rather the difference between district sizes. If we postulate that our apportionment method minimize the difference between district sizes between any pair of states, then inequality (11.5) must hold for every i and j. This implies that we must be using Dean's method. We have thus proved:

Theorem 11.2. *Dean's method is the unique apportionment method with the property that the difference between the sizes of districts in any two states cannot be made smaller by transferring a seat from the better represented state to the worse represented state.* □

Let us return to the example we considered earlier, with $p_1 = 10$, $p_2 = 25$, and $h = 5$. If the apportionment assigned $a_1 = 2$ and $a_2 = 3$, then the district sizes would be $10/2$ and $25/3$, respectively. If, on the other hand, the apportionment assigned $a_1 = 1$ and $a_2 = 4$, then the district sizes would be $10/1$ and $25/4$. In the first case, the difference is $25/3 - 10/2 = 10/3$ with the advantage going to the first state (smaller being better for district size), while, in the second case, the difference is $10/1 - 25/4 = 15/4$ in favor of the second state. Dean's method selects the first apportionment, because $10/3$ is smaller than $15/4$. Note that this is a different apportionment than Webster's method gives on these data. Measuring whether two numbers are close together is subtly but significantly different from measuring whether their reciprocals are close together.

The **Dean critical divisors** are seen from inequality (11.4) to be equal to

$$p_j(2a_j + 1)/2a_j(a_j + 1).$$

Whichever one of these numbers, as j is varied over all the possible states, is the largest dictates which state receives the subsequent seat. Notice that this formula requires a division by a_j, which is problematic if $a_j = 0$. To resolve this problem, we regard division by 0 as yielding the value infinity, which is thought of as being larger than any finite number. The Dean critical divisors are not finite until a_j is at least 1 for every j. This corresponds to the fact that the harmonic rounding never rounds a positive number down to 0. Hence Dean's method cannot be executed unless $h \geq n$, in which case the first n seats are doled out automatically, one to each state, before the finite Dean critical divisors come into play. The n infinite Dean critical divisors, the ones with $a_j = 0$ for the n possible values of j, correspond to each state's infinitely powerful claim to its first seat.

It can be shown that the geometric mean of two numbers is always between their arithmetic mean and their harmonic mean. This suggests that Hill's method should be a kind of compromise between the methods of Webster and Dean. If Webster's method is optimal in one sense and Dean's method is optimal in another, then one may ask if there is some other sense in which Hill's method is optimal. The answer is yes. We imitate the derivations above to discover the sense in which Hill's method minimizes the discrepancy between states.

11.3 Why Hill Is Best

We have seen that either Webster's method or Dean's method can be considered optimal from a certain point of view. We now consider a third viewpoint from which Hill's method rises to the top. To understand the sense in which Hill's method is optimal, we first consider different ways to compare two numbers to see how nearly equal they are. One way is to subtract the smaller number from the larger and see how close to 0 we get. But another approach is to divide the larger number by the smaller and see how close to 1 we get. These two approaches give rise to subtly different notions of nearness. The former regards 10 and 30 as more nearly equal than 100 and 200, while the latter regards 100 and 200 as more nearly equal than 10 and 30. Hill's method adopts this second approach.

Hill's method is the one that rounds the modified quota p_i/d down to a_i whenever

$$\frac{p_i}{d} < \sqrt{a_i(a_i + 1)}$$

and rounds the modified quota p_i/d up to a_i whenever

$$\sqrt{a_i(a_i - 1)} \leqslant \frac{p_i}{d}.$$

In other words,

$$\sqrt{a_i(a_i - 1)} \leqslant \frac{p_i}{d} < \sqrt{a_i(a_i + 1)}$$

for every i. (As for Dean's method, we know that a_i is nonzero for all i.)

Inverting these inequalities and multiplying through by p_i yields

$$\frac{p_i}{\sqrt{a_i(a_i+1)}} < d \leqslant \frac{p_i}{\sqrt{a_i(a_i-1)}},$$

provided $a_i > 1$. Since this holds for every i with $a_i > 1$, it is also true that for every i and j with $a_i > 1$ we have

$$\frac{p_j}{\sqrt{a_j(a_j+1)}} < d \leqslant \frac{p_i}{\sqrt{a_i(a_i-1)}}. \tag{11.6}$$

Eliminating d and squaring both sides yields

$$\frac{p_j^2}{a_j(a_j+1)} < \frac{p_i^2}{a_i(a_i-1)}.$$

Algebraic manipulations allow us to express this alternatively as

$$\frac{p_j/a_j}{p_i/a_i} < \frac{p_i/(a_i-1)}{p_j/(a_j+1)} \tag{11.7}$$

or equivalently as

$$\frac{a_i/p_i}{a_j/p_j} < \frac{(a_j+1)/p_j}{(a_i-1)/p_i} \tag{11.8}$$

for all i and j with $a_i > 1$. In fact, inequalities (11.7) and (11.8) hold even when $a_i = 1$, provided that we understand division by 0 as yielding infinity and that all numbers are less than infinity. Suppose that the district sizes in state j are greater than they are in state i, so that the advantage is with state i. Then the left side of inequality (11.7) is greater than 1. Inequality (11.7) tells us that the transfer of one seat from state i to state j will pass the advantage from state i to state j, and moreover that the ratio of district sizes will be even greater (will surpass 1 to an even larger degree) after such a reapportionment has been executed. In the form of inequality (11.8), where again the left side is arranged to be greater than 1, we see that the ratio of degrees of representation will be greater after such a reapportionment. Note how the emphasis this time is on ratios being small (that is, close to 1) rather than differences being small (that is, close to 0).

Suppose that we have assigned seats to states in such a way that the transfer of any seat from one state to another increases the ratio of district sizes between the two states. Then inequality (11.7) holds for all i and j. This indicates that the method is that of Hill. We have thus proved:

Theorem 11.3. *Hill's method is the unique apportionment method with* *the property that the ratio between the average sizes of districts in any two states (expressed as a number greater than or equal to 1) cannot be made smaller by transferring a seat from the better represented state to the worse represented state. Moreover the same assertion is true if the phrase "average sizes of districts" is replaced by the phrase "degrees of representation".* □

Let us take a third look at the example from above with $p_1 = 10$, $p_2 = 25$, and $h = 5$. As we contemplate the assignment $a_1 = 2$ and $a_2 = 3$ versus the assignment $a_1 = 1$ and $a_2 = 4$, we should be examining the district sizes $10/2$ and $25/3$ versus the sizes $10/1$ and $25/4$. In which case is the imbalance greater? If we measure this imbalance by computing the difference between these numbers, then choosing the smaller difference led us to the method of Dean and to choosing the first option. If instead we measure this imbalance by computing the ratio between these numbers, then choosing the smaller quotient leads us to the method of Hill. Adopting this approach, we measure the imbalance in the first case as $25/3 \div 10/2 = 5/3$ while in the second case as $10/1 \div 25/4 = 8/5$. Hill's method selects the second option, because $8/5$ is smaller than $5/3$. The methods of Dean and Hill disagree on these data, even though they can be understood as measuring the imbalance between the same pairs of numbers. One minimizes a difference while the other minimizes a quotient.

Inequality (11.6) shows that the numbers of the form

$$p_j \Big/ \sqrt{a_j(a_j + 1)}$$

are the **Hill critical divisors**. As one computes the Hill apportionment for successive values of h, one gives each subsequent seat to the state j for which the ratio $p_j/\sqrt{a_j(a_j + 1)}$ is largest. Often this is the way that Hill's method is described when the current United States apportionment method is explained.

Let us review our findings. It is certainly desirable, in doing apportionment, to keep district sizes in the various states as close as possible to one another. But there are different ways to measure closeness between numbers. One way is to subtract the smaller number from the larger and see how close to 0 you get. This notion of closeness leads to Dean's method. A second way is to subtract the reciprocal of the larger number from the reciprocal of the smaller, again seeing how close to 0 you get. This notion of closeness leads to Webster's method. A third way is to divide the larger number by the smaller, seeing how close to 1 you get. This approach to closeness leads to the method of Hill. It isn't easy on this basis to choose among these three methods, although this point of view represents a powerful argument to narrow the decision down to one of these three options.

One way to further distinguish between these methods is to compare their treatment of balance between small and large states. Generally, among the five historical divisor methods — Adams, Dean, Hill, Webster, and Jefferson — small states tend to receive better treatment as one moves to the left and large states tend to receive better treatment as one moves to the right. The methods of Adams and Jefferson at the ends are often dismissed as too biased, and large states customarily violate quota when either of these methods is used. Among the middle three, Dean's method is most friendly to small states, while Webster's method is least friendly to small states. Still, it isn't possible to say definitively which of these three methods has the "appropriate" balance between small states and large, because it isn't simple to articulate what represents an ideal or perfect balance.

Another approach to deciding among these three attractive apportionment methods is to attempt to discern which ones are most likely to violate quota on realistic data. We know that no divisor method can satisfy the quota rule, but it turns out that the methods of Dean, Hill, and Webster don't seem to violate quota when facing actual census data. So while violation of quota remains a theoretical possibility, as a practical matter it is not a major concern. Some experiments have suggested that Webster's method is the least likely of these three to violate quota on real-life data (however that is defined), and some have advocated on behalf of Webster's method via such arguments.

In the next chapter, we examine the rocky and colorful history of the American apportionment process, decade by decade. Congress decided in 1930 to make Hill's method the permanent rule for conducting apportionment in the United States. Why? To a large degree, it was serendipity. There were strong advocates for the method of Hill and the method of Webster. (There does not seem to have been at that time any strong advocate for the method of Dean.) Members of Congress were by and large unable to settle the debate, and Edward Huntington's ties to the National Academy of Science, which advises Congress, may have been the critical factor in the end. Or, as the next chapter suggests, it may have been simple politics. Some members of Congress were persuaded that the position of the method of Hill in the center of the list made it a good compromise position. If the methods of Dean and Webster on either side are optimal from two reciprocal points of view, then perhaps a position naturally between them that is less biased toward small states than Dean's method but less biased toward large states than Webster's method is the right choice in the end. This is not an utterly persuasive stance, however, and the issue of which method is really fairest is not entirely resolved today.

But at least in current US policy one principle of fair play does reign: We've picked some rules, we've announce them in advance, and we abide by them. They may not be the fairest rules, but they are our rules, and we are doubtlessly better off using the method of Hill than if apportionment was done in Congress, as it was done prior to 1930, via a political catfight.

11.4 Exercises and Problems

11.1. Suppose that State A has population 5,000,000, while State B has population 6,770,000, and suppose that they are to share 20 representatives between them.

(a) If State A gets 8 representatives and State B gets 12 representatives, which state has a more favorable average district size?
(b) If State A gets 9 representatives and State B gets 11 representatives, which state has a more favorable average district size?
(c) Of the two apportionments suggested in parts (a) and (b), which gives average district sizes that are closer together?
(d) Of the two apportionments suggested in parts (a) and (b), which gives degrees of representation that are closer together?

11.2. Suppose that State A has population 7,000,000, while State B has population 4,000,000, and suppose that they are to share 4 representatives between them.

(a) If State A gets 3 representatives and State B gets 1 representative, which state has a more favorable average district size?
(b) If State A gets 2 representatives and State B gets 2 representatives, which state has a more favorable average district size?
(c) Of the two apportionments suggested in parts (a) and (b), which gives average district sizes that are closer together?
(d) Of the two apportionments suggested in parts (a) and (b), which gives degrees of representation that are closer together?

11.3. Give an example of an apportionment problem with $n = 2$ states that gives one outcome for Dean's method but a different outcome for Hill's method.

11.4. Give an example of an apportionment problem with $n = 3$ states that gives three different outcomes for Dean's method, Hill's method, and Webster's method.

11.5. Suppose that $n = 3$, $p_1 = 140{,}000$, $p_2 = 341{,}000$, and $p_3 = 519{,}000$.

(a) What are the top ten Dean critical divisors? What is the Dean apportionment for $h = 10$?

(b) What is the Hamilton apportionment for $h = 10$?

11.6. Suppose that $n = 3$, $h = 7$, $p_1 = 134{,}000$, $p_2 = 235{,}000$, and $p_3 = 331{,}000$.

(a) What is the Hamilton apportionment?

(b) What is the Dean apportionment? By how much does the district size in the worst represented state exceed the district size in the best represented state?

(c) What is the Webster apportionment? By how much does the degree of representation in the best represented state exceed the degree of representation in the worst represented state?

11.7. Let p and q denote two positive real numbers with $p < q$.

(a) Show that $p < \sqrt{pq} < q$.

(b) Show that $\sqrt{pq} < (p + q)/2$. (Hint: Attempt to prove the inequality obtained from this one by squaring both sides.)

11.8. Obtain the apportionment populations from the most recent US census.

(a) Compute the Dean, Hill, and Webster critical divisors.

(b) Find the smallest value of h such that the Dean apportionment and the Hill apportionment differ.

(c) Find the smallest value of h such that the Dean apportionment and the Webster apportionment differ.

(d) Find the smallest value of h such that the Hill apportionment and the Webster apportionment differ.

11.9. Give an example of a census involving 3 states for which Dean's method gives a quota violation.

11.10. Give an example of a census involving 3 states for which Hill's method gives a quota violation.

11.11. Prove that Webster's method satisfies quota on every census involving 3 states.

11.12. Show that no divisor method can encounter both a lower quota violation and an upper quota violation on the same census.

11.13. Show that Adams's method is the one that minimizes the largest district size among the states.

11.14. Show that Jefferson's method is the one that maximizes the smallest district size among the states.

12

History of Apportionment in the United States

"Though force can protect in emergency, only justice, fairness, consideration and cooperation can finally lead men to the dawn of eternal peace."
— Dwight D. Eisenhower

12.0 Scenario

Obtain estimates via extrapolation for the apportionment populations of the 50 states at the next census. Compute the apportionment with $h = 435$ for these populations using Hill's method. What changes are in store for the various states?

Compute the apportionment for these populations using the methods of Hamilton, Webster, and Dean. Which states have an immediate stake in the apportionment method to be used next decade?

At present, the District of Columbia does not have senators or ordinary representation in the House of Representatives. (They have a so-called "delegate" in the House, but the delegate does not have the right to cast votes on the House floor.) This leaves the residents of DC — there are more of them than there are residents of Wyoming — with no voting representation in Congress. Congress has considered a voting rights act that would treat DC as a state for the purposes of apportionment and allot to DC a representative in the House. The political consequences of such a change are clear, since DC is an overwhelmingly Democratic jurisdiction. Thus voting rights for DC have been opposed by most Republican members of Congress. One 2009 proposal meant to address this opposition was to increase the size of the House not merely by 1 to 436 but by 2 to 437. This proposal was supported by both Republican senators from the state of Utah. Explain why.

How would an increase from $h = 435$ and $n = 50$ to $h = 437$ and $n = 51$ affect the apportionment after the next census, assuming that Hill's method is used?

12.1 The Fight for Representation

Our founding fathers had the wisdom to create a federal system of representative government, but they probably did not recognize that the apportionment requirement they established would lead to rancorous partisan congressional debate for more than 150 years. The history of apportionment in the United States is a story of self-serving appeals for representation couched in terms of high mathematical principles. Political infighting and partisanship emerge as influences on the debate far more significant than dispassionate reasoning and the principles of fair play. Whenever members of the House of Representatives have propounded a mathematical argument in favor of an apportionment method, it is a good bet that the method being suggested favored that member's state. Whenever members of the House have argued with passion against a method that would result in a loss of a seat for their state, a generous view will see them as fighting on behalf of their constituents' voice in government, but a more cynical eye will see them as fighting for their own job. It should not surprise us that they display great creativity in finding arguments to support their claims. Every apportionment method favors some state, so every method has its advocates. Naturally, those advocates appeal, whenever possible, to mathematics in defense of their preferred method, since mathematics offers the language of rigor to establish their claims irrefutably. Yet given that no method can satisfy every desirable criterion, it remains unsettled to this day which method is the best choice.

Article 1, Section 2, of the US Constitution says "Representatives ... shall be apportioned among the several States ... according to their respective Numbers. ... The actual Enumeration shall be made within three years after the first meeting of the Congress ..., and within every subsequent Term of ten Years, in such manner as they shall by Law direct. The Number of Representatives shall not exceed one for every thirty thousand, but each State shall have at least one Representative." The Constitution goes on to describe how the seats in the first House of Representatives were to be apportioned until the first census could be taken in 1790. Sixty-six seats were thus doled out to the 13 colonies, and the first Congress was seated in 1789.

This first Congress proposed a package of 12 amendments to the Constitution. The first of these was a clarification of the apportionment clauses in Article 1, Section 2. This amendment and another were not ratified by a sufficient number of states. The 10 amendments that remained were ratified and comprise what we now call the Bill of Rights.

The results of the first census were reported to Congress in 1791, and thus began the long and bitter debate in Congress, lasting 150 years (if not all the way to the present time), over what mathematical procedure to use to implement the requirements of the Constitution. In this chapter, we review by the decade the apportionment bills passed by Congress. It has been a surprisingly unsettled area of law.

1790s: The founding fathers anticipated the admission of new states to the union and an increase in the size of the House of Representatives. Thus the Constitution does not name a value for the number of states n or the size of the House h. In fact, the number of states had already increased from the original 13 to 15 by 1790 with the addition of Vermont and Kentucky. The House of Representatives passed an apportionment bill in 1791 adhering closely to the words of Article 1, Section 2. The divisor $d = 30{,}000$ was divided into the populations of the states and fractional parts of the resulting quotas were ignored (i.e., the numbers were rounded down). This is the largest number of representatives permitted by a literal interpretation of the phrase "shall not exceed one for every thirty thousand". This process gave a house size of $h = 112$. The Senate passed a competing bill, using the same method but choosing instead $d = 33{,}000$, yielding $h = 105$. A stalemate ensued.

Alexander Hamilton argued that the method being used by the two legislative chambers was inappropriate, because quota was violated. The House bill assigned 21 seats to Virginia, yet Virginia's fair share of 112 seats could be calculated to be a mere 19.53. Hamilton, himself a New Yorker, considered this to be unfair. He brokered a compromise bill between the two chambers employing the quota method that we now name after him. Hamilton used the value $h = 120$, the largest number less than the total population of the United States divided by 30,000. This number could then be interpreted to meet the Constitutional requirement that h "shall not exceed one for every thirty thousand". With this value of h, the standard divisor was 30,133, so the representation across all states would be 1 in 30,133, which does not exceed 1 in 30,000, as required.

Southern representatives were quick to notice that southern states fared worse under Hamilton's compromise bill than they did under the original House bill. With $h = 120$, there would be 8 more seats in the House than the House bill had offered. But 6 of those extra seats went to northern states, only 2 to southern states. Virginia's number of 21 was to remain unchanged. Representative Richard Henry Lee of Virginia referred to the Hamilton bill as "a certain arithmetico-political sophistry". James Madison, also a Virginian, argued against the compromise bill as well. One argument against it was Constitutional: Hamilton's apportionment assigned Delaware, with a population of 55,540 and a standard quota of 1.843, a total of 2 representatives. When looked at from the point of view of Delaware alone, the number of representatives did indeed exceed one in 30,000, apparently in violation of the Constitution. Seven other states had the same flaw.

Bills passed by Congress require the President's signature to become law. The compromise bill was presented to President George Washington for signature. Washington consulted his cabinet. Hamilton, who was Secretary of the Treasury, argued in favor, joined by Secretary of War Henry Knox of Massachusetts. Secretary of State Thomas Jefferson, from Virginia, naturally argued against, suggesting that if $h = 120$ was to be the size of the House, then a divisor would have to be found such that the quotas rounded down would sum to 120. This is what we call Jefferson's method today, where h is preset and the divisor is adjusted correspondingly. Jefferson argued that his method was the only one supported by the Constitution. Hamilton argued that the quota violations of Jefferson's method were unfair.

Jefferson seems to have been unaware of the most serious objection to Hamilton's method: the paradoxes. But his argument prevailed nonetheless. Washington exercised the first ever presidential veto in rejecting the compromise bill of Hamilton. An attempted override of the veto failed. Washington remanded the bill to Congress, where the original Senate bill was quickly approved by both chambers. Jefferson's method was employed with $h = 105$ and $d = 33{,}000$.

1800s: By this time, Tennessee had been admitted to the union and the number n of states had increased to 16. Jefferson's method was employed again with $d = 33{,}000$. The population of the United States had increased 35% since 1790, resulting in an increase in the size of the House of roughly 35% as well, from $h = 105$ to $h = 141$. Congress continued in the mode of debating the value of d and letting h fall as it may, rather than setting the value of h and determining a corresponding d. But eventually, the divisor of 33,000 would prove too small, since the population of the United States was growing rapidly.

1810s: The population increased by another 35% in the preceding decade. There were at this time $n = 17$ states. Jefferson's method was employed again, this time with d increased to 35,000, yielding $h = 181$.

1820s: The population and the number of states both continued increasing. By this time, the population of the United States was almost 9,000,000, two-and-a-half times what it had been at the time of the first census in 1790. Jefferson's method was used for the fourth time, with a divisor $d = 40{,}000$, yielding $h = 213$, but not without objection. The bias against small states and the quota violations were widely recognized. New York, by then the largest state, had a standard quota of 32.5 but got 34 seats. Pennsylvania, the second largest state, had a standard quota of 24.9 but got 26 seats. Meanwhile, Delaware, a smaller state, had a standard quota of 1.7 but got only 1 seat.

During the debate, Representative William Lowndes of South Carolina proposed a new method to combat the bias of Jefferson's method. Lowndes's method is a quota method that operates much like Hamilton's method, except that the extra seats are meted out to the states according to the quotient of the fractional part of the standard quota and the lower quota. Lowndes's rationale was that this quotient measures the damage per congressional district that the state would suffer if it were awarded only its lower quota. The states where this damage is greatest should be first in line to obtain an extra seat.

Lowndes's method addresses the bias toward large states that Jefferson's method displays, but it overcompensates. In fact, Lowndes's method can be regarded as openly partisan in favor of small states. In 1820, Lowndes's method would have given all the extra seats to the smallest states. Proposals on the House floor that disadvantage large states have little chance of passage, since votes against such proposals are plentiful. Lowndes's method was rejected by Congress and has not been considered since.

1830s: For weeks, Congress debated about what divisor to use for the implementation of Jefferson's method. Representatives were more than happy to suggest numbers; a divisor just smaller than their state's population divided by the number of seats they desired in the House would be ideal. With such a divisor, the modified quota has a small fractional part, and little is lost in the rounding down. For example, with a population of 130,419, the state of Missouri would have been delighted with a divisor of 43,000, because $130{,}419/43{,}000 = 3.033$, while they would have suffered with a divisor of 44,000, because $130{,}419/44{,}000 = 2.964$. Of course every choice of a divisor disadvantages some states, the fractional parts of whose modified quotas are just smaller than 1. Representatives were more than happy to suggest divisors that would benefit themselves and their friends while punishing their political enemies. The congressional debate was a free-for-all.

James Polk of Tennessee sponsored a bill using Jefferson's method with $d = 47{,}700$ yielding $h = 240$. John Quincy Adams, then a representative from Massachusetts and an ex-president, was a bitter political adversary of Polk. He noted that Polk's bill harmed New England, whose population and influence were in decline. Massachusetts, which had been the second largest state in 1790, was now eighth largest. Adams would later write in his memoirs "I was all night meditating in search of some device, if it were possible, to avert the heavy blow from the State of Massachusetts and from New England." The device he was looking for was what we now call Adams's method, which he proposed with a divisor

$d = 50{,}000$ and a resulting $h = 250$. This increased the House size by 10 over Polk's bill, and 3 of the extra seats would be given to New England states. Moreover, Adams's method, with $d = 52{,}300$ would yield a house size of $h = 240$ but with the apportionment of Polk's Tennessee lowered from 13 seats to 12. Ultimately, Polk's bill passed in the House, and the best Adams could do was submit his suggestion for a new method to his Senate colleague from Massachusetts, Daniel Webster.

Webster at this time was chair of the Senate committee to study the apportionment problem. He noted that Adams's method overcorrected for the quota violations of Jefferson's method. For example, New York, the largest state at this time, had a standard quota of 38.6. Jefferson's method assigned 40 seats to New York, violating upper quota. But Adams's method assigned just 37 seats to New York, violating lower quota. Webster's discovery of the method now named after him seems natural. If always rounding down leads to a method (Jefferson's) that is biased toward the large states, while always rounding up leads to a method (Adams's) that is biased against the large states, then rounding in the usual way — to the nearest whole number — should afford a compromise position with an appropriate balance.

At the same time, Webster received a proposal from a University of Vermont mathematics professor by the name of James Dean. Dean's method, like Webster's, afforded a compromise between Jefferson's method and Adams's methods, but one that favored small states somewhat more than did Webster's. The methods of Adams, Dean, Webster, and Jefferson accorded New York 37, 38, 39, and 40 seats, respectively. Ultimately, Polk's proposal was to prevail in the Senate, so Jefferson's method with $d = 47{,}700$ determined the new apportionment, with a house size of $h = 240$. But the bias and quota violations were now widely known, and this was to be the last decade in which Jefferson's method would be tolerated by Congress.

1840s: The debate in the House began with the usual free-for-all, with every representative offering his own choice of divisor for Jefferson's method. The Senate chose a divisor of $d = 71{,}000$ but used arithmetic rounding, which is to say the method of Webster, the Senate's most famous and arguably most influential member. John Quincy Adams was instrumental in persuading his House colleagues to agree to go along with Webster's method, although he was unable to persuade them to use a smaller divisor that would accord a few extra seats to the advantage of New England states. The resulting apportionment bill using Webster's method was signed into law. The increase in d for this decade yielded a decrease in h to 223. This would be the only time in United States history that the size of the House was set smaller than in the preceding decade.

1850s: Samuel Vinton, a representative from Ohio, rose on the House floor, before the census data was reported, in an attempt to avert the chaos for which the apportionment process was known. He spoke in favor of a permanent apportionment act that would prescribe a fixed rule to mete out the seats in the House, avoiding the usual political scramble. The method he proposed was Hamilton's method with $h = 233$. Vinton's proposal was enacted into law in 1850 but was never strictly followed. In 1852, Congress passed an apportionment bill using Hamilton's method with $h = 234$. The advantage of this house size was that Hamilton's method rounded up every state whose fractional part exceeded 0.5 and rounded down every state whose fractional part was less than 0.5. This implies that the Hamilton method gave the same result as the Webster method, which could be implemented with the standard divisor.

1860s: Relations between the North and the South were at this time bitterly polarized. Hamilton's method with $h = 233$ was first computed, in line with the 1850 law, but then a proposal was brazenly introduced to add 8 extra seats and give them to northern states. There were enough votes to pass the proposal, and the resulting apportionment bill was passed into law. It cannot be said that any method or principle at all was employed in the determination of these final numbers.

1870s: One way of quieting rancor over apportionment is to increase the House size h. When this is done, many states get additional seats and so the impression of greater representation is spread widely. If h is increased sufficiently, it can be arranged that no state loses seats in the House. This averts a potential outburst from the delegation of a state whose numbers are threatened and whose incumbents are therefore at risk.

At first, Congress agreed to an apportionment with $h = 283$. This number was chosen because the methods of Hamilton and Webster agreed. Later a bill passed adding 9 further seats to the House, leading to an *ad hoc* apportionment of a House of size 292 that did not agree with either Hamilton's or Webster's method.

This apportionment was in force in 1876, when Rutherford B. Hayes was elected President by an electoral vote margin of 185 to 184. Had Hamilton's method been used in the apportionment bill for that decade, as required by Vinton's act of 1850, the resulting Electoral College would have been different, and Hayes would have lost to his opponent, Samuel Tilden. This illustrates the high stakes of the apportionment game.

1880s: One way for Congress to defuse the debate between supporters of Hamilton's method and supporters of Webster's method was to settle on a house size h such that the two methods agreed. Congress called upon the Census Office to provide the Hamilton and Webster numbers for values of h between 275 and 350. C. W. Seaton, chief clerk of the Census Office, noticed an oddity with regard to the state of Alabama using Hamilton's method: Alabama's share of the House with $h = 299$ seats would be 8 seats, but its share of $h = 300$ seats would be only 7 seats. This anomaly became known as the Alabama paradox. It is a violation of what we now call house monotonicity. Seaton argued that this paradox demonstrated that Hamilton's method must be abandoned. To replace it, he proposed a method of his own, apparently unaware that it was nothing but the old method of Jefferson in disguise. In the end an apportionment bill was passed with $h = 325$. Although no method was indicated, the apportionment numbers agreed with both Hamilton's method and Webster's method.

1890s: The apportionment process in this decade resembled the one from the previous decade. Congress investigated all values of h between 332 and 375, hunting for one with the following two properties: First, they desired a value of h such that each state would get at least as many seats as it had in the previous decade. Second, they desired a value of h such that the Hamilton method rounded up standard quotas precisely when the fractional part was greater than 0.5. The number $h = 356$ was found, and a bill was passed. The numbers were consistent with the methods of both Hamilton and Webster.

1900s: In 1901, a report was submitted to Congress listing the Hamilton apportionments for every value of h between 350 and 400. The Alabama paradox was ubiquitous. Colorado would receive 3 seats for all values of h between 350 and 400 except for $h = 357$, for which they would receive but 2. Representative Albert Hopkins of Illinois was no friend of Colorado, and he proposed a bill employing Hamilton's method with (guess what?) $h = 357$. A political battle erupted. Representative John Littlefield of Maine argued against the Hopkins proposal, noting that Maine's representation would fall from 4 to 3 seats under Hopkins proposal and that, employing Hamilton's method, Maine's representation vacillated between 3 and 4 as h increased from 350 to 400. Littlefield attacked Hopkins and Hamilton's method in a speech on the House floor, ridiculing the method and the partisanship of the bill. Littlefield accused Hopkins of adopting $h = 357$ "because Populists happen to come from Colorado". "God help the State of Maine," he continued, "when mathematics reach for her and undertake to strike her down." Hopkins fired back. "It is true that under the majority bill Maine is entitled to only three Representatives, and ... the seat of the gentleman who addressed the House on Saturday last is the one in danger. In making this statement he takes a modest way to tell the House and the country how dependent the State of Maine is upon him. ... [I]f the gentleman's statement be true that Maine is to be crippled by this loss, then I can see much force in the prayer he uttered here when he said 'God help the State of Maine'." In light of the Alabama paradox, the debate over the proper size of h had become so politicized that Hamilton's method could no longer be endured by Congress. The Hopkins bill was voted down. Instead, Webster's method was used with $h = 386$.

In 1907, Oklahoma was added to the union and given 5 seats in the House, since their population was estimated to be 1,000,000 and the standard divisor at the time was approximately 193,000. This brought the total number of states to 46 and the house size to 391. Had Hamilton's method been run with the 1900 census data, but with Oklahoma added in with a population of 1,000,000, indeed Oklahoma would have received 5 seats in the House. But a curious and disturbing feature of Hamilton's method was noticed at this time. If Hamilton's method had been used before the addition of Oklahoma, then New York and Maine would have received 38 seats and 3 seats, respectively. However, after the addition of Oklahoma, they would have received 37 seats and 4 seats, respectively, using the very same population data. Since the populations of New York and Maine did not change between these two applications of Hamilton's method, there was no reason for New York to have to relinquish a seat to Maine. This paradox has come to be known as the Oklahoma paradox or the new states paradox. It can be seen as an instance of what we now call the population paradox, a more general phenomenon. Congress was by this point widely aware of the paradoxes associated with Hamilton's method, and Hamilton's method was never to be used again.

1910s: A philosophy professor from Cornell named Walter Wilcox submitted to the House Committee on the Census a proposal to adopt Webster's method. With his proposal, he submitted tables for various choices of h between 390 and 440. Meanwhile, Joseph Hill of the Census Bureau proposed a new idea that pointed toward a method that had not been considered before. Congress passed an apportionment bill in 1911 using Webster's method, as Wilcox had suggested, and choosing the value $h = 433$, this being the smallest house size that could be adopted without any state losing a representative from the 1900 apportionment. In this bill, Congress anticipated the admission of Arizona and New Mexico as states the following year and made provisions for these two states to get one representative each, bringing the total to $h = 435$, the present number. From 1912 to the present, the reapportionment each decade has held fixed at $h = 435$. Since then, every increase in the number of seats for some states has required a corresponding decrease in the number of seats for others.

1920s: Webster's method seemed to be the method of choice for this decade, but finding a value of h was a dilemma. The smallest value of h that would prevent any state from losing a seat was $h = 483$, a significant increase over the already unwieldy $h = 435$. A bill to reapportion according to Webster's method with $h = 483$ was voted down, as was a bill to use Webster with $h = 435$.

At this point, Edward V. Huntington, a mathematics professor from Harvard, entered the stage. He had been a college classmate of Joseph Hill, and during World War I, Huntington had worked at the Census Bureau with Hill. Huntington reviewed Hill's new idea for apportionment, and while he found Hill's idea compelling, he noticed that Hill had incorrectly described the algorithm by which the apportionment should be computed. Huntington corrected this error and became an impassioned advocate for the new method that we call Hill's method but is sometimes known as the Huntington–Hill method. Opposing this new development was Walter Wilcox, who became a just-as-impassioned advocate for the method of Webster. Both Huntington and Wilcox argued that the method they proposed was the unique method that avoided bias between small and large states, whatever that might mean.

As the decade wore on, the controversy between Wilcox and Huntington intensified, and Congress, understandably confused by the opposing claims, could not pass a reapportionment bill. The Speaker of the House asked for a ruling from the National Academy of Sciences, and a four-member committee of the National Academy was formed to study the matter and report back to Congress. Their report sided unequivocally with Huntington. Huntington gloated in an article in the journal *Science*: "All controversy concerning the mathematical aspects of the problem of reapportionment in Congress should be regarded as closed by the recent authoritative report of the National Academy." He added that this report was "particularly timely, since Congress has been in serious danger of being confused and misled by an erroneous theory".

But time had run out for an apportionment bill. No reapportionment ever transpired in the 1920s. Throughout the decade, the seats in the House remained distributed as they were in the 1910s. This is in clear violation of the Constitution.

1930s: The stakes for the reapportionment scheduled to occur in the 1930s were high. The migration from East to West and from rural to urban during the 1910s and 1920s significantly altered the relative populations of many states, so the changes in the 1930 reapportionment were to be substantial. In 1929, Congress had finally managed to pass a permanent bill regarding the apportionment process. This bill provided for an automatic reapportionment if Congress failed to act within a certain period. This guaranteed that the failure of the 1920s would not be repeated. The bill required that Congress be presented with the data for Webster's method and Hill's method with $h = 435$. If Congress could not act in time, the bill provided that the method that had been most recently used would automatically be in effect. In 1931, the numbers were kind, and the apportionment methods of Webster and Hill agreed, so a dispute was averted, and a reapportionment bill was passed.

1940s: Unfortunately, the methods of Webster and Hill did not agree on the census data from 1940. The lone difference was that Arkansas earned an extra seat with the method of Hill, while Michigan earned an extra seat with the method of Webster. Arkansas was a Democratic state, while Michigan leaned Republican. An extra seat in the House for a representative affiliated with the Democratic party was irresistible to the Democratic majority in Congress, and on such partisan and unscientific grounds, a bill passed in 1941 prescribing Hill's method and $h = 435$. Moreover, the bill provided that this would be the permanent choice for the United States. Hill's method automatically gives at least one representative to every state. Ironically, the authors of the 1941 apportionment bill felt the need to add the clause "no State to receive less than one Member", apparently unaware that this is a feature of the method and that therefore this phrase is a redundancy.

1950s: The country had grown to a population of 150,000,000. Hill's method with $h = 435$ was implemented again in accordance with the 1941 bill. In 1959, Alaska and Hawaii were admitted to the union, bringing the count of states to $n = 50$. A single representative from each of Alaska and Hawaii was added to the House, but with the understanding that the increase in house size to 437 would be temporary and that Congress would revert back to the value $h = 435$ in the subsequent decade.

1960s, 1970s, 1980s, 1990s: The population of the United States grew to 250,000,000, and California became the largest state (by a wide margin). Throughout this period, the House size remained fixed at $h = 435$ and Hill's method remained in force.

In 1975, Michel Balinski and Peyton Young discovered their new apportionment method, which has certain theoretical advantages over other methods. Balinski and Young did not advocate for their own method, however, favoring instead the method of Webster.

The state of Montana was accorded 2 seats in 1960, 1970, and 1980, but was relegated to 1 seat in 1990. The population of Montana in 1990 was approximately 800,000, and with only 1 seat, the state became the most populous congressional district in the United States. The loss of a congressional seat is unusually costly for a state with just 2 representatives, since the loss signifies a 50% decrease in representation. Montana challenged Hill's method and the 1990 apportionment in federal court, but their lawsuit failed and the apportionment was affirmed by the US Supreme Court.

2000s: The US apportionment population exceeded 280,000,000. Hill's method was used with $h = 435$ again. On October 17, 2006, the Census Bureau reported that the population of the United States had reached 300,000,000.

2010s: Hill's method with $h = 435$ was employed for the 8th consecutive decade. The US population grew to over 309,000,000, raising the average population of a congressional district to over 710,000. Belying global warming, the population growth was seen disproportionately in southern states, continuing a trend of several decades. Texas gained 4 seats, and Florida gained 2. By contrast, Pennsylvania lost a seat, earning 18 seats in the House, just half of what it had a century before. Had the Dean method been used in this decade in place of the Hill method, only two states would have been affected, Montana, which would have gained its second seat much as it argued in the 1990s, and California, which would have lost one of its 53 seats. Had the Webster method been used in this decade in place of the Hill method, again only two states would have been affected, this time with North Carolina gaining a seat and Rhode Island losing one.

12.2 Summary

The apportionment act of 1941 established $h = 435$ and Hill's method as the law of the land. That there is a law of the land is undoubtedly a good thing. But laws can be amended or repealed, and so the choice of apportionment methods remains an issue of interest in Congress. It seems likely that the upcoming census will result in a House of Representatives that has $n = 50$, $h = 435$, and whose apportionment is determined by Hill's method. But none of this is certain. In particular, voting rights activists in the District of Columbia advocate for $n = 51$ and $h = 436$ or perhaps 437. Meanwhile, activists in Montana continue to advocate for the method of Dean, while a growing consensus of theorists advocates for the method of Webster.

Table 12.1 summarizes the history. During the period from 1790 to 1830, the method of Jefferson was used, but with the method of Hamilton competing for consideration. In 1840, the method of Webster was used. From 1850 to 1900, the method of Hamilton was prescribed by law, but it was not strictly followed. In 1910 and 1930, the method of Webster was again used. From 1940 to the present, the method of Hill has been in force, but with the method of Webster competing for attention.

It cannot be said that reason and mathematics dictated the choice of

Table 12.1 US apportionments across the decades.

Year	h	**Method used**	**Comments**
1790	105	Jefferson	Hamilton's method vetoed.
1800	141	Jefferson	
1810	181	Jefferson	
1820	213	Jefferson	Lowndes's method considered.
1830	240	Jefferson	Adams's, Webster's, and Dean's methods considered.
1840	223	Webster	
1850	234	Hamilton	Webster's method agrees.
1860	241	No method	
1870	292	No method	Tilden loses presidential election.
1880	325	Hamilton-Webster	Alabama paradox noted.
1890	356	Hamilton-Webster	
1900	386	Webster	Oklahoma paradox noted.
1910	433	Webster	Hill's method considered.
1920		No reapportionment	Wilcox versus Huntington.
1930	435	Webster-Hill	
1940	435	Hill	Arkansas prevails over Michigan.
1950	435	Hill	
1960	435	Hill	First apportionment with 50 states.
1970	435	Hill	Balinski–Young method discovered.
1980	435	Hill	
1990	435	Hill	Montana sues.
2000	435	Hill	

the method that we use today in the United States. Rather, the designation of the method of Hill may be attributed to the charisma of Edward Huntington and to the partisan advantage that the method afforded one political party in 1941. So we find ourselves today uncomfortably uncertain if the arguments in favor of our current policy on apportionment are cogent, persuasive, or rigorous. In the end, it may be unreasonable to expect mathematicians to settle the debate over apportionment methods, since finding the right notion of fairness is a question of values and philosophy rather than of science. In any case, while we may wonder if we are using the ideal method, at least we are using *some* method. We are certainly better off than if the decisions were made not by rules at all but by the kind of nasty partisan debate that for many decades accompanied the apportionment process in the United States.

12.3 Exercises and Problems

12.1. Gather the census data from 1790 and determine the apportionment with $h = 120$ according to the methods of Hamilton, Jefferson, Adams, Dean, Hill, and Webster.

12.2. Obtain the US census data from 1870.
(a) Determine the Hamilton apportionment with $h = 292$.
(b) Determine the winner of the presidential election of 1876, assuming that the number of electoral votes assigned to each state is 2 more than the numbers computed in part (a) and assigning 3 electoral votes to Colorado.

12.3. According to the year 2010 census and the ensuing apportionment, which five states have congressional districts of smallest size? Which five states have congressional districts of largest size?

12.4. Montana sued the government in 1992 over the apportionment. What method did their lawyers advocate? How would Montana have fared under this method? Did Montana prevail in the lawsuit?

12.5. Using census data from the 2010 census, determine which state would have received an extra seat by Hill's method if h had been increased from 435 to 436. Answer the same question for the 2000 census.

12.6. Using census data from the 2010 census, determine which state would have received an extra seat by Hill's method if h were increased from 434 to 435, from 435 to 436, and from 436 to 437.

12.7. In the year 2100, let us imagine that the United States will have 54 states and the House of Representative will grow to 500 seats. Congress will still apportion these seats using Hill's method and they will use a divisor of 900,000. In this Congress, the state of Maryland will receive 12 seats. From this information, determine what possible populations Maryland might have in the year 2100.

12.8. Repeat Problem **12.7** using Webster's method in place of Hill's method.

12.9. Again in 2010, Montana got just one seat in the House of Representatives. How many additional residents would Montana have needed in order to have gotten two seats? Assume all other census parameters remain the same.

Notes on Part II

Although the US Constitution solves many of the practical problems of running a democracy, it is vague on the question of how to apportion the US House of Representatives. Instead, the Constitution leaves the details of this essentially mathematical problem to the political process. This omission has resulted in the tumultuous history described in Chapter 12. The lack of explicit instructions for apportionment seems like an oversight, since it is doubtful that the framers of the Constitution realized how difficult the problem is.

The apportionment problem is a case of what is called a fair division problem. One familiar type of fair division problem involves sharing a piece of cake between two people. In the well-known "divider-chooser" algorithm, one person divides the cake into two pieces, and the other person chooses which piece to take. Assuming the divider acts rationally, he will cut the cake into two pieces that he regards as equally valuable. He will do this because it guarantees him exactly half of the value of the cake. The chooser will select what she regards as the most valuable piece, so she will get a piece at least as valuable (to her) as the one she rejects. In particular, her piece will also be worth half (or more) of the value of the cake. This algorithm is often attributed to King Solomon, although the fair division problem that King Solomon solved in the Bible is actually of a different nature. But the idea is undoubtedly very old.

Cake division problems with more than two people are more difficult. The following method for three people is attributed to the Polish mathematician Hugo Steinhaus from around the time of World War II. Suppose Alice, Bob, and Carol want to share a piece of cake. Alice cuts the cake into three pieces, and Bob and Carol each say which pieces they consider acceptable. As long as they do not both want the same piece, each of them is guaranteed a piece worth at least $1/3$ of the cake. Alice then gets the piece that is left. If, however, Bob and Carol both want the same single piece, Alice is first given one of the two pieces they both reject. Then Bob and Carol divide the two remaining pieces using the two-person divider-chooser method.

This algorithm is certainly fair in the sense that all the players get pieces that they regard as worth at least $1/3$. But the algorithm fails to be what is called "envy-free". If Carol chooses before Bob, Bob gets an acceptable piece, but he might have preferred the one that Carol got. Fairness and "envy-free" are two criteria that we might want to impose on cake dividing methods. The two-player divider-chooser method is both fair and envy free, but it fails to satisfy another criterion with which we are already familiar: the Pareto criterion. Suppose the cake is half chocolate and half strawberry. Imagine that the cutter prefers chocolate and the chooser prefers strawberry. Not knowing the preference of the chooser, the cutter will cut two pieces with equal amounts of both flavors. Both players will end up with half chocolate and half strawberry, even though the cutter would have preferred all chocolate and the chooser would have preferred all strawberry.

Cake cutting differs from apportionment in several ways. First, the cake is infinitely divisible, whereas members of Congress (despite King Solomon) cannot be cut in half. Second, the cake sharers are generally assumed to have equal claims to shares of the cake, whereas different states have different claims on shares of the House according to their populations. On the other hand, all House seats are identical, while cake slices need not be equal to each other nor equally valued by cake eaters with different tastes.

Another realm of fair division problems involves the division of estates. This is closely related to cake cutting but with the additional complication that the various items of value in the estate, such as art, racehorses, sports cars, and mansions, may have sentimental value and not be divisible.

The Talmud (circa 200 AD) describes another kind of estate division problem. Three creditors with claims against a debtor equal to (say) \$100, \$200, and \$300 are left to divide the debtor's estate, which is worth less than \$600. If the estate is worth \$100, the Talmud advises that each creditor should get \$33.33. If the estate is worth \$200, it advises that the first creditor should get \$50 and the other two should each get \$75. But if the estate is worth \$300, the Talmud advises that first creditor should get \$50, the second \$100, and the third \$150. Talmudic scholars and mathematicians puzzled over the origin of these numbers for many years, until in 1985 they were shown by Nobel laureate Robert Aumann and Michael Maschler [3] to have a game-theoretic interpretation.

Fair division problems are a substantial subject in themselves. Good introductions include [12] and [9]. We have avoided an extended general discussion of fair division in this book, choosing instead to concentrate on the case of apportionment with its immediate connection to politics. The definitive text on apportionment is the 1975 masterpiece of Michel Balinski and Peyton Young [6]. Nearly all the material in Part II of this text can be found there, although our emphasis is quite different. We remain agnostic about which method is the best choice, while Balinski and Young advocate quite openly for a change to Webster's method. In particular, Balinski and Young cast doubt on Edward Huntington's interpretation of the relative merits of divisor methods [29], on which the choice of Hill's method for the apportionment in the United States is premised. Balinski and Young do not advocate for their own method, as they regard population monotonicity as essential.

A few recent papers on the theory of apportionment suggest some new methods. Agnew [1] introduces divisor methods based on the so-called identric and logarithmic means (see Problems **8.17** and **8.18**). Agnew argues that the identric mean method is superior to any of the five historic divisor methods. Meanwhile, mathematician and law professor Paul Edelman [21] studied Supreme Court decisions on the relative sizes of districts for elected officials. Most of these cases the court considered were not about the US House of Representatives, but rather about representative assemblies at the state and local level. In striving to implement the principal of "one person, one vote" the court has consistently looked at what Edelman calls the "total deviation", and Edelman argues that this is the test the court would likely apply to any cases about congressional apportionment. Edelman says that the Supreme Court tried to apply this test to the 1990 case *Montana v. US Department of Commerce*, but "got the math wrong". To find the total deviation, one looks at p_i/a_i, the size of the average congressional district in state i. The total deviation is then the difference obtained as the largest value of p_i/a_i over all states i, minus the smallest value of p_i/a_i. Edelman describes an elaborate apportionment method due to Gilbert and Schatz [26] that minimizes total deviation. It is implemented as a hybrid of Adams's and Jefferson's methods.

Methods of apportionment can be used in all fair division problems that have the same character as the apportionment problem. We mention in the text, for example, the allocation of buses to bus routes according to numbers of riders, and the allocation of teachers to schools within a school district according to numbers of students. Economist Vincy Fon [24] proposes an application of apportionment to replace the winner-takes-all system of choosing members to the Electoral College. Her new method would assign the seats from each state to the various candidates using what amounts to a new apportionment method.

Apportionment problems have a different political interpretation in parliamentary systems, where seats are typically doled out to parties rather than to districts. However, since a party receives seats in proportion to the percentage of the vote it receives, the mathematical problem is essentially the same. Most of the important apportionment methods were first developed in the United States, often by famous figures in the early history of the republic: Hamilton, Jefferson, (John Quincy)

Adams, and Webster. Many of these methods were rediscovered in other countries, and thus they often go by different names in those countries. In Europe, Jefferson's method is called the d'Hondt method. It is used, for example, for parliamentary elections in Argentina, Brazil, Croatia, Denmark, Estonia, Finland, Hungary, Israel, Japan, Montenegro, the Netherlands, Poland, Portugal, Spain, Turkey, and Venezuela. Similarly, Webster's method is sometimes called the Sainte-Laguë method, and is used, for example, for parliamentary elections in Bosnia, Germany, New Zealand, and Sweden. Hamilton's method, when applied in this context works very much like the single transferable vote (STV) method for parliamentary elections, which is discussed in the notes to Part I.

One fair division problem in US politics that has not yet received a complete mathematical treatment is the problem of redistricting. After each decennial census, each state is told the number a_i of House seats to which it is entitled. It is then the job of the state legislature to carve up the state into a_i districts. Besides the obvious requirement that districts have roughly the same populations, many considerations go into such decisions, including boundaries of cities and counties, geographical features within the state, the integrity of districts of sitting House members, and legal requirements based on the Voting Rights Act to preserve minority rights. However, the party that controls the state legislature when the reapportionment occurs has the ability to manipulate the process to its own advantage in a significant way, and often does. Besides being roughly equal in size (not having too large of a total deviation in size), courts have generally held that districts should be "compact", which in this context means not too unusually shaped. Compactness stands in contrast to a "Gerrymander", named after Massachusetts governor Elbridge Gerry, who in 1812 created a sprawling district said to be shaped like a salamander to favor his party of Democratic-Republicans. Compactness is a rather vague geometric concept, and there have been several proposals as to what its precise mathematical meaning should be. To complicate matters, districts that clearly violate any reasonable definition of compactness have been allowed, or even mandated, by the courts in order to satisfy other criteria, like racial equity. In conclusion, the redistricting process presently carried out by each state every 10 years resembles the pre-1940 apportionment process; it is more political than mathematical. But could a mathematical approach help?

Many political scientists feel that a mathematical approach to redistricting would not work in practice. They argue that no algorithm could address the many human factors that the current political redistricting system takes into account. If we could write out a list of desired criteria for a redistricting method, including geometric requirements and human factors, it is doubtful that we could find a method that would satisfy them all. (There may even be an impossibility theorem lurking in the redistricting problem.) Nevertheless, our experience with apportionment and voting suggest that perfect methods are not really necessary. Good results can be achieved using reasonable but imperfect methods. In voting, for example, we do not wait until the votes are tabulated before deciding what method to use. In close races, that would invite each candidate to advocate for the method that would make her win. The Electoral College method for electing the President, though deeply flawed, works because everyone knows in advance what is required to be elected President. Similarly, since 1940, the House has been capped at 435 members, and the apportionment has been conducted without incident using Hill's method. Collecting the data for these processes first and only later asking how those data will be used invites a stampede of corruption. The method should be established first. And yet redistricting is still done in reverse, with census and apportionment data provided first but no algorithm identified for computing the resulting districts. The result is the present notorious system of political processes that permit the cynical manipulation of district boundaries to advance partisan interests or incumbency.

Part III

Conflict

Introduction to Part III

If you believe in free will, then you believe that people are capable of making decisions. But *how* does a person make decisions? Sometimes, when presented with a decision problem, one can determine precisely what choices exist, ascertain what the outcome of every choice is, and evaluate these choices on a linear scale (by, for example, measuring the dollar value of each outcome). In such cases, decision-making is easy. One simply opts for the choice that maximizes benefit. But in the real world, decision-making is more difficult, because one faces uncertainty that may be of two types. The first type is uncertainty about nature. Nature may be thought of as random and unconcerned with our welfare. For example, to decide whether to carry an umbrella, we should factor in the probability of rain. The second type is uncertainty about people. A person, unlike nature, may be our friend or our adversary, and a person may have a stake in the decision process that is unrelated to ours. We assume that other people do not act randomly but rather act in their own self-interest. In other words, they act like we do. For example, to decide how to negotiate a treaty, we need to contemplate the point of view of other signatories.

In Part III, we introduce a context for making decisions in the face of uncertainty. To provide some structure for our model, we assume that the decision-makers know what choices are available to them and to the other decision-makers, and we assume that the outcomes of these decisions can be objectively evaluated on a linear scale. These things are regarded as certain. The uncertainty stems either from random nature or from savvy opponents.

We manage uncertainty about nature with probability theory. We manage uncertainty about people with game theory. In the context of the phrase "game theory", a game need not be entertaining like Monopoly or frivolous like tic-tac-toe. A game is simply a framework for modeling the decision-making process. In our games, several players — usually just two — simultaneously render a decision. The possible choices for all players are known to all players, and the values of all outcomes — often thought of in terms of dollar values — are also known to all. Once the decisions are revealed, the outcome of the game is completely determined, the players are paid their due, and the game ends. We discover that probability theory is an important paradigm for analyzing games, even when there does not seem to be any randomness involved in the rules of the game itself.

Uncertainty generally makes people uncomfortable. It is rare for politicians to admit to uncertainty, and yet uncertainty is present in all complex policy decisions. By contrast, weather forecasters are more frank about their uncertainty, admitting on a daily basis the degree of their uncertainty about the subsequent day's likelihood of rain. Yet an indifferent natural world and cunning adversaries lurk everywhere in the political realm, and therefore a rational analysis of decisions requires both probabilistic and game-theoretic reasoning. It would be refreshing to hear politicians more often admit to employing the type of rational analysis that we encounter here in Part III.

13

Strategies and Outcomes

"The score never interested me, only the game." — Mae West

13.0 Scenario

Let's play a game. We begin with a matrix of numbers, like this:

-12	5	10
6	-13	-7

Game 13.1 A simple matrix game.

There are two players in this game, and we call them "Row" and "Column". Both players know all of the entries in the matrix. To play the game, Row secretly chooses a row (there are two choices available to her), and Column secretly — and independently — chooses a column (there are three choices available to him). Then the two players simultaneously reveal their choices. This identifies one particular entry in the matrix. For example, if Row chooses row 2 and Column chooses column 1, the entry "6" in the lower left entry is identified. This entry represents the number of dollars that Column must pay to Row: On the other hand, if the entry were negative, like the -12 in the upper left, then Row would have to pay Column. If Row selects her first row and Column selects his first column, then Row must pay Column $12.

How should the two players make their decisions? For each player, which choice is the safest? Which choice is the boldest? Remember that the information in the matrix is public, so it is possible for the two players to put themselves in the shoes of their opponents. Both players are clever and want to win as much money as possible. But they each recognize that their opponent is like-minded.

13.1 Zero-Sum Games

In this chapter we introduce a type of game called a **two-person zero-sum game**. We start by looking at a familiar example: a children's game called Roshambo. Lest the reader feel that game theory is not serious, however, we also consider a game that models a naval battle from World War II.

The game of **Roshambo**, known more familiarly in the United States as Rock-Paper-Scissors, is played by children throughout the world. Recently, Roshambo has also become a popular tournament game in bars. To play Roshambo, two players face each other, and, on the count of three, each makes a finger gesture representing Rock, Paper, or Scissors. The gesture for Rock is a closed fist, the gesture for Paper is an open palm, and the gesture for Scissors is two fingers. Certain gestures defeat others: Rock beats Scissors (i.e., rock dulls scissors), Scissors beats Paper (scissors cut paper), and Paper beats Rock (paper covers rock). If both players make the same gesture, then the outcome is considered to be a tie.

The choices available to a player in a game are called the player's **strategies**. We generally call the two players Row and Column. For the purpose of choosing appropriate pronouns, we often imagine that Row is female and Column is male. In Roshambo, each player has three strategies from which to choose: Rock, Paper, and Scissors. A single round of play involves the players choosing strategies, and then, at a prescribed moment, simultaneously revealing their strategy choices. A strategy choice by each player determines an **outcome** of the game. For example, in Roshambo, if Row chooses Rock and Column chooses Paper, then the outcome is (Rock, Paper). Since Paper beats Rock, Column wins.

In some games we allow outcomes that are more complicated than just winning and losing. For this purpose, we institute a score-keeping system that awards a numerical score to each player after each round of play. Because this score is frequently a monetary payment, it is called a **payoff**. In Roshambo, for example, we might require the loser to pay \$1 to the winner (with a payoff of \$0 to both players in case of a tie). In other words, the winner's payoff is \$1 and the loser's payoff is −\$1. Notice that in this system of payoffs, the sum of the payoff to one player and the payoff to the other is always \$0. This is what we mean when we call Roshambo a zero-sum game.

Now we are ready to establish some notation for zero-sum games. Suppose we want to describe a zero-sum game in which Row has m

strategy choices and Column has n strategy choices. We denote such a game by an m-by-n matrix in which the rows correspond to Row's strategies and the columns correspond to Column's strategies. An m-by-n **matrix** is nothing but an array of numbers with m rows and n columns. The matrix entry that lies in row i and column j, which we denote $u_{i,j}$, is the payoff to Row for the outcome (row i, column j). We call this matrix the **payoff matrix**, and we refer to a game given in terms of a payoff matrix as a **matrix game**.

It is important to note that even though the payoff matrix shows only the payoffs to Row, it also tells us the payoffs to Column. Because we assume the game is a zero-sum game, the sum of payoffs to the two players is always 0. Thus if the payoff to Row for the outcome (row i, column j) is given by $u_{i,j}$, then the payoff to Column for the same outcome is $-u_{i,j}$. In particular, $u_{i,j} + (-u_{i,j}) = 0$ for all i and j.

A key observation about two-person zero-sum games is that we can always implement the payoffs by having one player pay the other. In other words, a zero-sum game involves a transfer of wealth between the two players. The sign of $u_{i,j}$ indicates which player pays and which player gets paid.

Here is the matrix for the game Roshambo.

	Rock	Paper	Scissors
Rock	0	−1	1
Paper	1	0	−1
Scissors	−1	1	0

Game 13.2 Roshambo

In this matrix the entry $u_{2,1} = 1$ indicates that if Row plays row 2 (Paper), and Column plays column 1 (Rock), then Column must pay Row \$1. Later we will return to the question of how one ought to play Roshambo. But first we consider another example.

In 1944, during World War II, the Japanese navy needed to re-supply their base on New Guinea. In order to do this, they needed to sail a convoy around the island of New Britain, passing either to the North or to the South of the island. The United States controlled New Britain and wanted to stop the Japanese convoy by bombing it. The resulting battle is known as the **Battle of the Bismarck Sea**.

While it was certain that the United States would be able to find the Japanese convoy, it was uncertain how long the bombardment would last. This would depend on whether the United States searched for the Japanese in the right place. In his diary, the American general, George

Kenny, gave estimates of the number of days of bombing under four possible scenarios. First he considered what would happen if the Japanese sailed north of New Britain. In that case, if the United States searched to the north, the United States would have two days of bombing, but if the United States searched to the south, the United States would have only one day. On the other hand, if the Japanese sailed south and the United States searched south the United States would have three days of bombing, but if the Japanese sailed south and the United States searched north, the United States would have two days of bombing.

To model this battle by a game, we make the United States the row player and make the Japanese the column player and put the number of days of bombing as the payoffs. Whereas the United States wants to maximize the number of days of bombing, the Japanese want to minimize it. This makes it reasonable to model this situation as a zero-sum game.

		Japan	
		North	South
United States	North	2	2
	South	1	3

Game 13.3 Battle of the Bismarck Sea

This matrix is an abstract version of the real naval battle. Modeling the battle this way allows us to analyze it in an objective manner. We no longer need to pay attention to the nations, the geography, or the weaponry. All of that is replaced by the two players, Row and Column, and the payoff matrix.

Although we focus here on zero-sum games, we can handle a slight generalization with little added effort. A game is called a **constant-sum game** if there is a constant K so that Row's and Column's payoffs always add up to K. Just as in a zero-sum game, we write a constant-sum game as a matrix whose entries $u_{i,j}$ represent the payoffs to Row. The payoff to Column for the outcome (row i, column j) is then given by $K - u_{i,j}$. Almost everything we say about zero-sum games is also true for constant-sum games. Thus, for the purpose of strategic analysis, we usually disregard the distinction between zero-sum and constant-sum games, and refer to both as matrix games.

13.2 The Naive and Prudent Strategies

In order to play a game, a player must decide which strategy to choose. **Game theory** is the branch of mathematics that attempts to help a player make this decision by providing various principles and methods.

By a **game-playing method** we mean a technique that a player can use to analyze a game and obtain a recommended strategy. More technically, a game-playing method is a function whose domain is the set of all zero-sum games and whose codomain should be the set of all possible strategies available to the player. Actually, we allow the output of a method to be a set of strategies to allow for the method to equivocate between equally recommended strategies. (Such "ties" should be familiar to the reader from the theory of voting methods.)

To illustrate these ideas, we first consider a simple game-playing method based on an obvious idea. Suppose we call the outcome that gives a player his or her best payoff the **primary outcome** for that player. The primary outcome is a player's favorite outcome — the outcome a player would most like to occur. For Row this is the outcome corresponding to the largest entry of the payoff matrix and for Column it is the outcome corresponding to the smallest entry.

Definition 13.1. To play a game using the **naive method**, a player chooses the strategy corresponding to his or her primary outcome. This is called the player's **naive strategy**. If there is more than one primary outcome, any strategy corresponding to a primary outcome is considered to be a naive strategy.

In the Battle of the Bismarck Sea, the Americans (Row) would clearly prefer 3 days of bombing while the Japanese (Column) would prefer 1 day. The naive strategies for the two players are the strategies that aim for these outcomes. We indicate the naive strategies of the two players with arrows:

		↓	
		North	South
	North	2	2
→	South	1	3

If both players play their naive strategies, the outcome is (row 2, column 1). This is called the **doubly naive outcome**. Notice that the Japanese would be happy with this outcome, but the Americans would not. The naive strategy is also sometimes called the **greedy strategy** or the **optimistic strategy**. The naive strategy is not a very sophisticated approach to paying games. Just hoping to attain the best outcome does not mean that it will happen. A player adopting the naive strategy plays without considering the opponent's viewpoint.

Next, we consider a slightly more sophisticated method for selecting a strategy. The mindset on which it is based is opposite to that of the naive strategy. The idea is for a player to choose the strategy with the worst-case payoff that is least bad. We call the worst payoff that a player can get by choosing a particular strategy the **guarantee** of that strategy. In other words, a player can be secure in the knowledge that the consequences of a certain strategy choice can be no worse than its guarantee. For Row, the guarantee of a row is the smallest entry in the row. For Column, it is the largest entry in the column.

Definition 13.2. To play a game using the **prudent method**, a player chooses the strategy with the best guarantee. This strategy is called a player's **prudent strategy**.

We denote the guarantee for Row's prudent strategy by r and the guarantee for Column's prudent strategy by c. The prudent method amounts to using an approach that is called a "worst-case analysis". One imagines the worst possible result for each strategy choice, and selects the one whose worst case is the least bad. For this reason, the prudent strategy is also known as the **pessimistic strategy**.

Let us demonstrate how to apply the prudent method to the Battle of the Bismarck Sea. For the Americans, the guarantee for the strategy North (row 1) is 2 days of bombing, and the guarantee for South (row 2) is 1 day of bombing. Since 2 days of bombing is better than 1 day, North is the American prudent strategy, and $r = 2$. For the Japanese, the guarantee for the strategy North (column 1) is 2 days of bombing, and the guarantee for South (column 2) is 3 days of bombing. For the Japanese 2 days of bombing is better than 3 days of bombing, so North (column 1) is the Japanese prudent strategy, and $c = 2$. Notice that, in this game, the Japanese prudent strategy coincides with the Japanese naive strategy, but the American prudent strategy differs from the American naive strategy.

We can efficiently perform the calculations used in the prudent method using a simple device that we call the **min-max diagram**. We begin by writing the guarantee for each strategy next to the corresponding row and column. The prudent strategy for Row is the row with the largest guarantee. Similarly, the prudent strategy for Column is the column with the smallest guarantee. We mark these rows and columns with arrows. Here is the min-max diagram for the Battle of the Bismarck Sea:

	North	South		
North	2	2	2	←
South	1	3	1	
	2	3		
	↑			

When both players in a game play their prudent strategies, we refer to the outcome as **doubly prudent**. In the Battle of the Bismarck Sea, this is (row 1, column 1), which means that the Japanese sail North and the Americans search North. Subsequently, the Japanese convoy is subjected to two days of bombing, which is, in fact, what happened in the real battle in 1944.

This analysis of the Battle of the Bismarck Sea using game theory is due to cold war game theorist and retired army Colonel O. G. Haywood. Haywood notes that the official United States military field manual requires a commander to determine the worst possible outcome for each battle plan and then to execute the plan that has the best worst-case outcome. According to this field manual, official US military policy mandates the use of the prudent method.

There is a certain sense in which the prudent strategy is the best possible strategy for a player to choose. We summarize that discussion in the next proposition.

Proposition 13.3. *If Row plays her prudent strategy, her payoff will be at least r, the guarantee of her prudent strategy. If she plays any strategy that is not prudent, there is the possibility she will get less than r. Similarly, if Column plays a prudent strategy, the payoff (i.e., his "pay-out") will be at most c, his prudent strategy guarantee, and if he plays any strategy that is not prudent, there is the possibility that he will pay more than c.* □

Corollary 13.4. *The guarantees r and c for the prudent strategies of Row and Column satisfy $r \leq c$.*

Proof. Suppose (row i, column j) is doubly prudent. Then by Proposition 13.3, $u_{i,j} \geq r$ and $u_{i,j} \leq c$. It follows that $r \leq u_{i,j} \leq c$. □

Here is the min-max diagram for Roshambo.

	Rock	Paper	Scissors		
Rock	0	−1	1	−1	←
Paper	1	0	−1	−1	←
Scissors	−1	1	0	−1	←
	1	1	1		
	↑	↑	↑		

Note that for both players, all three strategies are tied as prudent strategies. Like the naive method, the prudent method offers no practical advice on how to play Roshambo.

13.3 Best Response and Saddle Points

In real Roshambo games, good players try to figure out or guess what their opponent is going to do and respond to it. In any game, a player who knows the opponent's strategy choice can respond to it in the most advantageous way.

Definition 13.5. A strategy choice by one player is called the **best response** to a strategy of the opponent if it gives the best payoff against the opponent's strategy.

In Roshambo, Paper is the best response to Rock because Paper beats Rock. Similarly, Rock is the best response to Scissors, and Scissors is the best response to Paper.

The best response to an opponent's game-playing method is a best response to every strategy that the method recommends for the opponent. The best response to an opponent's naive strategy is called the **counter-naive strategy**. The best response to an opponent's prudent strategy is called the **counter-prudent strategy**. Of course, if a method recommends several different strategies, there may be no best response.

All the information about best responses in a game can be conveniently captured in a simple diagram called a **flow diagram**. To draw a flow diagram, draw vertical arrows pointing to the largest entry in each column of a game matrix. (If two entries are tied as the largest, draw the arrows pointing to both.) Similarly, in each row, draw horizontal arrows pointing to the smallest entry. These arrows indicate the best responses each player has to all the strategies of the opponent.

Here is the flow diagram for the Battle of the Bismarck Sea.

	North	South
North	2	2
South	1	3

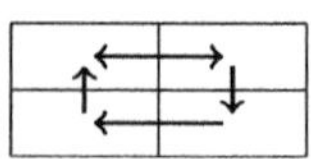

And here is the flow diagram for Roshambo.

0	−1	1
1	0	−1
−1	1	0

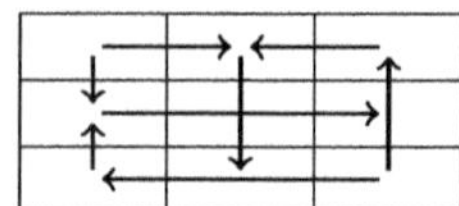

For example, the two arrows in the first row of the flow diagram for Roshambo point to the entry in the second column. This indicates that column 2 is the best response to row 1.

Suppose in a game of Roshambo that Row is about to play Paper, when she suddenly realizes that Column knows what she is going to do. Since Row assumes Column is rational, she concludes that he will take advantage of this knowledge and respond with Scissors. But now, since Row knows what Column will do, she can respond to Column by changing to Rock. Of course if Column is clever, he will anticipate this and counter with Paper. Such second guessing could continue indefinitely in a never-ending cascade of reasoning, anticipation, and indecision. This is why Roshambo seems like such a psychological battle.

Second guessing like this is called "backward induction". It can be modeled by following the arrows around the flow diagram. For example, suppose we start at the outcome (row 1, column 1) in Roshambo. The horizontal arrow out of that outcome shows that Column would have been better off choosing column 2 (Paper) instead of column 1 (Rock) against Row's row 1 (Rock). But were Column to switch to that, making the new outcome (row 1, column 2), Row would then have an incentive to change (following the arrow in column 2 down) to row 3. After that, Column might want to change back to column 1. In the case of Roshambo, this process never stops because every outcome has at least one arrow pointing out from it. Following the arrows in the flow diagram of Roshambo leads to an endless cycle. No matter what the outcome is in Roshambo, at least one of the two players can do better by playing a different strategy. In Battle of the Bismarck Sea, however, the situation is quite different. Following arrows in the flow diagram always leads to the one outcome (row 1, column 1), at which point there are no unidirectional arrows pointing out. For this outcome arrows point in from every other outcome in this row or in this column.

Definition 13.6. A **saddle point** is an outcome such that the strategy for each player is a best response to the strategy of the opponent.

In other words, an outcome is a saddle point if and only if, in the flow diagram, all the arrows in its row and column point in to it. Equivalently, the outcome (row k, column ℓ) is a saddle point if the corresponding entry $u_{k,l}$ is the smallest entry in its row and the largest entry in its column. This says

$$u_{k,\ell} \leq u_{k,j} \text{ for any column } j, \tag{13.1}$$

and

$$u_{k,\ell} \geq u_{i,\ell} \text{ for any row } i. \tag{13.2}$$

Looking at the flow diagram for Battle of the Bismarck Sea we see that (row 1, column 1) is a saddle point. On the other hand, the game Roshambo has no saddle points.

A saddle point is the simplest example of what is called a **Nash equilibrium** in game theory. Consider a situation in which both players in a game expect a certain outcome (perhaps they have negotiated it in advance). Now suppose that one or the other of the two players considers changing his or her strategy choice unilaterally. If the outcome is a saddle point, then neither player benefits from changing. Economists describe this situation as a "self-enforcing contract".

A **saddle point strategy** for a player is any strategy that corresponds to a saddle point outcome. In other words, if the outcome (row k, column ℓ) is a saddle point, then row k is a saddle point strategy for Row and column ℓ is a saddle point strategy for Column.

A common recommendation in game theory is for a player to always play a saddle point strategy when there is one. How can such a recommendation be justified? One justification observes that backward induction compels rational players to follow successive arrows in the flow diagram, and this process ends only at a saddle point. Rational players, so the argument goes, are thus inexorably drawn toward saddle points. Another, and perhaps more convincing justification is afforded by the following result.

Theorem 13.7. *A two-person zero-sum game has a saddle point if and only if $r = c$, in which case the saddle point is a doubly prudent outcome, and the payoff is $r = c$.*

Proof. Suppose $r = c$, and for convenience let $v = r = c$ denote this common value. Let row k and column ℓ be strategies that both have guarantee v, so that (row k, column ℓ) is a doubly prudent outcome. It follows that $v \leq u_{k,j}$ for all j and $u_{i,\ell} \leq v$ for all i. In particular, $v \leq u_{k,\ell} \leq v$, so $u_{k,\ell} = v$ is the smallest entry in its row and the largest entry in its column. Hence (row k, column ℓ) is a saddle point.

Conversely, suppose (row k, column ℓ) is a saddle point, and let $v = u_{k,\ell}$. Since $u_{k,\ell}$ is the smallest entry in its row and the largest entry in its column, the guarantees for both row k and column ℓ must equal v. But r is the largest guarantee among all the rows, so $v = u_{k,\ell} \leq r$, and c is the smallest guarantee among all the columns, so $v = u_{k,\ell} \geq c$. This shows that $c \leq r$, but Corollary 13.4 says that $r \leq c$. Thus $r = c = v = u_{k,\ell}$. This shows that row k and column ℓ are both prudent strategies. □

Theorem 13.7 tells us that a player who chooses a saddle point strategy in a zero-sum game is actually choosing a prudent strategy. Conversely, a player who chooses a prudent strategy in a zero-sum game that has a saddle point is automatically choosing a saddle point strategy.

The next result shows that if both players choose saddle point strategies, then the outcome will always be a saddle point. While this might seem obvious, the issue is rather subtle because the two players may be aiming for two different saddle points.

Corollary 13.8. *An outcome* (row k, column ℓ) *in a zero-sum game is a saddle point if and only if row k is a saddle point strategy and column ℓ is a saddle point strategy.*

Proof. Of course, row k and column ℓ are saddle point strategies whenever (row k, column ℓ) is a saddle point.

Now suppose row k and column ℓ are saddle point strategies. This means there is some saddle point, say (row k, column j) in row k and some saddle point (row i, column ℓ) in column ℓ. We prove that (row k, column ℓ) is a saddle point too.

Since this game has a saddle point we know from Theorem 13.7 that $r = c$, and that row k and column ℓ are prudent strategies. This implies that (row k, column ℓ) is a doubly prudent outcome, which by Theorem 13.7 (again) shows that (row k, column ℓ) is a saddle point. □

One practical consequence of Theorem 13.7 is that it shows how to use the min-max diagram to find saddle points. In particular, whenever $r = c$, a game has saddle points, and the saddle points lie at the intersections of all the prudent rows and prudent columns. Here is a 5-by-5 example showing how this works.

2	3	2	2	5	2	←
0	10	0	3	−9	−9	
−2	2	−1	2	7	−2	
2	10	2	2	2	2	←
0	4	1	0	−4	−4	
2	10	2	3	7		
↑		↑				

The diagram shows that each player has two prudent strategies. Since $r = c = 2$ in this game these prudent strategies are saddle point strategies, and there are four saddle points: (row 1, column 1), (row 1, column 3), (row 4, column 1), and (row 4, column 3). The reader can verify this by drawing the flow diagram.

We have seen that a player's saddle point strategy is optimal in the sense that it is prudent. But if Column plays a saddle point strategy, then Row's saddle point strategy is also the best response to the opponent.

So a saddle point strategy is optimal against prudent opponents, or in other words, it is counter-prudent. Moreover, Column has no incentive to attempt anything but the prudent strategy in response to Row's prudent strategy, since no other strategy can give Column a better payoff. This discussion in essence proves the following corollary.

Corollary 13.9. *A strategy in a two-person zero-sum game is a saddle point strategy if and only if it is both prudent and counter-prudent.* □

13.4 Dominance

Having described several types of desirable strategies, we now switch directions and describe a class of undesirable strategies.

We say that one row of a matrix **dominates** another if each entry of the one row is at least as large as the corresponding entry of the other, and at least one entry is strictly larger. To be precise, row k dominates row i if $u_{k,j} \geq u_{i,j}$ for each column j, and moreover, $u_{k,j} > u_{i,j}$ for at least one column j. We also say that row i is **dominated** by row k.

For example, in the game

1	2	4
5	3	6

row 2 dominates row 1 because all the entries of row 2 are larger than the corresponding entries of row 1. In such a situation, we say row 2 **strictly dominates** row 1. In the game

1	2	3
5	2	3

row 2 still dominates row 1, but not strictly. However, in the two games

1	2	3
5	1	6

1	2	3
1	2	3

neither row dominates the other.

In a similar way, we say that one column dominates another if each entry of the one column is no larger than the corresponding entry of the other, and at least one entry is strictly smaller. To be precise, column ℓ dominates column j if $u_{i,\ell} \leq u_{i,j}$ for each row i, and moreover, $u_{i,\ell} < u_{i,j}$ for at least one row i.

It is important to remember that "better" for Column means *smaller* entries in the payoff matrix. Thus smaller columns dominate larger ones. For example, in the game

1	2	4
5	4	6

columns 1 and 2 both strictly dominate column 3, but there is no dominance between column 1 and column 2.

A dominated strategy is, in all circumstances, at least as bad as the one that dominates it. It seems obvious that rational players should always avoid dominated strategies. Therefore, removing a dominated strategy should have no strategic impact on a game. Hence we may simplify a game by removing dominated strategies. This is called **reduction**. Sometimes, after removing a dominated strategy, some strategies that were previously not dominated become dominated. The process of simplifying a game by successively removing dominated strategies, one at a time, until the process can go no further is called **complete reduction**. As long as all the dominances encountered are strict, the result is unique.

Let us follow the reduction steps in the Battle of the Bismarck Sea.

		Japan	
		North	South
United States	North	2	2
	South	1	3

First notice that although neither row dominates the other, column 1 dominates column 2. In the first step, we remove column 2 (the Japanese option to sail South) to obtain the 2-by-1 game

	North
North	2
South	1

Once this is done, the United States acquires a dominated strategy: row 2 (South) is dominated by row 1 (North). Removing row 1 results in a completely reduced 1-by-1 game:

	North
North	2

Notice that the conclusion here is the same as the one we reached when considering flow diagrams. North is the prudent strategy for both players. The reductions in this game have a common-sense explanation. First, the Japanese conclude that sailing South would be foolish. The United States then becomes convinced that the Japanese will not sail South, and they soon realize that this means it would be foolish for them to search in the South. This leaves North as the only serious option for both players.

We conclude this introduction with a final example of a matrix game.

0	−1	−2	5	4
−3	1	2	3	6
−4	−5	−6	−7	7

In this game, Row has three strategies available to her, and Column has five strategies available to him. The matrix has both positive and negative entries, indicating that there are prospects for payoffs both to Row and to Column.

The naive strategy for Row is to choose the third row, aiming for her primary outcome 7 in the lower right corner. The naive strategy for Column is to choose the fourth column, aiming for his primary outcome −7 in the bottom row. Row finds her best guarantee by playing the first row, with guarantee −2, hence the first row is her prudent strategy. If she chooses row 1, her worst possible outcome will be the payoff −2, representing a payoff of $2 to Column. All other strategies for Row could lead to an even worse outcome. The best guarantee for Column is found in the first column, which is the only column that guarantees that Column will not have to pay a positive amount to Row. The first column is prudent for Column and has guarantee 0.

The counter-naive strategy for Column is to choose the fourth column, since this results in the payoff −7 against Row's naive strategy. Row's counter-naive strategy is the first row, obtaining a payoff 5 against Column's naive fourth column. Similarly, we see that the counter-prudent strategy for Column is the third column, where he receives a payoff 2 against Row's prudent strategy, while the first row is Row's counter-prudent strategy, where the payoff will be 0.

One can continue the best response analysis by asking what is Column's best response to Row's counter-prudent first row. It is the third column, which therefore is the counter-counter-prudent strategy for Column. What is Row's best response to that? It is her counter-counter-counter-prudent strategy, which is the second row, since the payoff 2 in the middle of the matrix is the largest entry in its column.

We can continue this backward induction as long as we wish, in effect following arrows in the associated flow diagram. Before producing the flow diagram, it is reasonable first to reduce the matrix by eliminating dominated rows and columns, which have no influence on optimal play in zero-sum games. No rows dominate any others in this 3-by-5 matrix, but the last column is dominated by the first column, the second column, and the third column. We reduce the matrix by eliminating the fifth column. This illustrates the naivete associated with Row's naive strategy (the third row). While the primary payoff 7 was tempting, there is no prospect of obtaining that payoff, because there is no prospect that a rational opponent will select the fifth column. Hence there is no justification for Row to play the third row. In fact, it becomes dominated by either of the other two rows once the fifth column is removed. Hence we reduce further by removing the third row. After that, the fourth column becomes dominated by the other columns, so we eliminate that as well. We finally reach the matrix here.

0	−1	−2
−3	1	2

At this point, no further reductions are possible. This is the completely reduced form of the original 3-by-5 game. The flow diagram for this matrix is as follows:

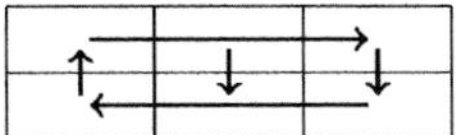

This 2-by-3 game has no saddle point, because from every one of the 6 outcomes there is a unidirectional arrow pointing outward, illustrating that at least one of the players has a better response to the opponents strategy than the one that leads to that outcome. Since the completely reduced form of the game has no saddle point, neither does the original 3-by-5 game. We can surmise that savvy players will choose only those strategies in the original game that correspond to some choice in the completely reduced game. But at this point we need to introduce probability theory to produce a more complete analysis of this example. We begin our journey into probability theory in the next chapter.

13.5 Exercises and Problems

13.1. Consider the games associated with the following matrices:

3	0
1	−1

(*i*)

3	−1
0	1

(*ii*)

0	−1
1	3

(*iii*)

For each game, draw the flow diagram, and use it to find any saddle points.

13.2. For each game in Problem **13.1**:

(a) Draw the min-max diagram and use it to find each player's prudent strategy.
(b) Find the prudent strategy guarantees r and c. Does $r = c$?
(c) Use the min-max diagram to find the saddle points in each game.
(d) Describe what happens if both players use their prudent strategies. Does either player do better than his or her prudent strategy guarantee?

13.3. For each game in Problem **13.1**:

(a) Find each player's naive strategy.
(b) Explain what happens when both players play their naive strategies, and say who the outcome favors.
(c) Find each player's counter-naive strategy, if there is one.
(d) Explain what happens when Row plays her counter-naive strategy instead of her naive strategy against Column's naive strategy.

13.4. For each game in Problem **13.1**:

(a) Find the prudent and counter-prudent strategies for both players.
(b) Describe what happens if a player plays his or her counter-prudent strategy against the opponent's prudent strategy.
(c) Find the outcome that occurs when both players play their counter-prudent strategies.

13.5. For each game in Problem **13.1**:

(a) Find all the dominant strategies for both Row and Column.
(b) Completely reduce each game by successively eliminating all dominated strategies.

13.6. In the game given below, use the min-max algorithm to find the prudent strategies for both players and to find all the saddle points. Draw the flow diagram. Find any dominated strategies and completely reduce the game.

2	0	0
1	0	−1
0	−1	−2

13.7. Repeat the analyses of Problem **13.6** for the game associated with the following matrix:

3	0	4	1	1
0	0	3	1	−2
−1	2	0	3	−1
3	1	1	1	3

13.8. Analyze the following game by successive reduction:

2	1	3	16
4	5	12	14
6	9	10	13
7	8	11	15

13.9. In the following game, draw the flow diagram and find the saddle point.

3	4	5
2	0	6
1	6	0

What does the flow diagram predict will happen if the players start at (row 2, column 2)? Will they ever reach the saddle point?

13.10. Row and Column play a simple game of cards. Each player is given two cards: 2♡ (i.e., "the 2 of hearts") and 8♣ for Row, and 3♡ and 5♣ for Column. To play, the players simultaneously show one card. Row wins if the players show the same suit (hearts or clubs), whereas Column wins if they show different suits. The loser has to pay the winner the number of dollars showing on the winner's card. Set up this card game as a two-person zero-sum game. Find the prudent and counter-prudent strategies for both players, and say whether there is a saddle point. Who wins if both players play their prudent strategies? Who wins if one player plays the counter-prudent strategy against the prudent strategy of the other player?

13.11. Terrorists are planning to attack one of two landmarks, A or B. The FBI is aware of the threat but needs to decide which landmark to defend. Neither the FBI nor the terrorists have the resources to cover both targets. The terrorists value a successful attack on B as only 80% as spectacular as a successful attack on A. They view an unsuccessful attack as worthless. The terrorists know that if they find their chosen target unguarded, their attack will succeed. Landmark A is easy to defend, and if the FBI guards it, they are certain to be able to stop any attack. Unfortunately, landmark B is harder to defend, and even if the FBI guards it, they have only a 60% chance of stopping an attempted terrorist attack. Set up this conflict as a two-person zero-sum game. Does it have a saddle point? What happens if both players play their naive strategies? What happens if both players play their prudent strategies?

13.12. Company A intends to buy 6 laptops while Company B plans to buy 4. Ms. Row represents Dell, which currently supplies both companies, while Mr. Column represents Apple. Each sales rep has enough time to visit only one company. If Ms. Row and Mr. Column visit different companies, each will get all the business from the company they visit. However, if they visit the same company they will split that company's business. As the representative of the current supplier, Ms. Row will get all the business of a company that is not visited by either rep. Set up this competition as a two-person constant-sum game ($K = 10$, the total number of laptops to be sold). Find the naive and the prudent strategies for both players, and say whether there is a saddle point. Who wins if both players play their naive strategies? Who wins if both players play their prudent strategies? What happens if one player plays the naive strategy against the other player's prudent strategy?

13.13. Consider the game-playing method in which Row chooses whichever row has the largest average entry. Give an example of a matrix for which this method, the naive method, and the prudent method yield three different strategies.

13.14. Show that if a game completely reduces to a 1-by-1 game, then that entry is a saddle point in the original game.

13.15. Is the converse to Problem **13.14** true? If a game has a saddle point, will it always completely reduce to a 1-by-1 game?

13.16. Create an example of a matrix in which the prudent strategy, the counter-counter-prudent, and the counter-counter-counter-counter-prudent strategy for Row are all different.

14

Chance and Expectation

"No victor believes in chance." — Friedrich Nietzsche

14.0 Scenario

In a casino, there is a sort of roulette wheel that randomly produces a number from 1 to 10, each with equal probability. The rules invite you to place any of the following kinds of bets:

(1) You can bet that the outcome will be an even number, receiving \$4 if you win, paying \$5 if you lose.

(2) You can bet that the outcome will be the number 7, receiving \$6 if you win, paying \$1 if you lose.

(3) You can bet that the outcome will be 2, 3, 4, 5, 6, 7, 8, or 9, receiving \$1 if you win, paying \$6 if you lose.

(4) You can bet that the outcome will be 3, 4, or 9, receiving \$5 if you win, paying \$2 if you lose.

(5) You can decline to place any of the bets above, but then you must pay 25 cents for the privilege of watching the game.

Which of these bets would you choose, and why? Which bet would a pessimist choose? Which bet would an optimist choose? Would it make any difference if you were planning to play just once or if you were planning to play repeatedly, thousands of times? Which bet is likely to be the most exciting?

14.1 Probability Theory

Gambling — playing games of chance — may be one of the oldest of all human activities, but it was not until the 16th century that the basic principles that underlie gambling were studied in a systematic way. This study eventually matured into the modern **theory of probability**. Ideas from probability now permeate such steadfast pursuits as life insurance, investing, and medicine. Yet at its core, probability theory is still about gambling.

Probability theory studies random processes. By a random process we simply mean something that occurs in which the end result is unpredictable. The process under consideration might be a roll of a die or a football game. It might be tomorrow's weather or the possibility of having an automobile accident. We refer to the different ways that the random process can turn out as its **outcomes**. Borrowing a term from statistics, we call the set of all possible outcomes the **sample space**. In this book, we always assume the sample space is finite.

One familiar example of a random process is a coin toss. The outcomes of this process — the elements of the sample space — are "heads" and "tails", and we might write the sample space as $\{h, t\}$.

Once the sample space is specified, we need to describe how likely each outcome is to occur. This is done by assigning a number to each outcome, called the **probability** of the outcome. If we number the outcomes from 1 to n then we denote the probabilities of the n outcomes by $p_1, p_2, \ldots, p_n$. Each of the probabilities p_i is required to satisfy $0 \leq p_i \leq 1$. The idea is that an outcome that is more likely to occur should have a higher probability, with 0 corresponding to an outcome that will never occur (i.e., that is impossible) and 1 corresponding to an outcome that will always occur (i.e., that is certain). Taken together, the probabilities are required to satisfy $p_1 + p_2 + \cdots + p_n = 1$, which corresponds to the assumption that, in the end, exactly one of the n outcomes will occur.

We often assemble all the probabilities together into a list $P = (p_1, p_2, \ldots, p_n)$, called a **probability distribution**. When $n = 2$, any probability distribution can be written in a useful special form $P = (1 - p, p)$, where the probability p of the second outcome satisfies $0 \leq p \leq 1$. In everyday language, it is common to express probabilities as percentages that lie between 0% and 100%. When probabilities are expressed this way, we sometimes use the word **chance**. For example, people often say something like "the chance of rain tomorrow is 50%" to describe a probability $p = 1/2$.

Consider the example of a coin toss. A coin is considered to be "fair" if, when flipped, it is equally likely to come up heads or tails. In this case, heads and tails each have probability $1/2$, so the probability distribution that governs this process is $P = (1/2, 1/2)$. More generally, if a random process has n equally likely outcomes, then the associated probability distribution is $P = (1/n, 1/n, \ldots, 1/n)$.

Consider, for example, a gambler who has 9 coins in his pocket, say 8 pennies and a dime. Suppose the gambler reaches into his pocket and picks out the first coin he touches. What is the probability that he will pick the dime? We can model this process by taking the sample space to be $\{\text{penny } 1, \text{penny } 2, \ldots, \text{penny } 8, \text{dime}\}$. Assuming that each of the coins is equally likely to be picked, each outcome has probability $p = 1/9$ (or a chance of around 11%). The probability distribution for this random process is given by $P = (1/9, 1/9, \ldots, 1/9)$.

14.2 All Outcomes Are Not Created Equal

Everyday speech often takes the phrase "at random" to mean both "at random" and "equally likely". Since we will adopt a broader usage, we will explicitly specify "equally likely" when that is what we mean. So far, the examples we have considered have equally likely outcomes, which allowed us to compute the probability of an outcome as the reciprocal of the total number of outcomes. We now move on to consider examples of random processes with outcomes that are not necessarily equally likely.

Suppose, once again, that a gambler has 9 coins in his pocket, this time 5 pennies, 3 nickels, and 1 dime. Regarding coins of the same type as indistinguishable, we take the sample space to be $\{\text{penny}, \text{nickel}, \text{dime}\}$, which consists of 3 rather than 9 outcomes. We want to find the probability distribution P on this sample space that governs the process of picking out a random coin. We do this by noting that $5/9$ of the coins are pennies, $3/9 = 1/3$ of the coins are nickels, and $1/9$ of the coins are dimes. In this way, we obtain the probability distribution $P = (5/9, 3/9, 1/9)$. More generally, if each element the sample space is obtained by grouping together equally likely outcomes, then the probability of each grouped outcome is the fraction of the equally likely outcomes that comprise it.

This idea can be extended to construct a random process with any given rational distribution. Suppose a player wishes to design a random process with probability distribution $P = (2/12, 3/12, 4/12, 3/12)$. A player can put 2 pennies, 3 nickels, 4 dimes and 3 quarters in her pocket. Another approach is to construct a sort of "wheel of fortune" as in Figure 14.1, which might be found in a county fair. After paying, one spins the arrow and wins the indicated prize. The sample space here is $\{\text{pie}, \text{cupcake}, \text{lose}\}$. Notice that the sector labeled "pie" measures 60 degrees, or $1/6$ of the wheel; the sector labeled "cupcake" measures 120 degrees, or $1/3$ of the wheel; and the sector labeled "lose" is $1/2$ of the wheel. It is reasonable to assume that the probability of each outcome equals its relative size. Thus we arrive at a probability distribution here with $P = (1/6, 1/3, 1/2)$. It is a simple matter to make a wheel of fortune that gives any desired probability distribution.

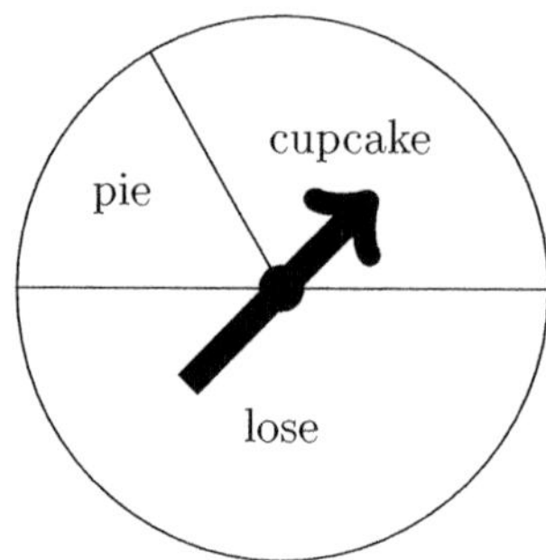

Figure 14.1 Wheel of fortune.

Where do probabilities come from? How does one decide that the probability of some event is the number p? There is no single paradigm that determines probabilities in all types of situations. So far, all of the examples of probabilities we have considered use some sort of mechanical device to produce random outcomes. We assign probabilities to the outcomes of these sorts of "gambling devices" by idealizing the process and drawing on laws of physics or laws of mathematics. We call the analysis of this kind of randomness **theoretical probability**. There are, however, several other situations in which randomness occurs, and these require different ideas to establish their probabilities.

Suppose that we want to assign probabilities to a random process that we have observed repeatedly through many trials. We can assign each outcome a probability equal to the fraction of the times it occurs. We refer to this kind of probability as **empirical probability**. Suppose, for example, that we have observed a repeated random process a large number N of times. If a particular outcome has occurred M times out of these N observations, then we declare the probability of that outcome to be equal to the fraction M/N. Empirical probability is useful in situations where there is no theoretical basis for setting a probability. Instead, we rely on historical data or randomized trials to assess how likely an outcome is. Of course, empirical probabilities are subject to revision as more data becomes available.

Consider, for example, the process of repeatedly flipping an *unfair* coin that is much more likely to come up heads than tails. Suppose, for example, that in a particular trial of 100 flips, 71 flips come up heads and 29 come up tails. In such a situation, the number $0.71{=}71/100$ may be assigned to the outcome "heads" as its empirical probability, and 0.29 may be assigned to the outcome "tails". Thus the empirical probability distribution for this series of trials is $P = (0.71, 0.29)$.

Empirical probability represents the long-term frequency with which we expect an outcome to occur. It makes sense in situations where it is possible to reproduce the random process repeatedly. For example, if we repeat the random process involved in a theoretical probability, we expect the empirical probability to be close to the theoretical probability. This is known as the law of large numbers.

Probability can sometimes even make sense when a process is not repeatable. A person's **subjective probability** is her best guess as to how likely an outcome is to occur. Of course, subjective probabilities can be different for different people; they depend on people's perspectives, backgrounds, and expertise. A good way to quantify someone's feeling about the likelihood of an outcome is to ask her how much she is willing to bet that the outcome will occur. For example, in a bet that pays \$1 for winning, we regard a person willing to risk up to \$1 as expressing her belief that winning has probability $1/2$. And if she is willing to pay up to \$1 for the privilege of winning $\$c$ if a particular outcome occurs, then this indicates that her subjective probability of that outcome is $1/c$.

One place where subjective probability is common is in weather forecasting. One forecaster might say that there is a 30% chance of rain tomorrow. Here, we may regard the sample space as $\{\text{rain}, \text{shine}\}$ and $P = (0.3, 0.7)$. However, another forecaster on the same day might say that there is a 40% chance of rain tomorrow, so for the second forecaster, the probability distribution is $P' = (0.4, 0.6)$. Which forecaster is correct? Although it might seem strange, we say that both are correct, whether it rains in the end or not. The forecasters are merely expressing their best guesses as to the likelihood of rain. The experiment cannot be repeated precisely. The weather forecasters are using their experience to deliver a subjective probability.

In many situations, the assignment of a probability to an outcome may be defended on the basis of all three philosophies. Consider the situation that arises when investing in stocks. Suppose that an investor buys a stock for \$100 and plans to sell it at the end of the quarter. An analysis of the stock's performance over the last several years, combined with an analysis of the fundamentals of the company, as well as the optimism of the investor lead him to estimate that its value will increase 40% with probability of 0.3, will increase 10% with a probability of 0.22, will decrease 10% with a probability of 0.31, and will decrease 40% with a probability of 0.17. Here, we might model this situation with the sample space $\{40\%, 10\%, -10\%, -40\%\}$ and the probability distribution $P = (0.30, 0.22, 0.31, 0.17)$.

Where do these probabilities come from? An economist might argue that they come from the immutable laws of economics and are therefore theoretical probabilities. A financial analyst might study the histories of similar stocks, and compute empirical probabilities. A stockbroker might forgo analysis, but nonetheless have a spectacular knack for predicting stock performance. For her, these numbers are subjective probabilities.

14.3 Random Variables and Expected Value

Consider a random process on a sample space with n outcomes and a probability distribution $P = (p_1, p_2, \ldots, p_n)$. A **random variable** X on the sample space is an assignment of a real number to each outcome. Mathematically speaking, a random variable is nothing but a function whose domain is the sample space and whose codomain is the set of all real numbers. For the random variable X, we usually write x_i for the number assigned to outcome i. Unlike the probabilities p_i, the numbers x_i are not restricted to being between 0 and 1. They can be arbitrary real numbers, positive or negative.

To understand the purpose of a random variable, one should imagine a player who gambles on the outcome of the random process. In such a scenario, the number x_i might represent the **payoff** the gambler will receive if outcome i occurs. The next idea provides a way to measure the worth of such a bet.

Definition 14.1. Consider a random process on a sample space with n outcomes, having the probability distribution $P = (p_1, p_2, \ldots, p_n)$. Let X denote the random variable that assigns the payoff x_i to outcome i. We define the **expected value** (or the **expected payoff**) E of X by

$$E = p_1 x_1 + p_2 x_2 + \cdots + p_n x_n.$$

In other words, to calculate the expected payoff of X, we multiply the payoff x_i of each outcome i by its probability p_i, and sum these numbers over all outcomes i.

For a random variable X that models the payoff on a bet, the expected value E models the average amount that the bettor can expect to win each time she places the bet. This idea makes the most sense if the bettor places the bet repeatedly. In a long sequence of N repetitions, the payoff x_i occurs with frequency p_i. Averaging the payoffs x_i over the N repetitions, we find that each outcome i contributes approximately $p_i x_i$ to the total. Altogether, the sum of all these contributions is therefore

$p_1x_1 + p_2x_2 + \cdots + p_nx_n$. So the expected payoff E for the bet is the rate at which money is paid, in units of dollars per trial.

Consider, for example, a state lottery that pays a grand prize of \$100,000,000 with probability 1/150,000,000. There are also several consolation prizes: \$200,000 with probability 1/3,000,000, \$10,000 with probability 1/150,000, and \$10 with probability 1/300. What would be the fair price for a ticket in this lottery?

We take $\{\text{Grand prize}, \text{2nd prize}, \text{3rd prize}, \text{4th prize}, \text{lose}\}$ as the sample space, and put $p_1 = 1/150{,}000{,}000$, $p_2 = 1/3{,}000{,}000$, $p_3 = 1/150{,}000$, $p_4 = 1/300$ and $p_5 = 1-(p_1+p_2+p_3+p_4) \approx 0.997$. Entering this data into a table, we obtain:

	Grand	2nd	3rd	4th	lose
Payoff:	\$100,000,000	\$200,000	\$10,000	\$10	\$0
Probability:	1/150,000,000	1/3,000,000	1/150,000	1/300	0.997

The expected value calculation is accomplished by multiplying each payoff by its associated probability (the number below the corresponding box), and then adding up these products. Therefore,

$$E = \frac{100{,}000{,}000}{150{,}000{,}000} + \frac{200{,}000}{3{,}000{,}000} + \frac{10{,}000}{150{,}000} + \frac{10}{300} + (0)(0.997) \approx 0.833.$$

This amount is in dollars, so the expected payoff is about 83 cents. What this means is that a lottery ticket will pay approximately 83 cents on the average. We call 83 cents the **fair price** of the ticket. If a gambler pays this much, then he or she will break even on average. In a similarly way, if the state charges the fair price, then it too will break even on average. However, bets are rarely sold for their fair value; they are usually marked up to provide a profit for the seller.

Imagine that the state sells the lottery tickets for \$1. The profit of the bettor is a new random variable obtained by subtracting \$1 (the cost of the ticket) from each of the payoffs. The data for this new random variable is as follows.

	Grand	2nd	3rd	4th	lose
Payoff:	\$99,999,999	\$199,999	\$9999	\$9	−\$1
Probability:	1/150,000,000	1/3,000,000	1/150,000	1/300	0.997

The new expected value is -0.167, which means that in the long run a bettor can expect to lose approximately 17 cents on each ticket. On the other hand, the state can expect to make a profit of 17 cents on each ticket.

Notice the similarity between this lottery and a zero-sum game. Taking into account the cost of the lottery ticket, either the bettor gives money to the state or the state gives money to the bettor. Between these two players, no money is gained or lost. We can model this lottery as a 1-by-5 matrix game in which the bettor is Row and the state is Column. As usual, the entries in the matrix represent the payoffs to Row.

99,999,999	199,999	9999	9	−1

Notice that Row has just one strategy choice (since we assume she has already bought the lottery ticket). So Row is nothing but an observer in this game, with no strategic role to play. On the other hand, Column (the state) appears to have five choices. However, by the rules of the lottery, the state is not allowed to behave as a rational player. Rather the state must choose an outcome at random. After all, the lottery is advertised as having certain odds of winning. The random process that the state uses to pick a winner is often televised to reassure the public that the state is not "fixing" the lottery.

We can make the lottery look more like a matrix game if we give Row two strategy choices.

99,999,999	199,999	9999	9	−1
0	0	0	0	0

Row 1 is the choice to buy a ticket and row 2 is the choice not to buy a ticket. If Row chooses not to buy, her payoffs are all \$0, regardless of what happens in the lottery. The two rows represent two random variables that record the different payoff schemes that will apply depending on what Row decides to do. There are now two corresponding expected payoffs: $E(\text{Bet}) = -0.167$ and $E(\text{Don't}) = 0$. Which option should Row choose?

Definition 14.2. Given a choice between two different random variables (i.e., payoff schemes) on two random processes, the **expected value principle** says that a rational player will (or should) always choose the one with the largest expected value.

Game theory and probability theory both tend to accept the expected value principle uncritically. However, there are some good reasons to remain a little skeptical. Under a literal interpretation of the expected value principle, a rational person would never buy a lottery ticket, or even automobile insurance. But these activities can make sense, and they can be explained by what is known as the theory of **utility**. The idea is that not all dollars are equally valuable. If one payoff is twice as large as another, it might not really mean it is exactly twice as good. For example, people who play the lottery may feel that the "pleasure" from winning \$100,000,000 (unimaginable wealth!) is actually more than

100,000,000 times the "pain" from losing \$1 (just spare change). The next example shows another instance of this principle.

Suppose that the price of an automobile collision insurance policy is \$1500 a year. Suppose the company pays off (on average) \$0 if there is no accident or a minor accident, \$2000 if there is a moderate accident, and \$16,000 if there is a major accident. Based on empirical probability, the company estimates probabilities of a customer having no accident, a minor accident, a moderate accident, and a major accident to be 0.5, 0.25, 0.2, and 0.05, respectively. Is the price of this policy fair?

To answer, we compute the expected payoff. Let X denote the random variable of payoffs, shown in the matrix here, with the probabilities of each payoff written below.

No	Minor	Moderate	Major
\$0	\$0	\$2,000	\$16,000
0.5	0.25	0.2	0.05

Then $E(X) = (0)(0.5)+(0)(0.25)+(2,000)(0.2)+(16,000)(0.05) = 1200$. In other words, the fair price for this policy would be \$1200. Even though the actual price, \$1500, exceeds the fair price by \$300, a consumer might still decide to purchase the policy. The consumer may regard the \$16,000 cost of a major accident to be catastrophic, whereas the \$1500 yearly insurance premium may be affordable. In effect, the consumer pays the insurance company \$300 to assume the risk of a major accident.

From now on, we will adopt the expected value principle. In doing this, we are assuming that the payoffs accurately model the good that comes from each outcome.

14.4 Mixed Strategies and Their Payoffs

Now that we have introduced some basic probability theory, we are ready to relate it back to game theory. Consider the game of Roshambo from Chapter 13.

	Rock	Paper	Scissors
Rock	0	-1	1
Paper	1	0	-1
Scissors	-1	1	0

We have not yet found any good methods for playing this game. Moreover, we face the prospect that some opponents may be good at guessing other players' intentions and then exploiting them. How is a player to defend against such an opponent? As it turns out, probability theory provides the answer.

In order to defend against an opponent who seems able to predict a player's moves, it is clear that a player should try to behave unpredictably. And what better way to behave unpredictably than to behave randomly? The idea of systematically playing random strategies was first suggested by the mathematician John von Neumann. In this context, the original strategy choices (the rows for Row and the columns for Column) are sometimes called **pure strategies** to avoid confusion.

Definition 14.3. Consider an m-by-n game. A **mixed strategy** for Row is a probability distribution $P = (p_1, p_2, \ldots, p_m)$ on her set of m pure strategies. In particular, Row uses a random process to choose one of her m pure strategies at random, with p_i being the probability of row i. Similarly, a mixed strategy for Column is a probability distribution $Q = (q_1, q_2, \ldots, q_n)$ on his set of n pure strategies.

As an example, consider the mixed strategy $Q = (1/4, 1/2, 1/4)$ for Column in Roshambo. This mixed strategy tells Column to randomly select and play a pure strategy, picking Rock 1/4 of the time, Paper 1/2 of the time, and Scissors 1/4 of the time. Column could implement this mixed strategy, for example, by choosing coins from his pocket or building a "wheel of fortune" spinner.

Once Column decides to use a mixed strategy, the game becomes a kind of lottery.

	Rock	Paper	Scissors
Rock	0	−1	1
Paper	1	0	−1
Scissors	−1	1	0
	1/4	1/2	1/4

Consider, for example, what happens if Row plays her pure strategy Rock against Column's mixed strategy Q. The payoffs in row 1 now define a random variable X, with Column's mixed strategy providing the probabilities. The expected value of this random variable is given by

$$E = (1/4)0 + (1/2)(-1) + (1/4)1 = -1/4. \tag{14.1}$$

In other words, if Row plays Rock against Q, she can on average expect to lose \$0.25 per game to Column. But Row is not entirely defenseless against Column. If she plays Scissors (instead of Rock) against Q, her expected payoff will be more favorable

$$E = (1/4)(-1) + (1/2)1 + (1/4)0 = 1/4, \tag{14.2}$$

or 25 cents.

One of the advantages of using mixed strategies is that they give a player many more options than pure strategies do. Instead of having to choose from among a finite number of rows or columns, a player has an infinite number of mixed strategies from which to choose. In fact, even a pure strategy can be considered to be special kind of mixed strategy.

Definition 14.4. A **basic mixed strategy** is a probability distribution with all but one probability equal to 0 (and with that one probability equal to 1). We denote by P_i the basic mixed strategy for Row that has the probability p_i of row i equal to 1. We denote by Q_j the basic mixed strategy for Column that has the probability q_j of column j equal to 1.

The basic mixed strategy P_i tells Row to play row i with probability 1, or in other words with certainty. So P_i is actually the same as the pure strategy row i. Similarly, the basic mixed strategy Q_j is the same as the pure strategy column j for Column.

Pure strategies, when played against each other, yield definite payoffs. However, once players start using mixed strategies, the payoffs become uncertain. Nevertheless, in accordance with the Expected Value Principle, we will evaluate the results of using mixed strategies by calculating their expected payoffs.

We begin by considering the case where one player plays a pure strategy against a mixed strategy of the opponent. When Row plays row i against Column's mixed strategy Q we denote the expected payoff by $E(P_i, Q)$. Similarly, when Column plays the pure strategy column j against Row's mixed strategy P we denote the expected payoff by $E(P, Q_j)$. We have already done two such calculations in the case of Roshambo. If Row plays Rock against the Column mixed strategy $Q = (1/4, 1/2, 1/4)$, then (14.1) shows that $E(P_1, Q) = -1/4$, and (14.2) shows that $E(P_3, Q) = 1/4$. The following lemma generalizes these calculations.

Lemma 14.5. *Consider an m-by-n matrix game. If Row plays row i against the Column mixed strategy* $Q = (q_1, q_2, \ldots, q_n)$, *then the (expected) payoff is*

$$E(P_i, Q) = q_1 u_{i,1} + q_2 u_{i,2} + \cdots + q_n u_{i,n}.$$

Similarly, if Column plays column j against the Row mixed strategy $P = (p_1, p_2, \ldots, p_m)$ *then the (expected) payoff is*

$$E(P, Q_j) = p_1 u_{1,j} + p_2 u_{2,j} + \cdots + p_m u_{m,j}. \quad \square$$

It is then easy to prove the following corollary.

Corollary 14.6. *Let P_i and Q_j be the basic mixed strategies corresponding to the pure strategies row i and column j. Then $E(P_i, Q_j) = u_{i,j}$.* □

In other words, when both players use pure strategies, the expected payoff is the same as the actual payoff.

14.5 Independent Processes

Suppose that we have two random processes, each with its own sample space and its own probability distribution. For any outcome of either process, the probability distribution tells us how likely that outcome is. But if the two processes take place simultaneously, we can think of this pair of occurrences as a single process with a sample space consisting of all pairs (i, j), where i is an outcome of the first random process, and j is an outcome of the second. How can we assign a probability to this compound outcome (i, j)?

The answer is that in general we can't. It depends on whether or not the two processes are related in some underlying way. This brings us to a key concept from probability theory: the notion of independence. When two independent random processes occur simultaneously, we can compute the probability that a certain pair of outcomes occurs.

Definition 14.7. Consider two random processes. Let p_i denote the probability of an outcome i in the first process, and let q_j be the probability of an outcome j in the second process. The two processes are said to be **independent** if the probability of each compound outcome (i, j) is given by $p_i q_j$.

In other words, two processes are independent if the probability of every compound outcome is obtained by multiplying the probabilities of its component outcomes. The intuition behind independence is that the outcome of one processes has no influence over the outcome of the other. It is customary to assume independence when two random processes are performed separately and have no connection to each other. Some examples will illustrate the concept.

Suppose a gambler tosses two fair coins. What is the probability that both coins show heads? We think of the compound process as having a four-outcome sample space: $\{(h, h), (h, t), (t, h), (t, t)\}$, where the first component of each pair indicates the result of the first coin and the second component indicates the result of the second coin. Although each coin is modeled by the probability distribution $P = (1/2, 1/2)$, we need some way to measure the probability of these compound outcomes. The assumption that the processes are independent provides the key to measuring this probability. When the two coin tosses are independent, each outcome in the compound process has probability equal to $1/2 \times 1/2 = 1/4$.

To illustrate that not all processes need to be independent, imagine that the same two fair coins are tossed, but that this time they are taped together so that when one lands "heads" the other lands "tails" and vice versa. Individually, each coin is still modeled by the probability distribution $P = (1/2, 1/2)$, but the independence between the two coins is now broken. In this version of the compound process, the probabilities of (h, h) and (t, t) are 0 while the probabilities of (h, t) and (t, h) are each $1/2$. The purpose of this curious example is to point out that there is a hidden assumption in the usual reading of "Suppose two fair coins are tossed." What is intended is "Suppose two fair coins are tossed *independently*."

When two processes are independent, knowing the outcome of one process does not change our estimate of the probabilities of the outcomes of the other. Informally, this is often expressed by saying that the two processes have "no effect" on each other. In practice, we often apply this in reverse. If it is obvious that two processes can have no effect on each other, then we conclude that the two processes are independent. For example, it seems obvious that the weather has no effect on the state lottery. So we conclude that two processes — one associated with tomorrow's weather and the other associated with tomorrow's lottery — are independent.

Now suppose that Row flips one biased coin that has a probability $2/3$ of showing "heads", and Column flips another biased coin that has a probability of $3/4$ of showing "heads". What is the probability that Row obtains "heads" and Column obtains "tails"? The probability that Row obtains "heads" is $2/3$, and the probability that Column obtains "tails" is $1 - 3/4 = 1/4$. Assuming that the coins are tossed independently, the probability in question is obtained by multiplying $2/3 \times 1/4 = 1/6$.

14.6 Expected Payoffs for Mixed Strategies

Suppose both players in a two-person zero-sum game play mixed strategies. What will the expected value of the payoff be?

Our fundamental assumption about how games are to be played is that the two players make their strategy choices alone and secretly. Assuming they both play mixed strategies, this allows us to assume that the random processes the players use to implement their mixed strategies are independent. To see how this works in practice, we consider an example.

Suppose in a game of Roshambo, Row plays the mixed strategy $P = (1/4, 1/2, 1/4)$ and Column plays the mixed strategy $Q = (1/6, 1/3, 1/2)$. We write these probabilities at the end of their respective rows and columns, as shown:

	Rock	Paper	Scissors	
Rock	0	−1	1	1/4
Paper	1	0	−1	1/2
Scissors	−1	1	0	1/4
	1/6	1/3	1/2	

Since the Row and Column mixed strategies are assumed to be independent, the probabilities of the outcomes in this game are obtained by multiplying the probabilities along the bottom of the matrix by the probabilities along the right side. These products are shown here:

1/24	1/12	1/8	1/4
1/12	1/6	1/4	1/2
1/24	1/12	1/8	1/4
1/6	1/3	1/2	

Once we have these probabilities, it is possible to compute the expected payoff, denoted $E(P, Q)$, by multiplying each payoff in the game matrix by the corresponding probability.

$$\begin{aligned} E(P,Q) &= (1/24)0 + (1/12)(-1) + (1/8)1 \\ &\quad +(1/12)1 + (1/6)0 + (1/4)(-1) \\ &\quad +(1/24)(-1) + (1/12)1 + (1/8)0 = -1/12. \end{aligned}$$

The conclusion is that each time Row and Column use the mixed strategies P and Q, the expected payoff to Column will be $1/12$ dollars, which is a little more than 8 cents. The following result generalizes this calculation.

Lemma 14.8. *Suppose in an m-by-n game with payoffs $u_{i,j}$ that Row plays the mixed strategy $P = (p_1, p_2, \ldots, p_m)$ and Column independently plays the mixed strategy $Q = (q_1, q_2, \ldots, q_n)$. Then the probability of the outcome* (row i, column j) *is given by $p_i q_j$, and the expected payoff $E(P, Q)$ is obtained by adding up the numbers $p_i q_j u_{i,j}$ over all i and j.* □

As a practical matter it is often easier to compute the expected payoff $E(P, Q)$ in a different way (starting with the numbers in Lemma 14.5).

Lemma 14.9. *Suppose in an m-by-n game that Row plays the mixed strategy $P = (p_1, p_2, \ldots, p_m)$ and Column independently plays the mixed strategy $Q = (q_1, q_2, \ldots, q_n)$. Then the (expected) payoff is given either by*

$$E(P, Q) = p_1 E(P_1, Q) + p_2 E(P_2, Q) + \cdots + p_m E(P_m, Q),$$

or alternatively by

$$E(P, Q) = q_1 E(P, Q_1) + q_2 E(P, Q_2) + \cdots + q_n E(P, Q_n).$$

Proof. We can expand these formulas using Lemma 14.5 to obtain the sum of the terms $p_i q_j u_{i,j}$. Since we assume Row and Column choose their strategies independently, Lemma 14.8 says this sum is $E(P, Q)$. □

14.7 Exercises and Problems

14.1. Simultaneously flip a penny, a nickel, a dime, and a quarter. The sample space for this random process consists of 16 outcomes. Assume independence.

(a) List the elements of the sample space.

(b) What is the probability of getting four heads? What is the probability of getting either four heads or four tails? What is the probability that all four coins do not come up the same?

(c) Which is more likely: two heads and two tails or three heads and one tail? Explain.

(d) Which is more likely: two heads and two tails or three of one and one of the other? Explain.

14.2. Suppose you have a pair of dice, one red and one green, each marked with the numbers 1 to 6. When the pair of dice is rolled, the sample space consists of 36 outcomes.

(a) Draw a diagram of this sample space as a 6-by-6 matrix.

(b) What is the probability of getting a pair of 6's?

(c) How many different ways can the dice come up as "doubles" (both dice show the same number). What is the probability of doubles?

(d) How many different ways can the sum of the two dice come up 7? What is the probability of getting a sum of 7?

(e) What has a higher probability: getting a sum of 4 or getting a sum of 6?

14.3. A huckster on a back street offers the following bet. You ante \$5, and he flips two coins. If the both come up the same, you get your money back, plus \$6 more if it is two heads or \$3 more if it is two tails. If the coins come up different, the huckster keeps your \$5. Analyze this bet using expected value and decide whether it is fair.

14.4. You discover your younger brother offering his friends the following bet. They pay him \$1 and he rolls a pair of dice. He pays back \$3 if it comes up either seven or doubles, otherwise he keeps the dollar. Is this bet fair?

14.5. You believe that the probability of rain today is 30%. A friend proposes a bet paying \$5 if it rains. How much should you pay for this bet? How much should you pay if you believed the probability of rain was some other value p?

14.6. A company offers a small city an economic opportunity: a new factory that will pump \$1,000,000 into the local economy. The only catch is a small chance that its dangerous product will leak, costing the city an estimated \$10,000,000 clean up. The mayor reasons that the city should accept the factory as long as the probability of an accident is not too high. According to the expected value principal, what is the highest acceptable probability of an accident?

14.7. Suppose in the 2-by-2 game

1	2
3	−1

Row uses the mixed strategy $P = (4/5, 1/5)$ and Column uses the mixed strategy $Q = (1/2, 1/2)$. Determine the expected payoff $E(P, Q)$ by first computing Row's expected payoffs against each of Column's pure strategies. What would happen if Column changed his mixed strategy Q?

14.8. Suppose that a variant of Roshambo is played in which rock-over-scissors is regarded as a double victory. The matrix governing this game is

0	−1	2
1	0	−1
−2	1	0

If Row uses the mixed strategy $(1/3, 1/3, 1/3)$ and Column uses the mixed strategy $(8/10, 1/10, 1/10)$, which player will be ahead in the long run?

14.9. If Row uses the mixed strategy $(2/5, 3/5)$ and Column uses the mixed strategy $(1/7, 4/7, 2/7)$, which player will be ahead in the long run in the following matrix game?

1	−2	7
−5	3	−6

14.10. A card is drawn at random from an ordinary deck of cards.

(a) What is the probability that the resulting card is a face card?

(b) What is the probability that the resulting card is not a face card?

(c) Suppose that a game pays \$1 for an ace, \$2 for a two, \$3 for a three, and so forth through \$10 for a ten, but charges \$17 for any face card. What is the expected payoff of this game? (Consider a charge to be a negative payment.)

(d) Suppose we alter the game so that the charge for drawing a face card is \$$C$. For what value of C is the game fair?

14.11. Gamblers use the terminology of **odds** to express probabilities. If the probability of an outcome is given by the fraction $p = a/b$, then the odds against the outcome are said to be "$b-a$ to a". For example, if the probability of winning a certain bet is $p = 2/7$ then the odds (against it) are 5 to 2.

(a) If the odds against a certain outcome are 100 to 1, what is the probability that the outcome will occur?

(b) If the probability of an outcome is 1/9 what are the odds against it?

(c) You believe the odds against rain tomorrow are 4 to 1. A friend bets you \$1 it will rain. For the bet to be fair, how much should you pay her if it doesn't rain?

14.12. What is the expected value of the roll of a single die? What is the expected value of the roll of a pair of dice? Look at the result of Problem **14.2**.

14.13. A complex bill is before Congress. One party argues that the bill will produce prosperity to the tune of \$10,000 in the pocket of every American. The opposing party argues that the bill will be a catastrophe, costing every American \$50,000. Experts all agree that no one actually knows what will happen if the bill is passed. A small group of experts attempts to quantify the uncertainty by measuring the probability that the prosperity will occur as planned or that the catastrophe will occur. They estimate that the probability of prosperity is 17%, the probability of catastrophe is 3%, and the probability that the bill will have no effect at all is 80%. If you were in Congress, would you vote for the bill?

14.14. Two gamblers Rico and Carla are playing five rounds of a game for a \$100 winner-takes-all prize. Each player has a 50% chance of winning each round. The winner of the prize is the gambler who wins the most rounds. After the third round, when the score is 2 wins for Rico and 1 win for Carla, the owner of the establishment tells the gamblers that he has received a tip that the police are coming and so they must stop the game and split the prize. How much of the $100 prize should each gambler get?

14.15. Assume that $P = (p_1, p_2, \ldots, p_m)$ and $Q = (q_1, q_2, \ldots, q_n)$ are probability distributions. Show that the mn probabilities $p_i q_j$ also form a probability distribution. (This is why the product formula for independence makes sense.)

14.16. You are visiting the home of an associate who you know has two children.

(a) What is the probability that the associate has two daughters?

(b) How would your answer to part (a) change if your associate told you that his older child was a girl?

(c) How would your answer to part (a) change if your associate told you that he did indeed have at least one daughter?

(d) How would your answer to part (a) change if your associate told you that he did indeed have at least one daughter named "Drizella"?

14.17. "Powerball" is the name of a multistate lottery game in which players bet on a random drawing of six numbered balls. Find out the various types of tickets that are awarded prizes, the payoff for each type of winning ticket, and the probability of each of these payoffs. This information including the probabilities can be found on the world-wide web, as can the value of the jackpot prize, which changes weekly. Compute the expected payoff of a Powerball ticket and compare that to the price of the ticket. How much would the jackpot have to be in order that the $2 price of the ticket is fair?

14.18. Repeat the previous exercise with the "Mega Millions" multistate lottery game. Which lottery, Powerball or Mega Millions, gives a better expected payoff for a ticket?

15

Solving Zero-Sum Games

"Young man, in mathematics you don't understand things. You just get used to them." — John von Neumann

15.0 Scenario

In *The Adventure of the Final Problem* by Sir Arthur Conan Doyle, Sherlock Holmes, aboard a train departing from London to Dover, sees Professor Moriarty on the platform and realizes that Moriarty has seen him. Holmes fears that Moriarty will take an express train to Dover and catch him. His best chance for escape is to get off at Canterbury, but Moriarty has that option too.

If Holmes and Moriarty end up at the same station it is certain that Holmes will be killed. If Holmes goes to Dover but Moriarty stops at Canterbury, Holmes is guaranteed to escape to France and survive. If Holmes stops at Canterbury but Moriarty goes on to Dover, Holmes will temporarily escape, but because he remains in England, he will have a 40% chance of eventually being killed. What should Holmes do?

Model the situation by a two-person constant-sum game with Moriarty as the row player. Check for saddle points and dominated strategies. What is Holmes's prudent strategy? Why is this strategy of little use to him? What is Holmes's counter-prudent strategy? Explore the possibility of Holmes using a mixed strategy. What are the advantages of such an approach?

15.1 The Best Response

In a two-person zero-sum game with a saddle point, a saddle point strategy is both prudent and counter-prudent. But when a game has no saddle point, every outcome provides the opportunity for at least one player to benefit from a unilateral strategy change. As we have seen, this may result in an endless cycle of mind changing and second guessing. What is a player to do?

To address this problem, John von Neumann proved that each player in a two-person zero-sum game always has a mixed strategy that, in a certain precise sense, is best or optimal. The change from pure strategies to mixed strategies comes with both advantages and disadvantages. While mixed strategies give the players many more strategy choices, the price of this freedom is a significant loss of certainty. In effect, a game played using mixed strategies becomes a lottery. The mixed strategy paradigm rests on the Expected Value Principle, and so carries with it any philosophical doubts we may have about expected values. These may be especially troubling if we intend to play the game only once. Nevertheless, we will accept the idea of mixed strategies and proceed bravely. From now on, we will understand a game in a new way. To play, Row and Column will each pick a mixed strategy. We regard this pair of chosen mixed strategies as an outcome, and we regard the resulting expected payoff as if it were the amount actually paid when the game is played.

Suppose that, in a game of Roshambo, Column plays the mixed strategy $Q = (2/7, 4/7, 1/7)$. What is Row's best response?

	Rock	Paper	Scissors
Rock	0	−1	1
Paper	1	0	−1
Scissors	−1	1	0
	2/7	4/7	1/7

Using Lemma 14.8, we see that if Row plays Rock, then

$$E(P_1, Q) = (2/7)(0) + (4/7)(-1) + (1/7)(1) = -3/7,$$

and if she plays Paper then,

$$E(P_2, Q) = (2/7)(1) + (4/7)(0) + (1/7)(-1) = 1/7$$

and if she plays Scissors then,

$$E(P_3, Q) = (2/7)(-1) + (4/7)(1) + (1/7)(0) = 2/7.$$

Thus Scissors is Row's best *pure strategy* response to Q because it gives her the largest expected payoff: 2/7. Could Row do better by playing a mixed strategy? The next result shows that the answer is no.

Lemma 15.1. *There is always a pure strategy among the best responses a player has to any pure or mixed strategy played by the opponent.*

Proof. Suppose P_k is Row's best pure strategy response to Column Q. Then

$$E(P_k, Q) \geq E(P_i, Q), \text{ for all rows } i.$$

and we have by Lemma 14.9 that

$$E(P, Q) = p_1 E(P_1, Q) + p_2 E(P_2, Q) + \cdots + p_m E(P_m, Q)$$

for any Row mixed strategy $P = (p_1, p_2, \ldots, p_m)$. Combining these two, we infer that

$$E(P, Q) \leq p_1 E(P_k, Q) + p_2 E(P_k, Q) + \cdots + p_m E(P_k, Q) = E(P_k, Q).$$

The last equality comes from the fact that $p_1 + p_2 + \cdots + p_m = 1$. It follows that $E(P_k, Q) \geq E(P, Q)$. A similar argument applies to Column. □

In short, the expected payoff for a mixed strategy is always a weighted average of pure strategy payoffs. And an average can never be larger than its largest component.

Now we come to one of the fundamental concepts of game theory.

Definition 15.2. A mixed strategy outcome (P, Q) in a zero-sum game is called an **equilibrium** (or **Nash equilibrium**) if P is a best response to Q, and Q is a best response to P. The corresponding strategies P and Q are called equilibrium strategies.

John von Neumann introduced the concept of an equilibrium for two-person zero-sum games in 1928. Later, in 1950, John Nash studied equilibria for games that are not necessarily zero-sum and that may have more than two players. Nash proved that any game always has at least one such equilibrium. This result, called the Nash Equilibrium Theorem, is the subject of Chapter 17. As a consequence of Nash's work, equilibria in game theory are now generally called Nash equilibria.

In terms of inequalities, a mixed strategy outcome (P, Q) is a Nash equilibrium if and only if

$$E(P, Q) \geq E(R, Q) \text{ for any Row mixed strategy } R, \tag{15.1}$$

and

$$E(P, Q) \leq E(P, S) \text{ for any Column mixed strategy } S. \tag{15.2}$$

When the outcome of a game is a Nash equilibrium, neither player has any incentive to unilaterally change his or her strategy.

The idea of a Nash equilibrium for mixed strategies generalizes the idea of a saddle point for pure strategies.

Lemma 15.3. *A pure strategy outcome (row k, column ℓ) is a saddle point if and only if the corresponding basic mixed strategy outcome (P_k, Q_ℓ) is a Nash equilibrium.*

Proof. Corollary 14.6 shows that $E(P_i, Q_j) = u_{i,j}$ for any row i and column j. Thus the inequality (13.1) holds if and only if the inequality(15.1) does, and the inequality (13.2) holds if and only if the inequality (15.2) does. □

Now let us consider Roshambo again. The calculations before Lemma 15.1 make it clear that the mixed strategy $Q = (2/7, 4/7, 1/7)$ is not a good choice for Column. If Column plays Q, then Row has two responses (Paper and Scissors) that provide her with positive expected payoffs.

But what if Column plays $Q' = (1/3, 1/3, 1/3)$ instead of Q? We can calculate Row's expected payoffs against Q' as follows.

$$E(P_1, Q') = (1/3)(0) + (1/3)(-1) + (1/3)(1) = 0,$$

$$E(P_2, Q') = (1/3)(1) + (1/3)(0) + (1/3)(-1) = 0,$$

and

$$E(P_3, Q') = (1/3)(-1) + (1/3)(1) + (1/3)(0) = 0.$$

Thus $E(P, Q') = 0$ for any Row strategy P. In short, once Column chooses the mixed strategy Q', he neutralizes Row's ability to choose a response that makes a difference in the game.

Definition 15.4. A mixed strategy is called a **neutralizing strategy** if the expected payoff is the same for every possible response by the opponent.

We call an outcome (P, Q) a **neutralizing outcome** if P and Q are both neutralizing strategies.

Lemma 15.5. *A neutralizing outcome in a zero-sum game is a Nash equilibrium.*

Proof. Let (P, Q) be a neutralizing outcome. Since P is a neutralizing strategy for Row, every Column strategy Q is a best response to P. In particular, Q is a best response to P. In the same way, P is a best response to Q. □

The following is von Neumann's main result about two-person zero-sum games.

Theorem 15.6. (***von Neumann's equilibrium theorem***) *Every two person zero-sum game has a Nash equilibrium.* □

We will give a proof of this theorem for the 2-by-2 case later in the chapter. However, for the general proof, we refer to the Nash equilibrium theorem, discussed in Chapter 17. Theorem 15.6 provides us with a new game-playing method called the equilibrium method.

Definition 15.7. The **equilibrium method** for a zero-sum game is the method in which players choose one of their equilibrium strategies.

Von Neumann argued that this method is the one players should generally use when playing a two-person zero-sum game. We will discuss von Neumann's reasons for this in the next section. But first, we apply this method to Roshambo.

Suppose in Roshambo that Row plays $P' = (1/3, 1/3, 1/3)$ and Column plays $Q' = (1/3, 1/3, 1/3)$. As we have seen, P' and Q' are neutralizing strategies, so according to Lemma 15.5, the outcome (P', Q') is a Nash equilibrium. To play Roshambo using the equilibrium method, Row should always play P' and Column should always play Q', which translates to both players choosing Rock, Paper, or Scissors uniformly at random. Playing at random in this way protects a player from an expert opponent who is good at guessing what the player is going to do. Although Roshambo has no saddle points, like all games it does have a Nash equilibrium. This mixed-strategy Nash equilibrium stands in for the nonexistent saddle point.

15.2 Prudent Mixed Strategies

In Chapter 13 we introduced the idea of the guarantee of a pure strategy as the worst case payoff that a player could get by playing it.

Definition 15.8. The **guarantee** of a mixed strategy is the (expected) payoff that results when the opponent plays his or her best response to it.

Lemma 15.1 makes it easy to find mixed strategy guarantees. We need only evaluate the expected payoff against each of the opponent's pure strategy responses and pick the worst case payoff.

Consider the case of Roshambo. We have seen that if Column plays $Q = (2/7, 4/7, 1/7)$, then

$$E(P_1, Q) = -3/7, \quad E(P_2, Q) = 1/7, \text{ and } E(P_3, Q) = 2/7,$$

and because $2/7$ is the largest expected payoff, Row's best response is row 3. It follows that the guarantee for Column using strategy Q is $2/7$ (or in other words, $2/7$ is Column's worst case expected payoff if he plays Q).

On the other hand, Column's guarantee for $Q' = (1/3, 1/3, 1/3)$ is 0. This is because, for each column i, $E(P_i, Q') = 0$. Clearly Q' is a more secure strategy choice for Column than Q is.

Definition 15.9. The **prudent mixed strategy** for a player is the mixed strategy that has the best guarantee.

For Row the prudent mixed strategy is the mixed strategy with the largest guarantee, whereas for Column, it is the one with the smallest guarantee. This idea seems simple enough, but it rests on a subtlety. To find a prudent mixed strategy, a player would need to compare the guarantees of infinitely many mixed strategies. Unfortunately such infinite optimization problems tend to be difficult.

For now, we shall simply assume that every game does have prudent mixed strategies for both players (we will prove this later). Thus we can define the **prudent mixed strategy method** to be the method in which a player plays his or her prudent mixed strategy.

Each player's prudent mixed strategy has a guarantee. We write $\overline{r}$ for Row's prudent mixed strategy guarantee, and we write $\overline{c}$ for Column's prudent mixed strategy guarantee.

Lemma 15.10. *The mixed strategies $P' = (1/3, 1/3, 1/3)$ and $Q' = (1/3, 1/3, 1/3)$ are (respectively) Row's and Column's unique prudent mixed strategies in Roshambo, and their guarantees are $\overline{r} = 0$ and $\overline{c} = 0$.*

Proof. Consider any Row mixed strategy, $P = (p_1, p_2, p_3)$ that is not equal to $P' = (1/3, 1/3, 1/3)$. Then at least one probability, p_1, p_2, or p_3, must be bigger than $1/3$. For example, in $P = (1/4, 1/2, 1/4)$, the probability of Paper satisfies $p_2 = 1/2 > 1/3$. Now Column's best response to Paper is Scissors, or Q_3. If he plays this against P the expected payoff is

$$E(P, Q) = (1/4)(1) + (1/2)(-1) + (1/4)(0) = -1/4 < 0.$$

This is not good for Row. It shows that her strategy P has a negative guarantee. In a similar way, we can show that any other Row strategy P that is not equal to P' has a negative guarantee. On the other hand, if Row plays P' then $E(P', Q) = 0$ for every column response Q. It follows that the guarantee of 0 for P' is better than the negative provided by every other choice P' that Row has. This shows that P' is Row's prudent mixed strategy, and $\overline{r} = 0$. The argument for Column is essentially the same. □

John von Neumann's fundamental insight into zero-sum games was that there is a correspondence between prudent mixed strategies and equilibria.

Theorem 15.11. (***von Neumann's Min-Max Theorem***) *In a two-person zero-sum game every Nash equilibrium* (P, Q) *is doubly prudent with* $\overline{r} = \overline{c}$. *In particular, every such game has prudent mixed strategies for both players. Conversely, every doubly prudent mixed strategy outcome* (P, Q) *is a Nash equilibrium.*

Proof. If (P, Q) is a Nash equilibrium, then we have (15.1) and (15.2), which show that $E(P, Q)$ is the guarantee of P and of Q. In other words, $\overline{r} = \overline{c} = E(P, Q)$. Thus both P and Q are prudent.

Conversely, suppose that (P, Q) is doubly prudent, then $E(P, Q) = \overline{r} = \overline{c}$. Now, since P is prudent, all the payoffs against P must be at least $\overline{r}$. Thus $E(P, S) \geq \overline{r} = E(P, Q)$ for all S, which is (15.2). In a similar way, we obtain inequality (15.1), and it follows that (P, Q) is a Nash equilibrium. □

Theorem 15.11 shows that $\overline{r} = \overline{c}$ in any two-person zero-sum game. We write this number as v, and call it the **value** of the game (so, $v = \overline{r} = \overline{c}$). The min-max theorem show that there is a certain precise sense in which an equilibrium strategy in a two-person zero-sum game is optimal. If Row plays her equilibrium strategy, she can be sure to win at least v on average, and if Column plays his equilibrium strategy, he can be sure to lose no more than v on average. No other strategies can guarantee the players better expected payoffs. An equilibrium strategy in a two-person zero-sum game is both prudent and counter-prudent.

In analyzing a two-person zero-sum game, the goal is often to find an equilibrium strategy for each player and to find the value v. We refer to this as a **solution** of the game. If a game has a saddle point (row i, column j), then the solution is the pure strategy P_i for Row, the pure strategy Q_j for Column, and $v = u_{i,j}$. A two-person zero-sum game with $v = 0$ is called a **fair** game.

The solution of Roshambo is given by $P' = (1/3, 1/3, 1/3) = Q'$, and $v = 0$. In particular, Roshambo is a fair game. This fairness explains why Roshambo is sometimes used in place of coin flipping.

15.3 An Application to Counterterrorism

In December 2002, after the fall of the Taliban, Osama bin Laden and the leadership of al-Qaeda escaped to the border region between Afghanistan and Pakistan. US Special Forces, knowing that al-Qaeda could hide in either country, had to choose where to search. Let us make some estimates about the chances of catching al-Qaeda. We will suppose that if al-Qaeda hid in Afghanistan, and the United States searched there, then the chance of capture would be 60%. But if the United States did not search there, then an escape was certain. On the other hand, if al-Qaeda hid in Pakistan, and the United States searched there, the chance of capture would be 40%. But even if the United States did not search there, we assume there is a 10% chance that Pakistani security forces would capture the leaders of al-Qaeda and turn them over to the United States. These numbers are made up, but at least they seem plausible. Finding such numbers is the job of military analysts.

We model this conflict with the matrix shown below, equating the probability of capture with the payoff to the United States from the game. We call this game **US versus al-Qaeda**.

		al-Qaeda	
		Afghanistan	Pakistan
United States	Afghanistan	60	10
	Pakistan	0	40

This game seems similar to the Battle of the Bismarck Sea from Chapter 13 but the game-theoretic analysis is rather different. Here is the flow diagram.

60	10
0	40

It shows that this game has no saddle points. Common sense tells us that it would be foolish for al-Qaeda to announce where they are hiding, or for the United States to announce where it is searching. But it isn't easy to see what the players should do. Our goal now is to solve and find the equilibrium strategies for both players.

Suppose that al-Qaeda uses the mixed strategy $Q = (1 - q, q)$ for some $0 \leq q \leq 1$. We calculate the expected payoff for each of the US pure strategy responses. The payoff for a search in Afghanistan (row 1) is

$$E(P_1, Q) = 60(1 - q) + 10q,$$

and for search in Pakistan is

$$E(P_2, Q) = 0(1 - q) + 40q.$$

Writing both left-hand sides as E, these equations simplify to

$$E = -50q + 60 \tag{15.3}$$

and

$$E = 40q, \tag{15.4}$$

which are the equations for two lines in the (q, E)-plane. The graphs of these lines are shown in Figure 15.1. The downward sloping line

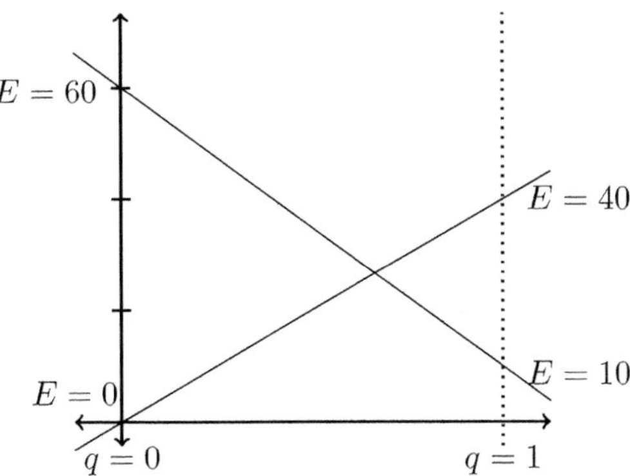

Figure 15.1 Payoffs to the United States for al-Qaeda's mixed strategies

$E = -50q + 60$ corresponds to the US Afghanistan strategy, while the upward sloping line $E = 40q$ corresponds to the US Pakistan strategy. The value of q where the lines cross corresponds to an al-Qaeda neutralizing strategy $Q = (1 - q, q)$. Since every neutralizing strategy is an equilibrium, this will be the equilibrium strategy we are looking for.

The crossing can be found using algebra. Equating the right-hand sides of (15.3) and (15.4), we obtain $-50q + 60 = 40q$, which yields $q = 2/3$. In other words $Q = (1/3, 2/3)$ is al-Qaeda's equilibrium strategy. This strategy advises al-Qaeda to hide in Afghanistan with probability 1/3 and to hide in Pakistan with probability 2/3. The fact that it is a neutralizing strategy means that no matter whether the United States searches in Pakistan or in Afghanistan, the chance of catching al-Qaeda will be the same. That chance, $\overline{c} = 80/3 \approx 26.7$, is the height of the intersection point. In other words, the strategy Q guarantees al-Qaeda that the United States has only a 26.7% chance of catching them. No strategy can make al-Qaeda any safer than that.

Now we start over and analyze the game from the US point of view. Like al-Qaeda, the United States uses a mixed strategy $P = (1-p, p)$. The expected payoffs against al-Qaeda's two pure strategies are $E(P, Q_1) = 60(1-p)+0p = -60p+60$, and $E(P, Q_2) = 10(1-p)+40p = 30p+10$. In this case, the value of p for the neutralizing strategy satisfies $-60p + 60 = 30p + 10$, which implies that $p = 5/9$. Thus $P = (4/9, 5/9)$

is the US equilibrium strategy, and $\overline{r} = 80/3 \approx 26.7$. This strategy guarantees the United States at least a 26.7% chance of catching al-Qaeda. Note that $\overline{r} = \overline{c}$ (as promised by Theorem 15.11). This common value $v = 26.7\%$ is the value of the game.

Osama bin Laden was found in Pakistan in 2011.

15.4 The 2-by-2 Case

We now generalize to an arbitrary 2-by-2 matrix.

Theorem 15.12. *Any 2-by-2 zero-sum game has a Nash equilibrium.*

Proof. Consider the game

	1	2
1	a	b
2	c	d

where we regard a, b, c, and d as given fixed numbers. This is the most general 2-by-2 matrix game.

If the game has a saddle point, then by Lemma 15.3 it has a Nash equilibrium, in which case the theorem is proved. Now consider the case where there is no saddle point. Without loss of generality, we assume that

$$a \geq c. \tag{15.5}$$

(In the opposite case, the proof is similar and left to the reader.) Since there are no saddle points, the flow diagram looks like this:

	1	2
1	a	b
2	c	d

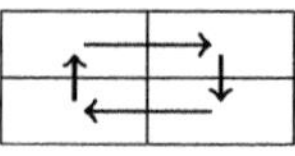

which indicates that

$$a > c \text{ and } d > b. \tag{15.6}$$

Moreover, the horizontal arrows in this flow diagram indicate that

$$b < a \text{ and } c < d. \tag{15.7}$$

Consider a Column mixed strategy $Q = (1 - q, q)$. The expected payoffs for pure strategies 1 and 2 of Row are

$$E(P_1, Q) = a(1 - q) + bq, \tag{15.8}$$

and

$$E(P_2, Q) = c(1 - q) + dq. \tag{15.9}$$

These are the equations of lines in the (q, E)-plane.

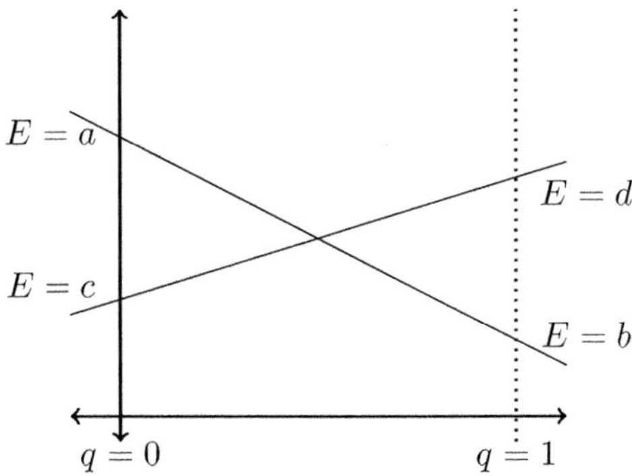

Figure 15.2 The first line goes from a on the E-axis (the line $q = 0$) to b on the vertical line $q = 1$. The second line goes from c on the E-axis to d on the vertical line $q = 1$.

The inequalities (15.6) and (15.7) show these two lines cross (more or less as drawn in Figure 15.2). This crossing occurs where $a(1-q)+bq = c(1 - q) + dq$. Solving for q, we obtain

$$q = \frac{a - c}{(a - c) + (d - b)}.$$

Plugging back into (15.8) and (15.9), we obtain

$$E(P_1, Q) = E(P_2, Q) = \frac{ad - bc}{(a - c) + (d - b)},$$

so Q is a neutralizing strategy for Column.

Repeating everything for the Row mixed strategy $P = (1 - p, p)$, we obtain

$$p = \frac{a - b}{(a - b) + (d - c)},$$

and

$$E(P, Q_1) = E(P, Q_2) = \frac{ad - bc}{(a - b) + (d - c)},$$

so that P is a neutralizing strategy for Row. Thus by Lemma 15.5, the outcome (P, Q) is a Nash equilibrium. □

Corollary 15.13. *The solution to a 2-by-2 zero-sum game without saddle points is the mixed strategy*

$$P = \left(\frac{d-c}{(a-b)+(d-c)}, \frac{a-b}{(a-b)+(d-c)} \right)$$

for Row and the mixed strategy

$$Q = \left(\frac{d-b}{(a-c)+(d-b)}, \frac{a-c}{(a-c)+(d-b)} \right)$$

for Column. The value of the game is

$$v = \frac{ad-bc}{(a-b)+(d-c)}.$$

□

There is an easy trick for applying the formulas in Corollary 15.13 that we refer to as the **cross-over algorithm**. The preliminary step is to verify that the game has no saddle point, because if a game does have a saddle point, the formulas in Corollary 15.13 do not apply. Next, swap the positions of a and d, and swap the positions of b and c. Then change the sign of b and c. We call the resulting matrix the **cross-over matrix**.

a	b
c	d

$\longrightarrow$

d	$-c$
$-b$	a

Now take the cross-over matrix and add the entries in each row and in each column. Write these numbers next to their respective rows and columns, being careful to preserve the signs (positives and negatives).

d	$-c$	$d-c$
$-b$	a	$a-b$
$d-b$	$a-c$	

Next, let N be the sum of the two numbers beside the two rows or below the two columns. Note that the two ways of computing this,

$$N = (d-c)+(a-b) \qquad \text{and} \qquad N = (d-b)+(a-c),$$

are actually the same (a fact implicit in Corollary 15.13). Dividing all the numbers outside the matrix by N gives the probabilities in the mixed strategies.

d	$-c$	$(d-c)/N$
$-b$	a	$(a-b)/N$
$(d-b)/N$	$(a-c)/N$	

Finally, calculate $\Delta = ad - bc$, a quantity called the **determinant** of the game matrix. Then the value of the game is given by $v = \Delta/N$. We emphasize that it is important to keep careful track of the signs throughout these calculations.

We illustrate the cross-over algorithm on the game US vs. al-Qaeda. Since we have already verified that this game has no saddle point, we proceed with our calculations. Crossing over and introducing minus signs yields

60	10
0	40

$\longrightarrow$

40	0
−10	60

.

Summing the rows and columns of the cross-over matrix gives

40	0	40
−10	60	50
30	60	

and $N = 30 + 60 = 90$. Dividing by N gives equilibrium strategies

40	0	40/90
−10	60	50/90
30/90	60/90	

or, in other words, $Q = (1/3, 2/3)$ for al-Qaeda and $P = (4/9, 5/9)$ for the United States. Finally $\Delta = 2400 - 0 = 2400$ so $v = 2400/90 = 80/3 \approx 26.7$.

There is an alternative way to interpret the US mixed strategy in the US vs. al-Qaeda game. Suppose al-Qaeda is concerned only with hiding Osama bin Laden, and also suppose that the United States will send in a large force to search for him. The conventional way to implement the mixed strategy $P = (4/9, 5/9)$ is either to send all the troops to Afghanistan or to send all the troops to Pakistan, with with respective probabilities 4/9 and 5/9. However, there is another interpretation. We may think of what orders to give to each soldier as a strategic decision. The mixed strategy suggests that a soldier should be sent to Afghanistan with probability 4/9 and to Pakistan with probability 5/9. So with a force, say, of 9000 to work with, the United States could send 4000 soldiers to Afghanistan and 5000 soldiers to Pakistan.

As a final example, consider the finger game called **finger gambling**. On the count of three, Millie and Al simultaneously extend a hand revealing one or two fingers. If both show one finger, Millie pays Al \$2. If both show two fingers, Millie pays Al \$8. If Millie shows one finger to Al's two, Al pays her \$3. If Millie shows two fingers to Al's one, Al pays her \$5. How should Millie and Al play? Who has the advantage in this game?

If we designate Millie as Row and Al as Column, then the game is given by

	One	Two
One	−2	3
Two	5	−8

The reader can check the flow diagram to see that neither player has a saddle point. Here is the cross-over matrix with the row and column sums

−8	−5	−13
−3	−2	−5
−11	−7	

which are all negative. Thus $N = -18$. Note that the probabilities for the optimal mixed strategies are positive (as nonzero probabilities must be) since they are quotients of two negative numbers.

−8	−5	13/18
−3	−2	5/18
11/18	7/18	

Finally $\Delta = 1$ so $v = 1/(-18) = -1/18$. The fact that v is negative, but small, means that this game is slightly favorable to Al.

15.5 Exercises and Problems

15.1. Solve the three games in Problem **13.1**.

15.2. Solve the game in Problem **13.6**.

15.3. For the game shown here, find a saddle point solution and a second solution made of mixed strategies that are not basic.

3	3	3
3	0	6
3	6	0

15.4. Solve the game in Problem **13.10**. Which player does this game favor?

15.5. Solve the game in Problem **13.11**. Which player does this game favor?

15.6. Solve the game in Problem **13.12**. Which player does this game favor?

15.7. Give a complete analysis of the 2-by-3 game with the following payoff matrix.

-1	7	-3
8	-12	5

15.8. Give a complete analysis of the 3-by-2 game with the following payoff matrix.

5	7
6	3
8	4

15.9. The day before a close presidential general election, the incumbent George Rose is leading his challenger John Collins by a small margin. In two large states, Florida with 29 electoral votes and Ohio with 18 electoral votes, the race is too close to call. Each candidate has enough time to visit only one of the two states. The pundits have said that if just one candidate visits a state, that state is a sure win for the candidate. On the other hand, if both candidates visit the same state, they will each have a 50% chance of winning there. If neither candidate visits Florida, the chance of Rose winning there is 75%. On the other hand, if neither candidate visits Ohio, the chances of Collins winning there is 75%. Set this up as a two-person constant-sum game. After analyzing the game, advise the candidates what to do. (Hint: There are four outcomes. For each outcome, first determine Rose's chance of winning Ohio and his chance of winning Florida. As the payoffs use the expected number of electoral votes a candidate will get.)

15.10. Suppose that Sarah Palin and Barack Obama are campaigning for the office of President in 2012. They each have time for one visit to either Alaska or Hawaii, which represent 3 and 4 electoral votes, respectively. Palin is favored in Alaska; Obama is favored in Hawaii. If neither candidate visits a state, or if only the favored candidate visit a state, the favored candidate is guaranteed to win. If both candidates visit the same state, the favored candidate is expected to win with probability 3/4. If only the unfavored candidate visits a state, each candidate has a probability of winning of 1/2. Model this situation as a two-person constant-sum game. Analyze the game. What does game theory predict?

15.11. Prove that a 2-by-2 zero-sum game with no saddle point is fair if and only if the determinant is zero. If a 2-by-2 zero-sum game with a saddle point is fair, does its matrix always have determinant zero? If a 2-by-2 matrix has determinant zero, is the game always fair?

15.12. Consider a matrix game

a	b
c	d

which we assume has value v. Construct a new game:

$-a$	$-c$
$-b$	$-d$

(a) What is the value of the new game?
(b) Prove that if the original game has a saddle point, the new game has one too.

15.13. Consider a matrix game

a	b
c	d

and construct a new game:

$-a$	$-b$
$-c$	$-d$

(a) Assume the original game has value v and no saddle point. Determine the value of the new game.
(b) Prove that if the original game has a saddle point, the new game has one too.
(c) Show that if the original game has a saddle point, the value of the new game may favor the same player as the original game or the other player.

15.14. Prove the inequality $r \leq \overline{r} = \overline{c} \leq c$ (between the prudent pure strategy and prudent mixed strategy guarantees). Explain, in words, what it is saying. Under what circumstances will one (or both) of the inequalities actually be equality?

15.15. Suppose a 2-by-3 or 3-by-2 game has no saddle point. A theorem says that the equilibrium strategy for the player with three strategies assigns a positive probability to exactly two of the three pure strategies. Consider, for example, the case below where Row has three strategies.

1	0
2	−3
−1	4

Solve this game as follows. First solve each of the three 2-by-2 games obtained by deleting one of Row's pure strategies, and find the Nash equilibrium (P, Q) and value v. In each of the three cases, play Column's equilibrium strategy Q against the pure strategy that Row deleted. In exactly one of the three cases, the payoff will be less than v. This Q is then Column's equilibrium strategy in the 3-by-2 game. The corresponding Row equilibrium strategy in the 3-by-2 game is P, interpreted as a strategy in the original game. The value of the game is v.

15.16. Solve the following game

−2	4
3	−1
−1	0

15.17. Solve the following game

−1	−1	−2
2	−1	−2
−1	0	3

15.18. Alex and Olaf are in an airport lounge waiting for their flights. "I know a good game" says Alex. "We point fingers at each other; either one or two. If we match with one finger, you buy me a mojito. If we match with two fingers, you buy me two mojitos. If we don't match, I'll let you pay me a dollar."

At first Olaf is not interested, but then he says "I'll play if you pay me four dollars before each game to help cover the cost of those five dollar mojitos."

"I'll pay you three-seventy-five," says Alex.

"Well okay," says Olaf. "You should really be paying more, but I suppose the game won't last long."

Model this scenario as a constant-sum game and explain the significance of the proposed \$4.00 payment and the \$3.75 payment that Alex agrees to make. Would you have advised Olaf to accept Alex's counter-offer?

16

Conflict and Cooperation

"It is our choices . . . that show what we truly are, far more than our abilities." — J. K. Rowling

16.0 Scenario

Rosie and Carl have just bought a new car and are ready to go for a ride, but each wants to be the first to drive. Each may either say "I want to drive" (the more aggressive stance) or "Please — you drive first" (the more polite stance). Rosie and Carl agree that the best outcomes are those that peacefully determine who gets to drive, even though such an outcome would mean that one of them would have to settle for being a passenger. Both agree that the *last* choice is fighting about who gets to drive. We call this game New Car.

Assign numerical scores to each of the four outcomes from Rosie's point of view. The numbers may be arbitrary, but higher scores should be assigned to the outcomes that Rosie regards as better. For example, you might want to award Rosie 10 points if she gets to Drive and Carl agrees to Ride, but assign her -10 points if they fight, etc. Put the scores into a 2-by-2 matrix like the one here.

	Drive	Ride
Drive		
Ride		

This is the game as Rosie sees it. We call this "Rosie's game". Next, in a similar matrix, write the numerical scores for the game as Carl sees it, again using larger numbers for outcomes that Carl prefers. In other words, obtain Carl's game.

Looking at Rosie's game, determine her naive strategy and prudent strategies. Also, in each column, draw the vertical arrow that points to Rosie's best outcome. These arrows show Rosie's best response to each of Carl's strategy choices. Now answer the same questions for Carl's game keeping in mind that in this game (unlike in zero-sum games) Carl prefers larger numbers.

Finally copy Rosie's vertical arrows and Carl's horizontal arrows, into a single flow diagram, and use this to find the two saddle points for the game New Car. Are these saddle points good outcomes? How could the players achieve them?

16.1 Bimatrix Games

In the previous chapters, we studied zero-sum games and constant-sum games. In this chapter we broaden our point of view and study a more general type of game. A **bimatrix** is a rectangular array in which two numbers appear in each cell. We can associate a two-person game with a

0, 3	−1, 2	0, −2
1, 0	0, 1	0, 0
2, −1	−1, −6	0, 3

Game 16.1 A bimatrix.

bimatrix in much the same way we did with matrices. As before, we call the two players "Row" and "Column." To play such a **bimatrix game**, Row secretly chooses a row and Column independently chooses a column. Then both players simultaneously reveal their choices, determining an outcome. The difference in the bimatrix case is that there are two payoffs corresponding to each outcome. The first payoff is the payoff to Row and the second payoff is the payoff to Column. We imagine that the money for these payoffs comes from a third party, like a banker.

We can extend the notation of Chapter 13 to the bimatrix case. Suppose Row chooses row i and Column chooses column j. Then, as before, we write $u_{i,j}$ for the payoff to Row, but in addition, we write $v_{i,j}$ for the payoff to Column.

The goal for each player in a bimatrix game is to obtain the largest possible payoff. In particular, Row wants to make $u_{i,j}$ as large as possible, and Column wants to make $v_{i,j}$ as large as possible. A key assumption is that the players do not care about the payoff that their opponent gets. This is what makes bimatrix games different from matrix games. In a matrix game, one player's gain always equals the other player's loss. In a bimatrix game, however, it possible for certain outcomes to be good for both players, or bad for both players, or anything in between.

Here is an example of how the payoffs work in Game 16.1. Suppose that Row chooses row 2 and Column chooses column 1, so that the outcome is (row 2, column 1). Then Row receives the payoff $u_{2,1} = 1$, and Column receives the payoff $v_{2,1} = 0$.

Note that when the outcome in Game 16.1 is (row 2, column 1), the sum of the payoffs to the two players is $u_{2,1} + v_{2,1} = 1$, which is not zero. This shows that Game 16.1 is not zero-sum. But also, for the outcome (row 1, column 1) the sum is $u_{1,1} + v_{1,1} = 3$, so the game is not constant-sum either. This is why it must be written as a bimatrix game. On the other hand, all zero-sum games and all constant-sum games (i.e., matrix games) can also be written as bimatrix games. For example, here is how the Battle of the Bismarck Sea game (from Chapter 13) looks as a bimatrix game.

$2, -2$	$2, -2$
$1, -1$	$3, -3$

Notice that $u_{i,j} + v_{i,j} = 0$, or equivalently, $v_{i,j} = -u_{i,j}$, for every row i and column j. And here is how the US versus al-Qaeda game (from Chapter 15) looks as a bimatrix game

$60, 40$	$10, 90$
$0, 100$	$40, 60$

Here, $u_{i,j} + v_{i,j} = K$ for all i and j, where $K = 100$.

16.2 Guarantees, Saddle Points, and All That Jazz

Many of the basic ideas that we discussed for zero-sum games extend to bimatrix games with only a few minor technical changes.

As before, the **guarantee** of a strategy is defined to be the worst payoff that a player can get by playing it. For row i, it is the smallest Row payoff $u_{i,j}$ in that row, and for column j, it is the smallest Column payoff $v_{i,j}$ in that column. For each player, the **prudent strategy** is the strategy that has the largest guarantee, and the **prudent strategy method** is the game-playing method in which a player plays his or her prudent strategy.

Much as in the zero-sum case, the **min-max diagram** provides a convenient way to find guarantees and prudent strategies for both players. Here is what we get for Game 16.1.

0, 3	−1, 2	0, −2	−1	
1, 0	0, 1	0, 0	0	←
2, −1	−1, −6	0, 3	−1	
−1	−6	−2		
↑				

Row's prudent strategy is row 2, which has a guarantee of $r = 0$, and Column's prudent strategy is column 1, which has a guarantee of $c = -1$.

Since a matrix game can also be written in bimatrix form there is some potential confusion here. We can illustrate the two approaches with the US versus al-Qaeda game by computing guarantees and prudent strategies both ways.

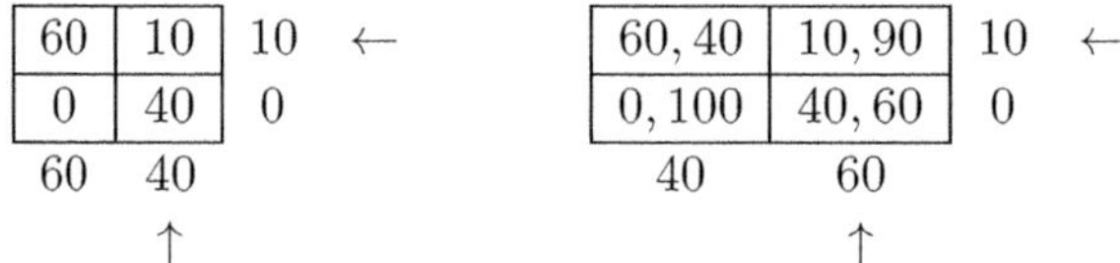

60	10	10	←
0	40	0	
60	40		
	↑		

60, 40	10, 90	10	←
0, 100	40, 60	0	
40	60		
	↑		

Figure 16.1 The matrix and bimatrix way of computing guarantees.

Notice that in Figure 16.1 we have $c = 40$ (a 40% chance al-Qaeda will be caught) in the matrix game version of the calculation, but $c = 60$ (a 60% chance that al-Qaeda will escape) in the bimatrix version. But these are just two different ways of expressing the same thing. Matrix games always frame the outcomes from Row's point of view. Bimatrix games treat Row and Column symmetrically. In the bimatrix world, Column has his own payoffs to maximize, while in the zero-sum matrix world, Column attempts to minimize his opponent's payoffs.

Next, we consider saddle points. As in the case of a matrix game, these are based on the **best response**. The best response to an opponent's strategy is the player's strategy that, when played against the opponent's strategy, gives the player the best payoff. As before, we can illustrate best responses using a **flow diagram**. Here is the flow diagram for Game 16.1.

0, 3	−1, 2	0, −2
1, 0	0, 1	0, 0
2, −1	−1, −6	0, 3

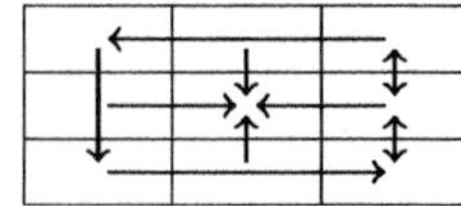

Notice that in the bimatrix case the vertical arrows point to the largest first number in each payoff pair, and the horizontal arrows point to the largest second number (when there is a tie, two sided arrows are drawn). For example, the down arrow in the column 1 shows that Row's best response to column 1 is row 3. The two arrows pointing in to the middle entry in row 2 show that Column's best response to row 2 is column 2.

As the reader might expect, a **saddle point** in a bimatrix game is an outcome for which each player's strategy is a best response to the opponent's strategy. Thus, just as before, an outcome is a saddle point if and only if, in the flow diagram, all the one-sided arrows in its row and in its column point toward it. Equivalently, an outcome (row k, column ℓ) is a saddle point if and only if $u_{k,\ell} \geq u_{k,j}$ for each column j, and $v_{k,\ell} \geq v_{i,\ell}$ for all rows i.

We can see from the flow diagram that Game 16.1 has two saddle point outcomes (row 2, column 2) and (row 3, column 3). Thus the Row strategies row 2 and row 3, and the Column strategies column 2 and column 3 are saddle point strategies (because they correspond to saddle points).

For zero-sum games, saddle points are both prudent and counter-prudent. As we will see, the same is not necessarily true for non-zero-sum games. Thus, although saddle points are still important, it is less clear that they are always a good strategy choice.

Finally we consider dominance. We say row i dominates row j if each row payoff in row i is at least as large as the corresponding row payoff in row j and one or more is strictly larger. In particular, row i dominates row j if $u_{i,k} \geq u_{j,k}$ for all k, and $u_{i,k} > u_{j,k}$ for at least one k. This is almost exactly the same as for zero-sum games. However, for Column the criterion is a little different than in the zero-sum case. In particular, column k dominates column ℓ if $v_{i,k} \geq v_{i,\ell}$ for all i and $v_{i,k} > v_{i,\ell}$ for at least one i. In other words, Column should look at his own payoffs, rather than Row's, when deciding whether one of his strategies dominates another. But even though the procedure is a little different, the interpretation is the same. It seems reasonable that neither player should choose a dominated strategy because in all cases the strategy that dominates it is at least as good, and in at least one case it is better.

16.3 Common Interests

To what extent do the players in a game have interests in common? In this section we study the balance between conflict and cooperation. For each player in a game, we can rank the outcomes from best to worst according to the payoffs they earn. For example, Row's payoffs in Game 16.1, written in descending order are 2, 1, 0 and -1. We rank these as her 1st, 2nd, 3rd, and 4th choices. Similarly, Column's payoffs are 3, 2, 1, 0, −1, −2 and −6, and we rank them as his 1st, 2nd, 3rd, 4th, 5th , 6th, and 7th choices. Numbers like 1st, 2nd, and 3rd, which are used for describing the order of things, are called **ordinal numbers**. When we write ordinal numbers in a bimatrix, we obtain what is called an **ordinal game**. Here is the ordinal version of Game 16.1.

0, 3	−1, 2	0, −2
1, 0	0, 1	0, 0
2, −1	−1, −6	0, 3

3rd, 1st	4th, 2nd	3rd, 6th
2nd, 4th	3rd, 3rd	3rd, 4th
1st, 5th	4th, 7th	3rd, 1st

In terms of the corresponding ordinal games, we classify bimatrix games into three types.

Definition 16.1. A **coordination game** is a bimatrix game which, when written as an ordinal game, the two players' preference orders are exactly the same. In other words, if Row prefers one outcome to another, then so will Column. A **strictly competitive game** is a bimatrix game which, when written as an ordinal game, the players' preference orders are exact opposites. In other words, if Row prefers one outcome to another, then Column will prefer the other to the one. Finally, a **mixed motive** game is any game that is neither a coordination game nor a strictly competitive game.

It is easy to see that Game 16.1 is a mixed motive game. On the other hand, if we look at the ordinal version of the Battle of the Bismarck Sea from Chapter 13

2, −2	2, −2
1, −1	3, −3

2nd, 2nd	2nd, 2nd
3rd, 1st	1st, 3rd

we see that it is strictly competitive. This is a feature of all zero-sum games.

Proposition 16.2. *Any two-person zero-sum or constant-sum game is strictly competitive.*

Proof. Since the game is constant sum, $u_{i,j} + v_{i,j} = u_{k,\ell} + v_{k,\ell}$ for any two outcomes (row i, column j) and (row k, column ℓ). Equivalently $u_{i,j} - u_{k,\ell} = v_{k,\ell} - v_{i,j}$. If Row prefers (row i, column j) to (row k, column ℓ) then $u_{i,j} \geq u_{k,\ell}$ or equivalently $u_{i,j} - u_{k,\ell} \geq 0$. It follows that $v_{k,\ell} - v_{i,j} \geq 0$ which says $v_{k,\ell} \geq v_{i,j}$, or in other words, Column prefers (row k, column ℓ) to (row i, column j). □

Let us now consider an example of a coordination game. The following game, called **Meet in New York** is adapted from the work of economist Thomas Schelling, who won the Nobel Prize in Economics in 2005 for his work in game theory.

Two friends plan to meet in New York, but have neglected to discuss where to meet. To simplify the story, we imagine there are just three possible meeting places: Times Square, Grand Central Station, and the Empire State Building. Assume that cell phone communication is impossible.

For each of the friends, the first choice is a successful rendezvous, no matter where it occurs. The second (i.e., last) choice is to end up at different locations. We assign point values to these outcomes: 1 for a meeting and 0 for a failure to meet. The game we get is

	Times Square	Grand Central	Empire State
Times Square	1, 1	0, 0	0, 0
Grand Central	0, 0	1, 1	0, 0
Empire State	0, 0	0, 0	1, 1

Game 16.2 Meet in New York

Definition 16.3. Two outcomes are **equivalent** if both players agree that they are equally good. For a bimatrix game with Row payoffs Row $u_{i,j}$ and Column payoffs $v_{i,j}$, (row i, column j) is equivalent to (row k, column ℓ) provided $u_{i,j} = u_{k,\ell}$ and $v_{i,j} = v_{k,\ell}$. The set of all outcomes that are equivalent to a given outcome is called an **equivalence class**.

In the Meet in New York game, there are two equivalence classes. One consists of the three good outcomes in which the two friends meet. The other consists of the remaining six outcomes in which they do not meet. By drawing the flow diagram we see that the good outcomes are all saddle points.

1, 1	0, 0	0, 0
0, 0	1, 1	0, 0
0, 0	0, 0	1, 1

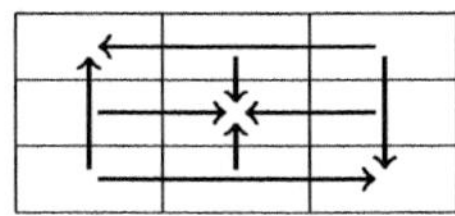

Sometimes the best way to play a coordination game is to use the naive strategy method. Recall that an outcome is called a player's **primary outcome** if it is that player's first choice. A player's **naive strategy** is the strategy corresponding to the player's primary outcome. In a coordination game, a primary outcome for one player will also be a primary outcome for the other. We call such an outcome **doubly primary**. These are the good outcomes in Meet in New York. Unfortunately, the naive approach does not help in Meet in New York because all strategy choices are naive. The difficulty here is a problem of coordination. If a player incorrectly guesses what strategy the other player has chosen, the players will not meet.

Coordination problems are common in many situations. For example, if two people are walking quickly toward each other along a narrow corridor they need to decide whether to swerve to the left or to the right. This is a coordination problem. A coordination problem was also the basis of a 1960s TV game show called the Newlywed Game, in which couples were separated and asked questions testing how well they knew their spouses (e.g., "How many children will your wife say she wants to have?").

Over the years, Schelling asked his students to play Meet in New York. The most common choice was "Grand Central Station". Schelling called outcomes like this **focal points**. He observed that focal points exist in many cities (e.g., the Eiffel Tower in Paris) but not in all cities (e.g., none in Washington, DC). Focal points also exist in some other real life coordination games. For example, in the game where two people approach each other walking down a corridor, most people instinctively swerve to the right, as suggested by the rules of driving. (A problem occurs when one player is from England or Japan, where cars drive on the left.) Similarly, social conventions like "ladies first" exist, at least in part, to avoid coordination problems.

We can model Schelling's focal point in Meet in New York by agreeing that a meeting at Grand Central Station is more preferable then meeting in the other two locations and score it a little higher as a result. This changes the game to:

	Times Square	Grand Central	Empire State
Times Square	1, 1	0, 0	0, 0
Grand Central	0, 0	2, 2	0, 0
Empire State	0, 0	0, 0	1, 1

Game 16.3 Focused Meet in New York.

In this modified version of the game there is a unique doubly-primary outcome. The naive strategy has now become the obvious choice for both players.

Sometimes, even if the doubly primary outcome is not unique, the game still presents some obviously good strategy choices. An example of such a situation is illustrated in Game 16.4. There, the four equivalent doubly primary outcomes are arranged in such a way that if each

2,2	0,0	2,2
0,0	1,1	0,0
2,2	0,0	2,2

Game 16.4 Four interchangeable primary outcomes.

player chooses a corresponding strategy, the outcome is guaranteed to be doubly primary. In this case we say the doubly-primary outcomes are interchangeable.

Definition 16.4. A set S of outcomes of a game is called **interchangeable** if whenever each player chooses a strategy corresponding to one of the outcomes, the resulting outcome is guaranteed to be in the set S.

The trouble with the original version of Meet in New York is that the three equivalent primary outcomes are not interchangeable. In fact, this is the crux of the coordination problem. The language of equivalent and interchangeable outcomes gives us a new way to express one of the crucial ideas from Chapters 13 concerning saddle points in a matrix game.

Proposition 16.5. *In a two-person zero-sum or constant-sum game, the collection S of all saddle point outcomes is interchangeable and any two outcomes in S are equivalent. For a non-zero-sum game, both of these statements may be false.*

Proof. The equivalence of the set of saddle point outcomes of a zero-sum game follows from Corollary 13.8. Their interchangeability follows from Corollary 13.9. Game 16.1 shows that both of these properties may fail for non-zero-sum games. □

The challenge in playing a coordination game comes from the rule that players may not confer or negotiate prior to playing a game. Although the players may try to infer what their opponent is going to do, they cannot know. In zero-sum games, there was never any need to enforce this restriction because in purely competitive situations there is no advantage to be gained from compromise or negotiation.

16.4 Some Famous Games

Now that we have discussed strictly competitive games and coordination games, it is time to consider some examples of mixed motive games.

The following game, called **Battle of the Sexes**, goes back to the 1950s. Row and Column want to go out on a date, but unfortunately, they have somewhat different tastes. Row prefers to go to the ballet while Column prefers to go to a hockey game. Nevertheless, both Row and Column would rather go out together than to go out apart — this is their highest priority. On the day before their date, Row and Column discuss a hockey game and a ballet, but they neglect to finalize their plans. At the last minute, each has to choose which of the two events to attend.

We model this scenario as a bimatrix game with "Ballet" and "Hockey" as the two strategy choices for both players. To begin with, we describe each of the four outcomes in words

	Hockey	Ballet
Ballet	Go alone to own event.	Go together to Row's event.
Hockey	Go together to Column's event.	Go alone to other's event.

Each player's 1st choice is to go — together with their date — to their own favorite event. Each player's 2nd choice is to go together with their date to the other's favorite event. Of course, both players would prefer to go to their own favorite event alone (the 3rd choice) than to go alone to the other's favorite event (the 4th choice). This results in the following ordinal game:

	Hockey	Ballet
Ballet	3rd, 3rd	1st, 2nd
Hockey	2nd, 1st	4th, 4th

To make a bimatrix game out of this, we assign payoffs to these outcomes. Since going out together brings at least some degree of happiness, we assign these outcomes positive payoffs: 10 points for a favorite event, 5 points for the 2nd choice. Let's say that going out alone is neutral (0 points), as long as a player goes to his or her favorite event, but going out alone to the other player's favorite event results in unhappiness (-5 points).

Both players agree that two of the outcomes in this game are better than the other two. Yet it is not quite a coordination game because they disagree about which is best and which is only second best.

From the flow diagram, we see that the two good outcomes are both saddle points.

	Hockey	Ballet
Ballet	0, 0	10, 5
Hockey	5, 10	−5, −5

Game 16.5 Battle of the Sexes.

	Hockey	Ballet
Ballet	0, 0	10, 5
Hockey	5, 10	−5, −5

↓ ⟶	↑
⟵	

One of these saddle points is Row's primary outcome and the other is Column's primary outcome. They are not equivalent — the players value them differently — even though both are reasonably good for both players. Because of their diagonal locations, it is also the case that these two saddle points are not interchangeable. In fact, this game has a coordination problem, and that problem is the central conflict in the game.

How should the players play a game like this? One approach the players could try is their naive strategies. For Row this is Ballet and for Column it is Hockey. But if both players play their naive strategies, Row goes to ballet and Column goes to hockey. In other words, they end up alone.

		↓	
		Hockey	Ballet
→	Ballet	0, 0	10, 5
	Hockey	5, 10	−5, −5

Another approach would be to try their prudent strategies. However, using the min-max diagram, we see that the prudent strategy is the same as the naive strategy in this game.

	Hockey	Ballet		
Ballet	0, 0	10, 5	0	←
Hockey	5, 10	−5, −5	−5	
	0	−5		
	↑			

Perhaps this outcome — each player going alone to their favorite event — should be considered a reasonable compromise. The players are, however, not likely to consider this outcome to be acceptable. After all, both players agree that either one of the two saddle points is preferable. Both would rather go out together than apart, even if it means going to the other's event. In other words, switching to a saddle point would increase the payoff to each player, although they would not agree about which saddle point is best.

There is a simple diagram, called the **payoff polygon**, that can reveal critical information about a bimatrix game. To draw this diagram, we plot each outcome as a point in the (p, q)-plane, using Row's payoff as the p-coordinate of a point, and Column's payoff as the q-coordinate. Once we have plotted these points, we draw the smallest possible convex polygon that contains all of them. The payoff polygon consists of the set of payoffs for all possible outcomes of the game. These outcomes include both pure and mixed strategy choices by the players, as well as some situations (that we will discuss later) in which the players do not choose their strategies independently.

In the case of Battle of the Sexes, the points to be plotted are $(0, 0)$, $(10, 5)$, $(5, 10)$, and $(-5, -5)$. The corresponding payoff polygon, which is a triangle, is shown in Figure 16.2.

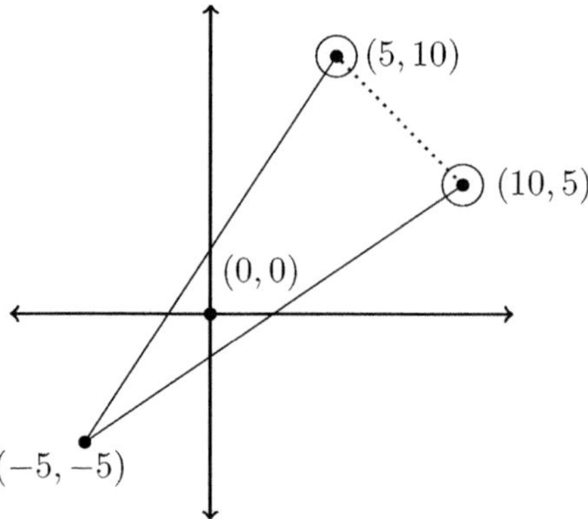

Figure 16.2 The payoff polygon for Battle of the Sexes with saddle points circled. The point $(0, 0)$ is inside the convex polygon.

Row prefers outcomes that have higher values of p. Informally, we say she prefers points in the payoff polygon that lie further north. Row is indifferent about points that lie along the same horizontal line. Similarly, since Column favors outcomes with higher q values, he prefers points that lie further east, and is indifferent about points that lie on the same vertical line. An outcome is called **Pareto inferior** if there is some other outcome that both players prefer, or that one player prefers and the other is indifferent about. In terms of the payoff polygon, an outcome is Pareto inferior if there is some other outcome that can be reached by going north, east, or any direction between north and east. An outcome is called **Pareto optimal** if it is not Pareto inferior to any other outcome.

No point in the interior of the payoff polygon (i.e., not on an edge) can be Pareto optimal since there will always be another point in the polygon that lies to its north and east. We call those edges of the payoff polygon that correspond to Pareto optimal outcomes the **Pareto optimal set**. Moving clockwise around the outside of the payoff polygon, the Pareto optimal set consists of the segments that go both to the right and down. Informally, we call these the northeastern facing segments. Clearly the players should never settle for a Pareto inferior outcome, since at no expense they can move to a Pareto optimal one that at least one player will prefer. This is called the **Pareto principle**. Like the closely related Pareto criterion from voting theory, it is named after the Italian economist Vilfredo Pareto (1848–1923). Recall that in Chapter 3 we argued that a voting method should not elect a candidate if there is a different candidate that every voter prefers.

The Pareto optimal set in Battle of the Sexes is the line segment from $(5, 10)$ to $(10, 5)$. It is shown as a dotted line in Figure 16.2. Note that the two saddle points are the Pareto optimal, but the doubly prudent outcome, with payoffs $(0, 0)$ is Pareto inferior to both saddle points, as is the last-choice outcome, with payoffs $(-5, -5)$. This shows that the two saddle points are the only Pareto optimal pure strategy outcomes. Thus, according to the Pareto principle, we should reject the doubly prudent outcome (as well as the last-choice outcome).

It is interesting to consider what happens for zero-sum and constant-sum games.

Theorem 16.6. *In a zero-sum or constant-sum game, every outcome is Pareto optimal.*

Proof. Since a zero-sum or constant-sum game is strictly competitive, an increase in the payoff of one player automatically means a decrease in the payoff of the other. □

A different way to understand Theorem 16.6 is by drawing the payoff polygon. Figure 16.3 shows the payoff polygon for the bimatrix version of Battle of the Bismarck Sea. In this case, the payoff polygon is "degenerate", which is to say that it is actually a line segment. Since the game is zero-sum, all the payoffs (p, q) lie along the line $p + q = 0$, which has slope -1, confirming the conclusion of Theorem 16.6.

Unfortunately, dismissing the doubly prudent outcome in Battle of the Sexes, leaves us to face a coordination problem. As is often the case with non-zero-sum games, there may actually be no really good resolution to this dilemma. In other words, the conflict cannot be solved by

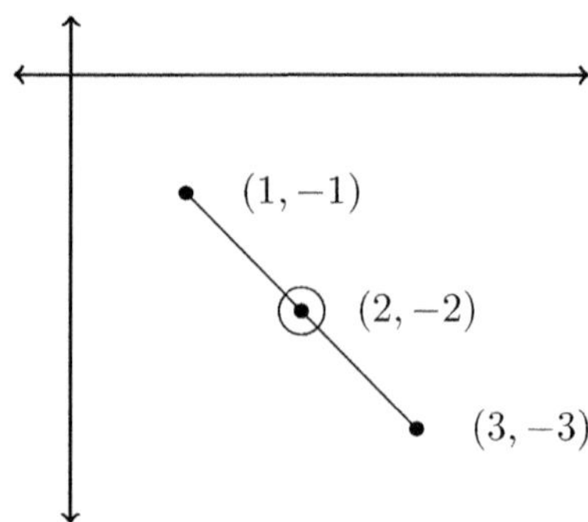

Figure 16.3 The payoff "polygon" for Battle of the Bismarck Sea is not really a polygon. It is collapsed into a line segment that lies entirely along the Pareto optimal set. The saddle point $(2, -2)$ at (North, North) is circled.

mathematics alone, and there is no "solution" to the game like the solutions we developed for zero-sum games. There are, however, some ways one or both players can improve their chances for a better resolution.

One way for a player to prevail (admittedly, by breaking the rules) is to announce her intended move in advance. This is called **preemption**. For example, Row might tell Column, "I am going to the ballet no matter what you do." Then, in order to maximize his payoff, Column would have to go to the ballet too. However, common sense tells us that preemption will sometimes fail to produce the desired result. Rather than responding as intended, Column may be angry and decide to go to the hockey game by himself. Although we assume players behave in a rational way, this may not always be the case in reality.

Another idea that Row and Column could use to resolve the conflict is to negotiate (which again breaks the rules). For example, they could decide whether to go to ballet or hockey by a coin toss. Or they might agree to alternate between ballet and hockey on different nights. These compromises both correspond to the point $(7.5, 7.5)$ on the payoff polygon. This point is in the Pareto optimal set, half way between the two saddle points. Note, however, that this point cannot be reached if the two players choose their moves independently.

There is a large branch of game theory, called cooperative game theory that studies games under the assumption that players are allowed to negotiate. Sometimes players are permitted to share their payoffs. Suppose we think of the payoffs in Battle of the Sexes as monetary and allow players to share them. In this case, the players should aim for the **maximum-sum outcome**, which is defined as the outcome that gives the two players, together, the largest amount of money. In Battle of the Sexes, the two saddle points are each worth \$15, so each is a maximum-sum outcome. In this interpretation, the point $(7.5, 7.5)$ on the payoff polygon corresponds to the equally shared prize. But it could also mean that Row had to pay Column \$2.50 to get him to go to the opera. The maximum-sum outcome is any outcome in the payoff polygon that lies out the furthest out in the direction exactly half-way between north and east.

We next consider the game called **Chicken**, in which two teenage drivers engage in risky behavior. The goal of the game is for the two drivers to drive directly toward each other on a dark one-lane country road, trying to avoid being the first driver to swerve. The driver who swerves first is the "chicken", and the other driver is "cool". If both drivers swerve, they each suffer some minor disrespect from their friends, but this is not nearly as bad as being the chicken. The worst outcome is what happens if neither driver swerves. In this case, there is a head-on collision.

Let us write down a description of the outcomes from the point of view of Row.

	Don't swerve	Swerve
Don't swerve	I'm dead.	I'm cool.
Swerve	I'm chicken.	I'm sensible.

The next step is to assign payoffs. These are based on the player's preferences. We assign 10 points to a player who is "cool" (the first choice), 0 points if both players are sensible and swerve. We assign -5 points for being the "chicken" and -10 points for what we assume is the worst outcome: being dead.

	Don't swerve	Swerve
Don't swerve	$-10, -10$	$10, -5$
Swerve	$-5, 10$	$0, 0$

This game has two saddle points, which can be found by drawing the flow diagram. The payoff polygon is shown in Figure 16.4.

At first glance, this game seems to resemble Battle of the Sexes, but a closer examination reveals it to be far more sinister. Even though the prudent method gives a player the sound advice to "swerve", and avoid the possibility of death, the naive method implores the player not to swerve. Although not swerving gives a player the possibility of being "cool", it leads to two deaths if both players do it. The imprudent preemptive threat by an opponent — "I will never swerve!" — carries a clear danger for a player who does not choose the best response and

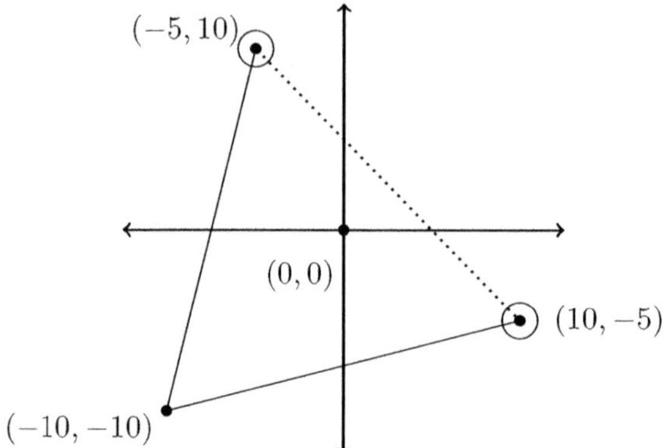

Figure 16.4 The payoff polygon for Chicken. The saddle points are circled, and the Pareto optimal set is dotted.

swerve. Such a player may well wonder if their opponent is crazy enough to go through with the threat.

In 1962, during the **Cuban Missile Crisis**, US spy planes observed Soviet troops assembling nuclear missiles in Cuba. The close proximity of the missiles to the United States was considered unacceptable by the Kennedy Administration, and in discussion with his advisors, President Kennedy considered two options. One option, the more aggressive of the two, was to invade Cuba. The other, less aggressive, option was a naval blockade. For the Soviets, there were also essentially only two possible responses: to maintain the missiles or to withdraw them. The table below, with the United States as Row and the Soviets as Column, describes the outcomes corresponding to all the strategy combinations.

	Maintain	Withdraw
Invade	Nuclear war	US victory
Blockade	Soviet victory	Cold War

Clearly, each side viewed its own victory as the best outcome. Each side also viewed a Cold War (essentially a stalemate) as preferable to a victory by their opponent. The key assumption, however, is that both sides considered a nuclear war to be the worst outcome. If we use the same scores as Game 16.3, we see that this game is essentially Chicken. In 1962, it was ultimately the doubly prudent outcome that prevailed. (The Cold War lasted another 25 years.)

An interesting feature of this model is the way it illustrates how one side's threat to choose the more aggressive strategy might be believable. For example, if the United States believed that the Soviets would find a nuclear war acceptable, then a Soviet threat to choose their aggressive strategy would have to be taken more seriously. This could have easily led to a Soviet victory. Even the appearance of recklessness can be an asset in a game like this.

The following game is called the **Prisoner's Dilemma**. Two partners in crime are captured by the police and interrogated separately. Each prisoner is offered freedom for a confession if it helps the police convict the other. The prisoner who does not confess and is convicted on the basis of the other's confession will receive a sentence of 15 years in jail. But if both prisoners confess, each will get 10 years. If neither prisoner agrees to confess the police will have to settle for convicting both prisoners on a lesser charge, and they will each get off with a sentence of only 1 year. We write the jail sentences as negative numbers in the game bimatrix.

	Confess	Keep quiet
Confess	$-10, -10$	$0, -15$
Keep quiet	$-15, 0$	$-1, -1$

The outcome where both prisoners keep quiet is in some sense jointly best for them. This outcome is Pareto optimal, and it carries the smallest total jail time (i.e., it is the worst case for the police). Keeping quiet is probably even something the prisoners discussed before committing the crime and agreed to do if they were caught. But will they keep their promise to each other?

The flow diagram for the Prisoner's Dilemma is show below.

$-10, -10$	$0, -15$
$-15, 0$	$-1, -1$

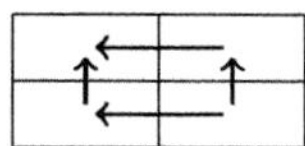

Here is the min-max diagram.

$-10, -10$	$0, -15$	-10	$\leftarrow$
$-15, 0$	$-1, -1$	-15	
-10	-15		
$\uparrow$			

Finally, Figure 16.5 shows the payoff polygon for the Prisoner's Dilemma. The saddle point is circled, and the Pareto optimal set is dotted.

Notice that the strategy "Confess" is both naive and prudent. In particular, a prisoner who fails to confess faces the possibility of a 15-year sentence, whereas the worst sentence for a prisoner who confesses is only 10 years. Moreover, the outcome where both prisoners confess is a saddle point. Even more striking is the fact that for both players,

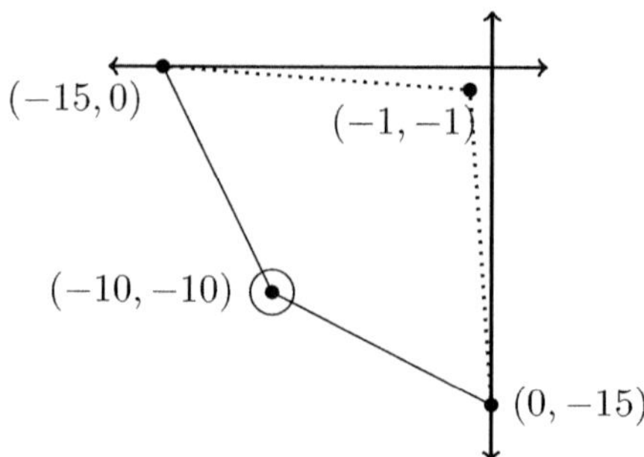

Figure 16.5 The payoff polygon for the Prisoner's Dilemma.

the strategy "Confess" dominates the strategy "Keep quiet". If Row confesses, then her payoff is better (her sentence is shorter) than if she keeps quiet, no matter what Column chooses to do.

What possible reason, then, could either player possibly have to keep quiet? Whether or not the other prisoner keeps quiet, a prisoner reduces his or her own sentence by confessing. Yet if both prisoners follow this reasoning, they both end up spending 10 years in jail. This is in spite of the fact that both would vastly prefer spending only 1 year in jail, which is what would happen if they could both keep quiet. This is the prisoner's dilemma. We study it in detail in Chapter 18.

16.5 Exercises and Problems

16.1. Consider the following game.

	A	B
A	1, −3	3, −1
B	−2, 4	−4, 2

(a) Find the guarantee for each row and each column. Find the prudent strategies for Row and for Column.
(b) Draw the flow diagram, and circle any saddle points.
(c) Draw the payoff polygon. Mark the Pareto optimal set, and circle any saddle points.
(d) Find any dominated strategies. Reduce the game by deleting the dominated strategies. Repeat this process as many times as possible.

What does the result suggest about how the players should play this game?

16.2. Consider the following game.

	A	B
A	1, 1	2, 0
B	−2, −1	0, −2
C	0, 3	0, 0

(a) Find the guarantee for each row and each column. Find the prudent strategies for Row and for Column.
(b) Draw the flow diagram, and circle any saddle points.
(c) Draw the payoff polygon. Mark the Pareto optimal set, and circle any saddle points.
(d) Find any dominated strategies. Reduce the game by deleting the dominated strategies. Repeat this process as many times as possible. What does the result suggest about how the players should play this game?

16.3. Consider the following game.

	A	B	C
A	3, 3	9, 4	6, 6
B	4, 5	8, 8	5, 9
C	2, 1	7, 7	1, 2

(a) Find the guarantee for each row and each column. Find the prudent strategies for Row and for Column.
(b) Draw the flow diagram, and circle any saddle points.
(c) Find any dominated strategies. Reduce the game by deleting the dominated strategies. Repeat this process as many times as possible. What does the result suggest about how the players should play this game?

16.4. Classify each of the following games as a coordination game, a strictly competitive game, or a mixed motive game.

−2, 2	0, 1
1, −1	2, −3

(a)

−2, −3	0, 0
1, 1	2, 3

(b)

−2, 3	0, 0
1, 1	2, −3

(c)

In each case, find the saddle points and draw the payoff polygon to find the Pareto optimal set.

16.5. A parent (Row) and child (Column), who are at the beach without cell-phones, get separated. There are two locations at which they might rendezvous: at the North lifeguard station or at the South lifeguard station. For each story below, model this as a 2-by-2 bimatrix game.

(a) The parent and child both want to find each other. Use a payoff of 1 for a good outcome and a payoff 0 for a not good outcome.
(b) The parent wants to find the child, but the child does not want to be found. Again, use the payoffs of 0 and 1.
(c) A video game arcade is located near the South lifeguard station. The child's most pressing desire is to visit the arcade, but the child does want to be found (eventually). The parent, on the other hand, just wants to find the child. Use 1 point and 0 points for the parent's payoffs. Use 0, 1, 2, and 3 as payoffs for the child. How will this game likely play out?
(d) For each case, say whether the game is strictly competitive, a coordination game, or a mixed motive game.

16.6. From the last will and testament of Josiah P. Withers: "Should my two ungrateful heirs, Roland and Calista, refrain from contesting my will, they may split my $10 million fortune evenly. But should one of my heirs contest my will, give the contester $8 million, give $2 million to my loyal cat Fleabuss, leaving nothing for the other (hmmph!). And if they both contest my will (as I fully expect them to do!), then donate $8 million to the Humane Society, leaving each of those lazy bums a mere $1 million. Each heir will have one chance in a private meeting with my attorney, Mr. Crum, to reveal whether or not to contest." Set this up as a (non-zero-sum) game, analyze it, and discuss the pros and cons of different strategies. What do you expect to happen?

16.7. Two fishing villages lie on opposite sides of a small lagoon and never communicate due to cultural and language barriers. Each village has one boat and can catch enough fish for the village to survive (50 pounds per week each). But both villages are considering building new boats, and if they do, the two new boats would be finished at roughly the same time. If three boats fish the lagoon, each boat will catch 40 pounds of fish. But with four boats fishing, the fish population will be decimated and (within a few weeks) each boat will catch only 15 pounds. Model this scenario as a (non-zero-sum) game. What is the prudent strategy for a village? Which outcomes are saddle points? Are the saddle points Pareto optimal? What is the nature of the conflict here? Explain why the saddle points are not satisfactory "solutions" to this conflict. How might a negotiator resolve this conflict?

16.8. In a TV game show called "Go For It!", a $1000 bill is placed on the table in front of two contestants. The players have to decide what to do before a buzzer sounds. If one player grabs the bill, that player gets to keep it. But if both players grab, nobody gets the \$1000. Instead both have \$500 deducted from their winnings. Model this as a bimatrix game and analyze it, finding prudent and naive strategies, saddle points, and Pareto optimal outcomes.

16.9. In the Unemployment Game, the government can either Pay or Withhold unemployment benefits from an unemployed worker, who can either Look for a new job, or Stop looking. The government's preferred outcome is to aide workers who are looking for work. Unfortunately, a common outcome is for the government to deny benefits to a worker who has stopped looking — the government's second choice. The government believes that the worst outcomes are not paying benefits to a worker who is looking for work (because it provides a negative incentive), or even worse, paying benefits to a worker who is not looking (voters hate this). The worker, on the other hand, would like to continue to receive benefits without having to look for a job, but if necessary, he will look to keep the benefits coming in. Once benefits get cut, however, he will reluctantly look for work, because otherwise there is no income at all. Model this situation as a 2-by-2 bimatrix game. Draw the flow diagram, and find the doubly naive, doubly prudent, and Pareto optimal outcomes. Interpret all these in the context of unemployment benefits.

16.10. Two members of the local school board want to be elected Chair, but each would settle for serving as Vice Chair. If the two can agree on who should be Chair and Vice Chair, the rest of the Board will go along. If neither campaigns hard to be Chair, the Board will eventually choose one, but the delay in a resolution makes this a third choice for everyone. Worse still, if the two fight over the Chair job, the outgoing Chair and Vice Chair will stay on, and the two fighters will face a difficult reelection campaign. Model this situation as a 2-by-2 bimatrix game. Analyze the game to find the naive and prudent strategies for both players. Find the saddle points and Pareto optimal set. Interpret your analysis in political terms.

16.11. The **maximum-sum outcome** of a bimatrix game is the outcome (row i, column j) with the largest value of $u_{i,j} + v_{j,i}$.

(a) What are the maximum-sum outcomes in Chicken, Battle of the Sexes, and the Prisoner's Dilemma?
(b) What are the maximum-sum outcomes in a zero-sum game?
(c) Show that a maximum-sum outcome is always Pareto optimal.
(d) Find an example of a game in which the maximum-sum outcome is a saddle point, and find an example of a game in which the maximum-sum outcome is not a saddle point.

16.12. Given any bimatrix game, we can make a new game, called the **coordination part** of the game, by giving both players the payoff $\frac{1}{2}(u_{i,j} + v_{i,j})$.

(a) Compute the coordination part of Battle of the Sexes, Chicken, the Prisoner's Dilemma, and Battle of the Bismarck Sea.

(b) Suppose the payoffs in the original game are monetary. Give an interpretation of what it means to play the coordination part of a game instead of the original game.

(c) For which games above would a player be better off (or worse off) playing the coordination part rather than the original game?

16.13. Given any bimatrix game, we make a new game, called the **zero-sum part** of the game. It is the matrix game with matrix entries (i.e., Row payoffs) $\frac{1}{2}(u_{i,j} - v_{i,j})$.

(a) Compute the zero-sum part of Battle of the Sexes, Chicken, the Prisoner's Dilemma, and Battle of the Bismarck Sea.

(b) For each of the games in part (a), comment on whether you would rather play the zero-sum part or the original game.

(c) Sometimes people playing a non-zero-sum game mistakenly play the zero-sum part instead. What misconception leads to this mistake?

16.14. Call a game **symmetric** if $u_{i,j} = v_{j,i}$ for all i and j.

(a) Show that the Battle of the Sexes, Chicken, and the Prisoner's Dilemma are all symmetric.

(b) Show that Roshambo, when written as a bimatrix game, is symmetric, but Battle of the Bismarck Sea is not.

(c) What equation should the payoffs $u_{i,j}$ of a matrix game satisfy in order for the game to be symmetric.

(d) Explain why a symmetric game is a game that is the same for both players.

(e) Prove that any symmetric zero-sum game is fair.

17

Nash Equilibria

"I'm afraid sometimes you'll play lonely games too, games you can't win because you'll play against you." — Dr. Seuss

17.0 Scenario

In a remote forest live two species of highly territorial birds, a more agressive species of hawk and a less agressive species of dove. When two birds meet, they compete over a territory, and their behaviors are determined by which of the two species they belong to. We assign scores in this competition in a way that models how likely a bird is to survive the season and reproduce. When a hawk meets a dove, the dove flees and the hawk wins the territory. Consequently, we award the hawk 10 fitness points and award 0 to the dove. When two doves meet, they both initially flee. However, seeing the territory empty, one of the doves inevitably returns and successfully occupies it. To model this, we award each of the doves 5 points, indicating that on average each dove has a 50% chance of earning 10 fitness points for occupying the territory. Finally, when a hawk meets a hawk, the fight that ensues is likely to injure both birds. These injuries are so severe that we award each of the hawks -10 points.

Model the encounter between two birds by a 2-by-2 bimatrix game, with strategies "hawk" and "dove". Note that individual birds do not actually play this game, since each bird's strategy choice is determined by its genes. However, in some situations, nature seems to exhibit an uncanny rationality. Let $P = (1-p, p)$ denote the probability distribution that describes the percentages of birds that are hawks and that are doves. It turns out that this population distribution P, viewed as a mixed strategy in the "Territory game", is the best response to itself. That is, nature chooses a population distribution P that results in a bird population with, collectively, the greatest fitness. Still another way to describe this is to say that the outcome (P, P) is a mixed strategy Nash equilibrium.

Analyze the Territory game and determine what percentage of birds will be hawks and what percentage will be doves.

17.1 Mixed Strategies

In Chapter 16, we discussed how to play bimatrix games using pure strategies. In this chapter, we discuss how to play bimatrix games using mixed strategies.

Suppose that two players are playing an m-by-n bimatrix game. In such a game, Row has m strategy choices and Column has n strategy choices. A mixed strategy for Row is defined to be a probability distribution $P = (p_1, p_2, ..., p_m)$, and a mixed strategy for Column is defined to be a probability distribution $Q = (q_1, q_2, ..., q_n)$.

Because we are no longer confined to the zero-sum case, there are two separate sets of payoffs: the payoffs $u_{i,j}$ to Row and the payoffs $v_{i,j}$ to Column. As a result, we need to consider two separate sets of expected payoffs. When Row plays the mixed strategy P and Column plays the mixed strategy Q, we denote the expected payoff to Row by $E(P, Q)$ and denote the expected payoff to Column by $F(P, Q)$.

The formulas for $E(P, Q)$ that we developed in Lemmas 14.8 and 14.9 remain valid in the bimatrix case. Moreover, Column's expected payoffs $F(P, Q)$ satisfy similar formulas, but with Column's payoff $v_{i,j}$ substituted for each occurrence of Row's payoff $u_{i,j}$. For example, $F(P, Q)$ is the sum of the numbers $p_i q_j v_{i,j}$ over all the outcomes (row i, column j).

How should Row respond if she knows that Column is going to play a certain mixed strategy Q? We say that the Row mixed strategy P is a **best response** to the Column mixed strategy Q if

$$E(P, Q) \geq E(R, Q) \text{ for all } R. \tag{17.1}$$

In other words, the response P for Row against Column's strategy Q gives Row a payoff that is better than (or at least as good as) any other response R. In a similar way, we say that a Column mixed strategy Q is a best response to P if

$$F(P, Q) \geq F(P, S) \text{ for all } S. \tag{17.2}$$

Often, a player has more than one best response to the opponent's strategy choice. The next lemma, which generalizes Lemma 15.1 to the non-zero-sum case, says that, among all the best responses, at least one of them is a pure strategy. The proof, which is the same as the proof of Lemma 15.1, is omitted.

Lemma 17.1. *A Row mixed strategy P is a best response to the Column strategy Q if and only if $E(P, Q) \geq E(P_i, Q)$ for every row i. Similarly, a Column mixed strategy Q is a best response to the Row strategy P if and only if $F(P, Q) \geq F(P, Q_j)$ for every column j.* □

In other words, the set of all best responses to any strategy of the opponent always includes at least one pure strategy.

Definition 17.2. A mixed strategy outcome (P, Q) is called a **Nash equilibrium** if P is a best response to Q and Q is a best response to P. Equivalently, (P, Q) is a Nash equilibrium if and only if both (17.1) and (17.2) are satisfied.

Although some of the results about Nash equilibria for matrix games extend to the bimatrix game case, others do not. This is one of the key reasons that non-zero-sum game theory is more difficult than zero-sum game theory. We begin by discussing some results that hold for all games. The first is a result on saddle points, which is proved the same way as Lemma 15.3.

Lemma 17.3. *In a bimatrix game, a pure strategy outcome* (row i, column j) *is a saddle point if and only if the corresponding basic mixed strategy outcome (P_i, Q_j) is a Nash equilibrium.* □

As in the zero-sum case, we call a strategy a **neutralizing strategy** if the payoff to the opponent is the same no matter what strategy the opponent uses. A **neutralizing outcome** is an outcome (P, Q) in which each player uses a neutralizing strategy. If changing strategies results in no change in payoff, then in particular (and obviously) changing strategies results in no improvement in payoff. Hence we obtain the following non-zero-sum version of Lemma 15.5.

Lemma 17.4. *Any neutralizing outcome is a Nash equilibrium.* □

Saddle points and neutralizing outcomes are the two simplest kinds of Nash equilibria. While other types of Nash equilibria exist, most of the equilibria we encounter in practice will be of one of these two types.

The next result is John Nash's famous equilibrium theorem, published in his 1950 Ph.D. dissertation. In 1994, Nash was awarded the Nobel Prize in Economics, largely on the basis of this theorem. We will provide a general proof of this theorem in Section 17.3. In the next section, we give the much simpler proof in the 2-by-2 case.

Theorem 17.5. (***Nash's Equilibrium Theorem***) *Every two person bimatrix game has at least one Nash equilibrium.*

17.2 The 2-by-2 Case

Let us consider any 2-by-2 bimatrix game.

a, e	b, f
c, g	d, h

Our goal here is to find all the pure and mixed-strategy Nash equilibria for such a game.

We call a game **decisive** if each pure strategy has a unique pure strategy best response by the opponent. A game that is not decisive is called **indecisive**. A game is decisive if and only if its flow diagram has no double arrows (i.e., no arrows pointing both ways).

Lemma 17.6. *If a 2-by-2 bimatrix game is indecisive, then it has a saddle point.*

Proof. For simplicity, suppose that the vertical arrow in column 1 is a double arrow (a similar argument holds in the three other cases). This means that row 1 and row 2 are equally good responses to column 1, or in other words, $a = c$. If either of the horizontal arrows point to the left, then there is a saddle point in the first column, and the lemma is proved.

Thus, suppose both horizontal arrows point to the right. The vertical arrow in column 2 either points up or down (or is a double arrow). So one of the entries in column 2 must be a saddle point. □

From now on in this section we consider only decisive games. The possible flow diagrams for decisive 2-by-2 bimatrix games can be classified into four **flow diagram types**, depending on the pattern of arrows. These types are illustrated in Figure 17.1.

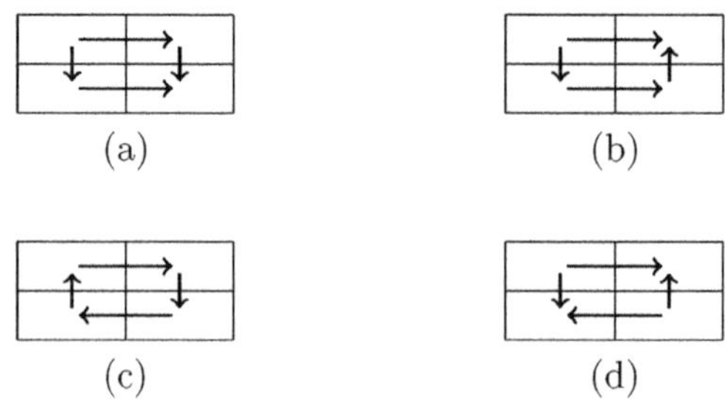

Figure 17.1 The four flow diagram types for decisive 2-by-2 bimatrix games: (a) doubly dominant, (b) singly dominant, (c) pursuit, and (d) coordination.

Two flow diagrams are considered to be of the same type if one can be changed into the other by a series of interchanges of rows and columns, or by an exchange of the roles of the two players. A game has a **doubly dominant** flow diagram if it has a dominant strategy for each player. A game has a **singly dominant** flow diagram if one player has a dominant strategy, but the other does not. The next lemma takes care of the doubly dominant and singly dominant cases. We leave the proof as an exercise.

Lemma 17.7. *Suppose a 2-by-2 bimatrix game is decisive and has a doubly or singly dominant flow diagram. Then the game has a unique saddle point and no other Nash equilibria.* □

The remaining two types of flow diagrams are called **pursuit** and **coordination**. A flow diagram of pursuit type has a circular flow pattern. This type of flow diagram is characteristic of games of pursuit, like the US vs. al-Qaeda from Chapter 15. Games of this type have no saddle points. A flow diagram of coordination type is characteristic of coordination games. The games of Chicken and Battle of the Sexes both have this type of flow diagram. Coordination type flow diagrams do not appear in the world of zero-sum and constant-sum games. A game with a coordination type flow diagram has two saddle points at opposite corners of the 2-by-2 matrix.

Proposition 17.8. *Any decisive 2-by-2 game with either a pursuit or a coordination type flow diagram has a neutralizing outcome. In particular, it has a non-basic mixed-strategy Nash equilibrium.*

Proof. We start by taking apart a bimatrix into the two matrices that contain the payoffs to Row and to Column.

a	b
c	d

e	f
g	h

We call these **Row's game** and **Column's game.**

First consider Row's game. Since the original bimatrix game was assumed to be decisive, we know that either $a > c$ or $a < c$. For concreteness, let us assume that $a > c$. Then it follows from the type of the flow diagram (we assumed it was either pursuit or coordination), that $b < d$. This is precisely the situation described in the proof of Theorem 15.12 that guarantees that Column has a neutralizing strategy. Note that we do not require $a > b$ and $c < d$ as we did in Theorem 15.12. Thus, if Column plays the mixed strategy $Q = (1 - q, q)$, where

$$q = \frac{a - c}{(a - c) + (d - b)},$$

then no matter what mixed strategy P Row chooses, she will receive an expected payoff of

$$E(P,Q) = \frac{ad - bc}{(a - c) + (d - b)}.$$

It follows that Q is a neutralizing strategy for Column. These formulas remain true if we assume $a < c$, in which case $b > d$, so they are true for any decisive bimatrix game with either a pursuit or a coordination flow diagram type.

Looking next at Column's game, the assumption of pursuit or coordination type means that either $e > f$ and $g < h$, or the opposite. In either case, if Row plays the mixed strategy $P = (1 - p, p)$, where

$$p = \frac{e - f}{(e - f) + (h - g)},$$

then no matter what strategy Q Column chooses, he will receive an expected payoff of

$$F(P,Q) = \frac{eh - fg}{(e - f) + (h - g)}.$$

It follows that P is a neutralizing strategy for Row. Thus (P, Q) is a neutralizing outcome, and so by Lemma 17.4, it is a Nash equilibrium. □

We are now ready to apply these results to some examples. Suppose we are given a decisive 2-by-2 bimatrix game and want to find all the Nash equilibria. The first step is to analyze the flow diagram. If the type of the flow diagram is doubly or singly dominant, then the game has a unique saddle point, which is the only Nash equilibrium. If the type of the flow diagram is pursuit, then the game has a unique mixed-strategy Nash equilibrium. The most complicated case is when the flow diagram is of coordination type. In this case, it turns out that the game has two saddle points, plus a mixed-strategy Nash equilibrium.

The last four equations in the proof of Proposition 17.8 show that, in the pursuit and coordination cases, we can use the cross-over algorithm from Chapter 15 to find the mixed-strategy Nash equilibrium. To do this, we split the game into Row's game and Column's game, and we have both players apply the crossover algorithm to their opponent's game.

Suppose we want to find all the Nash equilibria in the following bimatrix game, which we call **Chase**.

$3, 3$	$5, 7$
$2, 5$	$7, 2$

This game is decisive and has a pursuit type flow diagram:

3, 3	5, 7
2, 5	7, 2

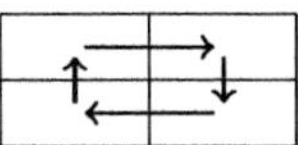

It follows that Chase has no saddle points. But it does have a mixed-strategy Nash equilibrium. To find it, we divide Chase into Row's game and Column's game.

3	5
2	7

3	7
5	2

We think of these as two separate matrix games, the games that each player would play if she or he were to simply ignore the payoffs of the opponent. However, ignoring the opponent is not what the players should do in order to find their Nash equilibrium strategies. Instead, each player should apply the crossover algorithm to the *other* player's game.

When Row applies the cross-over algorithm to Column's game, she gets the following.

2	−5	−3	3/7
−7	3	−4	4/7

This indicates that $P = (3/7, 4/7)$ is Row's equilibrium strategy. In particular, this strategy neutralizes Column. No matter how he responds his expected payoff is $F = (-29)/(-7) = 29/7 \approx 4.1$.

When Column applies the cross-over algorithm to Row's game, he gets the following.

7	−2
−5	3
2	1
2/3	1/3

This indicates that $Q = (2/3, 1/3)$ is the Column strategy that neutralizes Row, guaranteeing that she will receive an expected payoff $E = 11/3 \approx 3.7$. It follows that the outcome (P, Q) is a Nash equilibrium, and we think of the pair $(E, F) = (11/3, 29/7) \approx (3.7, 4.1)$ as the payoff pair for this equilibrium. We can illustrate this payoff by drawing the payoff polygon and plotting this point.

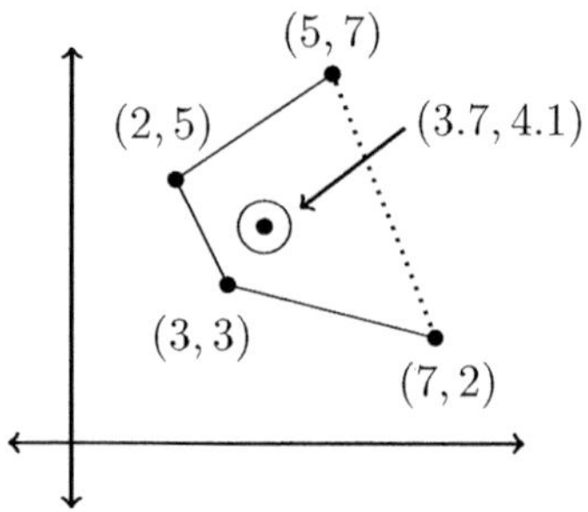

Figure 17.2 Payoff polygon for the game of Chase.

In order to find the Nash equilibrium, each player was asked to play their opponent's game (in effect, to ignore their own payoffs). What would happen if, instead, the players played their own games (ignoring their opponent's payoffs)? In particular, each player could pretend that his or her own game is a zero-sum game, and play the corresponding equilibrium strategy. By Theorem 15.6, the equilibrium strategy in a zero-sum game is prudent. It comes with a guaranteed worst case expected payoff. We call the strategy a player obtains by this method the player's **prudent mixed strategy**.

To find their prudent mixed strategies, both players solve their own games in the zero-sum sense. The first step in doing this is to look at the flow diagram. The flow diagrams for the two players' games in the game of Chase are shown below. The diagram for Row's game

3	5
2	7

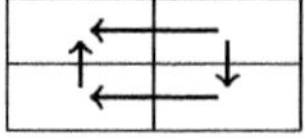

shows that Row's game has a saddle point in row 1, so Row's prudent mixed strategy is her saddle point strategy, row 1. This game has value $v = 3$ (the payoff for the saddle point). We say that Row's **mixed-strategy guarantee** in this game is $\overline{r} = 3$.

Next, we look at the flow diagram for Column's game:

3	7
5	2

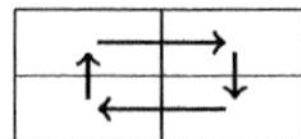

Notice that we have reversed the arrows compared to Chapter 14. This is because the numbers in this matrix represent payoffs *to* Column (rather than the usual payoffs to Row). Since Column's game has no saddle point, we look for a mixed-strategy Nash equilibrium using the cross-over algorithm

2	−5
−7	3
−5	−2
5/7	2/7

and obtain $Q = (5/7, 2/7)$. Column's mixed-strategy guarantee (which equals the value of Column's game) is, according to (17.2), $\overline{c} = v = -29/-7 \approx 4.1$, the same as the expected payoff of F that Column gets from the Nash equilibrium. Note that Column's prudent pure strategy is column 1, which with it's guarantee of $c = 3$ is worse.

We plot in Figure 17.3 the mixed-strategy guarantees on the payoff polygon as the vertical line $E = 3$ and the horizontal line $F = 29/7$. By playing her prudent strategy, Row is guaranteed a payoff of at least $E = 3$. In order for Row to get her equilibrium payoff, Column would

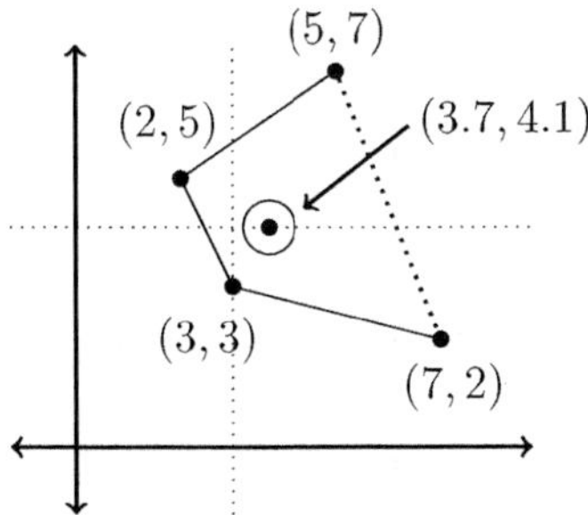

Figure 17.3 Payoff polygon and guarantees for the game of Chase.

have to play his equilibrium strategy too. Yet while this is Column's best response to Row's equilibrium strategy, there can be no guarantee that Column will play it. In the worst case, if Column plays column 1 against Row's equilibrium strategy, then Row gets only $E = 17/7$, which is 4/7 less than her guarantee. So while the equilibrium strategy offers the best average (or expected) payoff, the prudent strategy offers the better guarantee.

Next we consider the game of Chicken. This game is decisive, and the flow diagram is of coordination type.

−10, −10	10, −5
−5, 10	0, 0

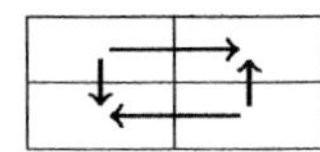

This shows, as we have already seen, that Chicken has two saddle points. So Chicken has two pure strategy Nash equilibria. But Proposition 17.8 tells us that Chicken has also a third Nash equilibrium that is comprised of non-basic mixed strategies. To find it, we separate the bimatrix for Chicken into Row's and Column's games.

−10	10
−5	0

−10	−5
10	0

Then we find mixed strategies for Row and Column by having each of them apply the cross-over algorithm to the game of their opponent.

0	−10	−10	2/3
5	−10	−5	1/3

0	5
−10	−10
−10	−5
2/3	1/3

The equilibrium strategies for both players are the same: $P = Q = (2/3, 1/3)$ (because the game is symmetric), and both players get the same expected payoff $E = F = -50/15 \approx -3.3$.

In the context of the story we proposed for this game (a dare involving cars), this mixed-strategy equilibrium outcome is not at all comforting. The table below shows the probabilities, under this mixed strategy outcome, of each of the four pure strategy outcomes.

	Don't swerve	Swerve
Don't swerve	4/9	2/9
Swerve	2/9	1/9

Notice that with a probability of 4/9, or about 44% of the time, the Nash equilibrium results in the death of both players!

Perhaps a safer approach for both players would be to play their prudent mixed strategies. However, the flow diagram for Row

−10	10
−5	0

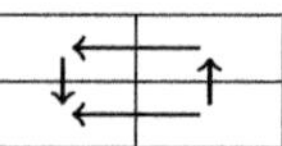

and the flow diagram for Column

−10	−5
10	0

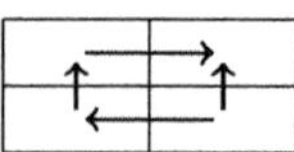

both have saddle points. Thus the prudent mixed strategies for both players are their prudent pure strategies: row 2 for Row and column 2 for Column. In other words, for each player the prudent strategy is to swerve. The prudent strategy guarantees are $r = c = -5$. This seems worse than the Nash equilibrium payoff (-5 rather than $-10/3 \approx -3.3$), but at least it avoids the possibility of death.

17.3 The Proof of Nash's Theorem

The following proof of Theorem 17.5 is based on a proof that Nash published in 1950. It uses a result from pure mathematics called the Brouwer fixed point theorem, which we will not prove, but if one is willing to accept the Brouwer theorem, the proof is not too difficult to follow. We note that Nash proved the equilibrium theorem in the more general setting of n-person games, but for simplicity, our explication of Nash's proof is restricted to the two-person case.

Nash's proof involves a function that we denote by the letter N. Both the domain and the co-domain for this function are the set of all possible pairs (P, Q), where P is a mixed row strategy and Q is a mixed column strategy.

In order to explain how the function N is computed, let us fix a two-person m-by-n bimatrix game, with payoffs $u_{i,j}$ to Row and $v_{i,j}$ to Column. Let $P = (p_1, p_2, \ldots, p_m)$ denote a mixed strategy for Row and $Q = (q_1, q_2, \ldots, q_n)$ denote a mixed strategy for Column. The Nash function N takes as its input a pair (P, Q), and gives as its output another pair (P', Q'). We now specify precisely how P' and Q' are construct from P and Q.

For each $i = 1, \ldots, m$, we let

$$a_i = \max\left(0, E(P_i, Q) - E(P, Q)\right), \tag{17.3}$$

and we let

$$p_i' = \frac{p_i + a_i}{1 + a_1 + a_2 + \cdots + a_m}. \tag{17.4}$$

In (17.3), P_i is the basic mixed strategy corresponding to row i, and the "max" means that the quantity a_i is set equal to $E(P_i, Q) - E(P, Q)$ as long as this quantity is positive, but is set to zero otherwise. We put $P' = (p_1', p_2', \ldots, p_m')$. In a similar way, we let

$$b_j = \max\left(0, F(P, Q_j) - F(P, Q)\right), \tag{17.5}$$

for each $j = 1, \ldots, n$, and we let

$$q_j' = \frac{q_j + b_j}{1 + b_1 + b_2 + \cdots + b_n}. \tag{17.6}$$

We put $Q' = (q_1', q_2', \ldots, q_n')$.

This is how the Nash function N is defined. For an input (P, Q), the output is $(P', Q') = N(P, Q)$, where P' and Q' are determined by (17.4) and (17.6).

The idea of the Nash function N is this. Think of the output P' and Q' of the Nash function as a pair of recommended strategies for the two players who might have been contemplating playing the strategies P and Q instead. If some pure strategy does better than the mixed strategy a player is contemplating, the Nash function advises the player to alter the mixed strategy by increasing the probability of the superior pure strategy.

Lemma 17.9. *If P and Q are probability distributions, then so are P' and Q'. Moreover, (P,Q) is a Nash equilibrium if and only if $(P',Q') = (P,Q)$.*

Proof. We first prove that P' is a probability distribution. Since by (17.3), $a_i \geq 0$ for all i, the denominator in (17.4) is not zero. Since $p_i \geq 0$, it follows that $p'_i \geq 0$. Also, since $p_1 + \cdots + p_m = 1$,

$$\begin{aligned} p'_1 + p'_2 + \cdots + p'_m &= \frac{p_1 + a_1}{1 + a_1 + a_2 + \cdots + a_m} + \\ &\quad \frac{p_2 + a_2}{1 + a_1 + a_2 + \cdots + a_m} + \cdots \\ &\quad + \frac{p_m + a_m}{1 + a_1 + a_2 + \cdots + a_m} \\ &= \frac{(p_1 + p_2 + \cdots + p_m) + (a_1 + a_2 + \cdots + a_m)}{1 + a_1 + a_2 + \cdots + a_m} \\ &= 1. \end{aligned}$$

Thus P' is a probability distribution, and by a similar calculation so is Q'.

Now suppose that (P,Q) is a Nash equilibrium. Then Lemma 17.1 implies that $E(P,Q) \geq E(P_i,Q)$ for every row i. It follows that $E(P_i,Q) - E(P,Q) \leq 0$, and since this is not positive, (17.3) says that $a_i = 0$ for every i. In this case, (17.4) tells us that $p'_i = p_i$ for all i. It follows that $P' = P$, and by a similar calculation $Q' = Q$.

Conversely, suppose $(P,Q) = (P',Q')$. That $P = P'$ implies that

$$p_i = \frac{p_i + a_i}{1 + a_1 + a_2 + \cdots + a_m} \tag{17.7}$$

for all i. By the first equation in Lemma 14.9, the payoff $E(P,Q)$ is a weighted average of the payoffs $E(P_i,Q)$. In fact, only those rows i for which $p_i > 0$ contribute to this average. Since an average can never be smaller than all of its components, there must be some row k with $E(P,Q) \geq E(P_k,Q)$, and row k must also be a row that has $p_k > 0$. For this row k, we have that $E(P_k,Q) - E(P,Q) \leq 0$, and since this difference is not positive, (17.3) says that $a_k = 0$. This fact, combined with (17.7) implies that

$$p_k = \frac{p_k}{1 + a_1 + a_2 + \cdots + a_m}.$$

But $p_k > 0$ implies that the denominator is 1, which means that $a_1 + a_2 + \cdots + a_n = 0$. Since we know that $a_i \geq 0$ for all i, it must actually be the case that $a_i = 0$ for all i. This implies that $E(P_i, Q) - E(P, Q) \leq 0$ for all i, and therefore, by Lemma 17.1, it follows that P is a best response to Q.

Repeating the argument (using the expected payoff F for Column instead of the expected payoff E for Row), we see that Q is also a best response to P. Since P is a best response to Q, and Q is a best response to P, it follows that (P, Q) is a Nash equilibrium. □

To complete the proof of Nash's Theorem, it remains to show that there is always at least one outcome (P, Q) so that $(P', Q') = (P, Q)$. In other words, it remains to show that there is an outcome (P, Q) so that $(P, Q) = N(P, Q)$ for the Nash function N. We call such a (P, Q) a **fixed point** for the Nash function.

To gain some perspective on fixed points, let us temporarily replace the Nash function N with a more familiar example: the squaring function f from elementary algebra. This is the function that takes as its input any real number and outputs the number's square. The squaring function f is defined by the formula $y = f(x) = x^2$. Now, a fixed point for the squaring function f would be a real number x that satisfies $x = f(x)$. That is, the number x would be its own square. Such a number would solve the equation $x^2 - x = 0$. This equation has exactly two solutions: $x = 0$ and $x = 1$. These are fixed points of f because $0 = f(0) = 0^2$ and $1 = f(1) = 1^2$.

The squaring function has an important feature in common with the Nash function: both are **continuous functions**. For the squaring function f, continuity means that if we square two numbers that are close together, their squares will also be close. The fact that the Nash function N is continuous follows from the four formulas that define it. These formulas involve only the four basic operations of arithmetic, addition, subtraction, multiplication and division, together with the "max" function. Any function made out of these five operations, as long as division by zero never occurs, must be continuous.

The squaring function f can take any real number x as its input, so its domain is the set of all real numbers. As the codomain of f we can also take the set of real numbers (even though the square of a number is never negative). In order to speak of fixed points, we require a function whose codomain is the same as its domain. For the Nash function N, the domain is the set of all pairs (P, Q), where $P = (p_1, p_2, \ldots, p_m)$ and $Q = (q_1, q_2, \ldots, q_n)$ are probability distributions. The importance of the first assertion in Lemma 17.9 is that it tells us that we can take the codomain of N to be equal to the domain.

In the case of a 2-by-2 game, each player has exactly two pure strategies. Let us now consider the set of all mixed strategies. For Row, these consist of all probability distributions of the form $P = (1 - p, p)$. Since such a probability distribution is completely determined by the choice of p, where p satisfies $0 \leq p \leq 1$, we can picture the set of all the Row mixed strategies as a line segment.

$p = 0$ $p = 1/3$ $p = 1$

For example, the point on this segment corresponding to the mixed strategy $P = (2/3, 1/3)$ is located at $p = 1/3$. Note that the endpoints of this interval, $p = 0$ and $p = 1$ correspond to the basic mixed strategies P_1 and P_2, which are simply the pure strategies row 1 and row 2.

In a similar way, all of Column's mixed strategies $Q = (1 - q, q)$ can also be depicted as a line segment.

$q = 0$ $q = 1/2$ $q = 1$

Here, the point $q = 1/2$ corresponds to the mixed strategy $Q = (1/2, 1/2)$.

An outcome (P, Q) is a pair of strategy choices, P for Row and Q for Column. Drawing an outcome requires drawing two line segments, but since that pair (P, Q) represents a single input to the Nash function N, we want to think of it as a single point in the domain of N. To accomplish this, we draw the two line segments perpendicular to each other, obtaining a square.

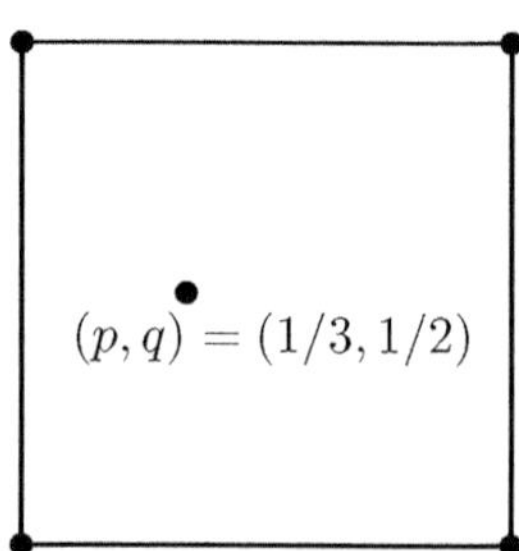

For example, in this drawing, the outcome (P, Q), where $P = (2/3, 1/3)$ and $Q = (1/2, 1/2)$, is represented by the point $(p, q) = (1/3, 1/2)$ in the square.

Now consider a game where a player has three pure strategies. In this case, we picture the space of all possible mixed-strategy choices as an equilateral triangle with sides 1. The vertices of this triangle represent the three pure strategies, while points inside the triangle represent mixed strategies. The mixed strategy $P = (p_1, p_2, p_3)$ (where $p_1 + p_2 + p_3 = 1$) is identified with the point in the triangle that is p_1 units from one edge, p_2 units from a second edge, and p_3 units from the third. The numbers p_1, p_2, and p_3 are said to be the **barycentric coordinates** of the point in the triangle.

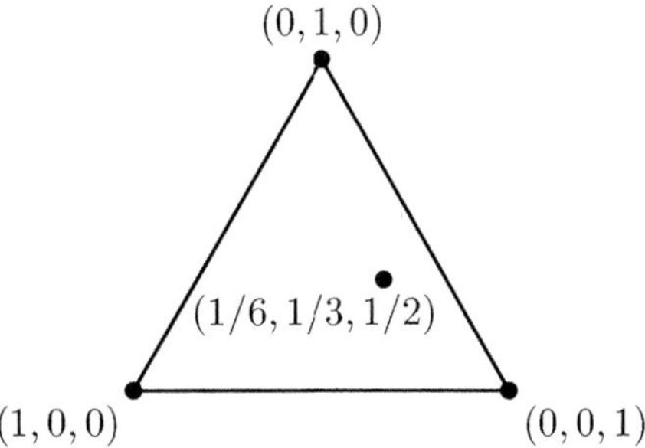

If the other player also has three strategies, the space of all their strategy choices (what was the square in the case of a 2-by-2 game) would require four dimensions to draw (two dimensions for each player). The picture would be even more complicated for larger games. However, even if we do not draw the domain of the Nash function N, we can conclude that it does share some properties with the square.

First, the domain of N is always completely enclosed by edges, which are themselves included in the domain. In mathematical terms, we say that the domain is **compact**. Second, for any two mixed-strategy outcomes in the domain, there is a line segment through the domain connecting the outcomes. Here is an illustration of this property in the case the square.

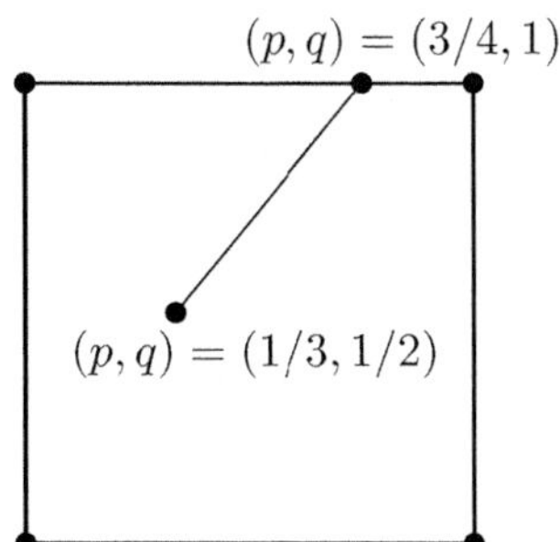

In mathematical terms, we say a domain with this property is **convex**.

Theorem 17.10. (***Brouwer's Fixed Point Theorem***) *If N is a continuous function with a compact and convex domain and the codomain is the same as the domain, then N has at least one fixed point.* □

As we said at the beginning of the section, the proof of this theorem goes beyond the scope of this book. We use the theorem nonetheless to complete the proof of Nash's Theorem.

Proof of Theorem 17.5. Consider a bimatrix game, and let N be the corresponding Nash function. Then N satisfies all the hypotheses of Brouwer's fixed point theorem, so by Theorem 17.10, there exists a fixed point $(P, Q) = N(P, Q)$. By Lemma 17.9, this outcome (P, Q) must be a Nash equilibrium. Thus every bimatrix game has a Nash equilibrium. □

17.4 Exercises and Problems

17.1. Consider the following game.

$0, -1$	$3, 4$
$2, 1$	$-1, -2$

(a) Draw the flow diagram, and classify it by its type. Find all saddle points.
(b) Determine whether there are any mixed-strategy Nash equilibria, and if there are, find them.
(c) Draw the payoff polygon, and plot the payoffs for all the Nash equilibria.
(d) Say whether the Nash equilibria are Pareto optimal.

17.2. Consider the following games:

$2, 1$	$0, 3$
$0, 2$	$1, 0$

(i)

$0, 4$	$3, -1$
$2, -2$	$-1, 1$

(ii)

$2, 1$	$0, 0$
$0, 0$	$1, 2$

(iii)

Answer the questions from Problem **17.1**.

17.3. For the game in Problem **17.1** do the following.
(a) Find the prudent pure strategies for both players and their guarantees.
(b) Find the prudent mixed strategies for both players and their mixed-strategy guarantees.

17.4. For each of the games in Problem **17.2** answer the questions in Problem **17.3**.

17.5. Explain why the Prisoner's Dilemma has a unique Nash equilibrium and it is a saddle point.

17.6. Find the Nash equilibrium in the Battle of the Sexes and the associated expected payoffs E and F for each player. Is this outcome Pareto optimal? (Hint: Plot the point (E, F) on the payoff polygon.) In the context of the story, describe what happens if both players play their equilibrium strategies. What are the probabilities of each of the four outcomes in this case?

17.7. Using your solution to the Unemployment game from Problem **16.9**, find the Nash equilibrium and the associated payoffs (E, F) for each player. Is this equilibrium Pareto optimal? In the context of the story, describe what it means for both players to play their equilibrium strategies.

17.8. Let D be the unit square in the plane with vertices $(0,0)$, $(1,0)$, $(0,1)$, and $(1,1)$. Find the fixed points for each of the following functions.
(a) Rotation of the square 90 degrees clockwise.
(b) $f(x,y) = (x/2, y/2)$.
(c) $f(x,y) = (1/3, 1/7)$.
(d) $f(x,y) = (1-x, y)$.
(e) $f(x,y) = (y, x)$.

17.9. Let I be the interval on the real line consisting of all numbers strictly between 0 and 1. Let f be defined with domain I by $f(x) = x/2$. Show that f has no fixed points. Which hypothesis of Theorem 17.10 is violated in this example?

17.10. Let D be the annulus in the plane consisting of all points whose distance from the origin is at least 1 but no greater than 2. Let f be defined with domain D by setting $f(x,y)$ equal to the result of starting at the point (x,y) and rotating around the origin by 90 degrees. Show that f has no fixed points. Which hypothesis of Theorem 17.10 is violated in this example?

17.11. Let I be the interval on the real line consisting of all numbers between 0 and 1, inclusive. Let f be defined with domain I by $f(x) = x + 1/2$ if $x \leq 1/2$ and $f(x) = x - 1/2$ if $x > 1/2$. Show that f has no fixed points. Which hypothesis of Theorem 17.10 is violated in this example?

17.12. Consider the zero-sum game associated with the matrix

−1	2
4	−3

.

Let $P = (1/5, 4/5)$ and $Q = (4/7, 3/7)$. Compute $N(P, Q)$.

17.13. Consider the game associated with the bimatrix

2, 6	8, 3
4, 1	5, 7

.

Let $P = (1/5, 4/5)$ and $Q = (4/7, 3/7)$. Compute $N(P, Q)$.

17.14. Consider the game of Chicken associated with the bimatrix

−10, −10	10, 0
0, 10	5, 5

.

Let $P = (1/5, 4/5)$ and $Q = (4/7, 3/7)$. Compute $N(P, Q)$.

17.15. Consider the zero-sum game associated with the 2-by-3 matrix

−1	2	1
4	−3	−1

.

Let $P = (1/5, 4/5)$ and $Q = (4/7, 2/7, 1/7)$. Compute $N(P, Q)$.

18

The Prisoner's Dilemma

"Power consists in one's capacity to link his will with the purpose of others, to lead by reason and a gift of cooperation." — Woodrow Wilson

18.0 Scenario

You and an opponent play a game in which you together count down "one-two-three-shoot!" and, when you say "shoot!", you extend your hand showing either one or two fingers. If one player shows two fingers while the other shows one finger, the player who shows two fingers gets \$10 (from the referee) and the opponent gets nothing. If both players show two fingers, both players receive \$1. If both players show one finger, both players receive \$9. Each player is assumed to be utterly indifferent to the fortune of the opponent. The object is to acquire as much money as possible, not to defeat the opponent.

Consider playing this game once with a stranger. Assume that negotiation in advance of playing is forbidden. How should you play? What is the prudent strategy? What is the equilibrium strategy? Is there a dominated strategy? Assuming both players employ a method that avoids dominated strategies, or that is prudent, or that is an equilibrium strategy, what outcome can be expected?

Now consider the possibility of consulting with the opponent in advance. Does that possibility open the door to an outcome that may be preferable to both players?

Next consider the possibility of playing the game repeatedly. Does the prospect of facing the opponent over and over suggest a different approach to playing the game? Would it make a difference if you knew in advance how many games you would play? If so, which approach would apply when you play the final iteration of the game?

Try playing this game with a variety of opponents, sometimes playing once, sometimes playing repeatedly, sometimes playing without prior consultation, and sometimes playing with prior consultation. How did you do?

18.1 Criteria and Impossibility

John von Neumann's min-max theorem (Theorem 15.11) suggests that the optimal way to play a zero-sum game is to use an equilibrium strategy. Here "optimal" simply means that no other strategy can promise a better expected outcome. But is this the promise we really want? Is the equilibrium strategy really the best way to play? In this section, we present a brief theory that can be used to evaluate the "reasonableness" or "appropriateness" of game-playing methods, leading to a simple impossibility theorem.

Recall from Chapter 13 that a game-playing method is a function whose domain is the set of all possible games and whose codomain is the set of mixed strategies for a player of that game. Of course pure strategies may also be among these in the form of basic mixed strategies. Roughly, a method inputs a game and outputs a strategy; it tells a player what to do. Every method is required to satisfy the universal domain property: for any game that the method is given as input, it is required to make at least one strategy recommendation. A method is, however, allowed to recommend more than one strategy. In practice, when this happens, a player could randomly choose one of the recommended strategies to play. We will consider two kinds of methods: those that work only for zero-sum games (i.e., for matrix games), and those that work for games in general (i.e., for bimatrix games).

We have already encountered three examples of game-playing methods. In the **prudent method**, a player chooses his or her prudent pure strategy. In the **naive method**, a player chooses the strategy corresponding to his or her primary outcome. There is also the **equilibrium method** in which a player chooses an equilibrium strategy. For zero-sum games, the fact that a player always has an equilibrium strategy follows from von Neumann's Equilibrium Theorem (Theorem 15.6). For bimatrix games, it follows from Nash's Equilibrium Theorem (Theorem 17.5).

Any game also has a prudent mixed strategy for each player. In a bimatrix game, a player finds his or her prudent mixed strategy by ignoring the opponent's payoffs and regarding the resulting game as a matrix game. In this matrix game, Theorem 15.6 tells us that the player has an equilibrium strategy, and by Theorem 15.11, this strategy is prudent. We call the method that selects this strategy the **prudent**

mixed method. In the case of a zero-sum game, the prudent mixed strategy is actually the equilibrium strategy (Theorem 15.11), so in this case the prudent mixed method is the same as the equilibrium method. However, for non-zero-sum games these two methods are different.

It is easy to come up with other methods, especially if one is willing to consider methods of no possible strategic value. There is, for example, the method that, when given a game, always outputs the first (pure) strategy. Slightly better is a method proposed by the 18th century mathematician Pierre-Simon Laplace. **Laplace's method**, when given a game, always outputs the probability distribution that makes every strategy equally likely. The idea behind Laplace's method is that, when faced with uncertain options — without any evidence to the contrary — one should always choose among the options at random and with equal probabilities. At the very least, this makes it difficult for an opponent to predict one's move.

Some of the ideas we have studied for playing games turn out not to be methods at all. For example, since not every game has a saddle point, there is no "saddle-point method". Just as we did when studying voting methods and apportionment methods, we insist that a game-playing method satisfy the universal domain property.

What makes a game-playing method reasonable, appropriate, or strategic? How can we decide which game-playing method to use? Our approach to this question is similar to the approach we followed in voting and apportionment: We establish criteria that identify those methods that lead to good strategic decisions and good outcomes. In game theory, this is a more subtle matter than it first appears, because games involve trade-offs, unpredictable opponents, and random play (i.e., mixed strategies).

We evaluate methods and criteria from two points of view. First, we ask how well they work for zero-sum (i.e., matrix) games. Second, we ask how well they work for games in general (i.e., for bimatrix games). Some methods may fare well on zero-sum games but fare poorly on non-zero-sum games, where examples display more varied and erratic behavior.

Our first criterion forbids a method from forcing a player to consider a dominated strategy.

Definition 18.1. A game-playing method satisfies the **dominance criterion** if at least one of the recommended strategies assigns probability zero to every dominated (pure) strategy.

In other words, the probability vector associated with at least one strategy recommended by the method must have zeroes in all of the entries corresponding to dominated strategies. Most people would regard choosing a dominated strategy as unwise, and would therefore regard the dominance criterion as compelling. What possible reason could a player have for selecting a dominated strategy? If a player implements a mixed strategy that assigns a positive probability to a dominated strategy, there is some chance that the player will actually play the dominated strategy. Surely this cannot be a good idea. Fortunately, many methods do satisfy the dominance criterion.

Proposition 18.2. *For a matrix game or a bimatrix game, the prudent method satisfies the dominance criterion.*

Proof. Consider a game with a unique prudent pure strategy for Row with guarantee r. Then every other strategy has some outcome that results in a payoff of less than r to Row. This implies that none of the other strategies can dominate the prudent strategy.

Now suppose there are a number of strategies, all with the same greatest guarantee r. Then the same argument shows that none of these strategies can be dominated by an imprudent one. But there must also always be some strategy among the prudent ones that is not dominated by any other prudent strategies. To see this, note that the dominance relation is transitive, which is to say that if strategy i dominates strategy j and strategy j dominates strategy k, then strategy i dominates strategy k. This implies that there can be no cycles of strategies in which each strategy is dominated by the next strategy around the cycle. Working backwards, we can find at least one prudent strategy that is not dominated by any other prudent strategy. □

Proposition 18.3. *For a matrix game or a bimatrix game, the equilibrium method satisfies the dominance criterion.*

For the proof of this theorem, we need the following lemma.

Lemma 18.4. *Suppose we are given an m-by-n game in which row $m-1$ is dominated by row m. Consider the new $(m-1)$-by-n game obtained by deleting row m from the original game. If (P, Q) is a Nash equilibrium for the new game, where $P = (p_1, \ldots, p_{m-1})$, then (P^0, Q) is a Nash equilibrium for the original game, where $P^0 = (p_1, \ldots, p_{m-1}, 0)$.*

Proof. To say that (P, Q) is a Nash equilibrium for the new game means that both (17.1) and (17.2) hold. Since $F(P^0, S) = F(P, S)$ for all S, it follows that

$$F(P^0, Q) = F(P, Q) \geq F(P, S) = F(P^0, S),$$

which is (17.2) for the original game. This is the first half of what it means for (P^0, Q) to be a Nash equilibrium in the original game. It remains to prove (17.1).

Let $R = (r_1, \ldots, r_{m-1}, r_m)$ be an arbitrary mixed Row strategy in the original game. Let $R' = (r_1, \ldots, r_{m-1} + r_m, 0)$ be the mixed strategy in the same game, obtained from R by moving all of the probability from row m to row $m-1$, and let $R'' = (r_1, \ldots, r_{m-1} + r_m)$ be the corresponding strategy in the new game. Note that $E(R', Q) = E(R'', Q)$ and $E(P^0, Q) = E(P, Q)$. Therefore

$$E(R', Q) = E(R'', Q) \leq E(P, Q) = E(P^0, Q), \tag{18.1}$$

where the inequality follows from the fact that (P, Q) is a Nash equilibrium in the new game. It remains to show that

$$E(R, Q) \leq E(R', Q), \tag{18.2}$$

since (18.1) combined with (18.2) yields (17.1).

Now, in the original game, row $m-1$ dominates row m, so $u_{m-1,j} \geq u_{m,j}$, or in other words, $u_{m-1,j} - u_{m,j} \geq 0$, for all j. Thus by Lemma 14.9,

$$\begin{aligned} &E(P_m, Q) - E(P_{m-1}, Q) \\ &= q_1(u_{m,1} - u_{m-1,1}) + \cdots + q_n(u_{m,n} - u_{m-1,n}) \leq 0. \end{aligned} \tag{18.3}$$

Using Lemma 14.9 again, we obtain

$$E(R, Q) = r_1 E(P_1, Q) + \cdots + r_{m-1} E(P_{m-1}, Q) + r_m E(P_m, Q)$$

and

$$E(R', Q) = r_1 E(P_1, Q) + \cdots + (r_{m-1} + r_m) E(P_{m-1}, Q).$$

Thus

$$E(R, Q) - E(R', Q) = r_m (E(P_m, Q) - E(P_{m-1}, Q)).$$

By (18.3) this is less than or equal to 0, and (18.2) follows.

□

Now it is an easy matter to complete the proof of Proposition 18.3.

Proof of Theorem 18.3. Make a new game by deleting all the rows that are dominated by another row, and all the columns that are dominated by another column. By Nash's Equilibrium Theorem (Theorem 17.5) the new game has a Nash equilibrium. By repeated application of Lemma 18.4 there is a corresponding Nash equilibrium of the original game that assigns probability zero to all the dominated strategies. □

Players who are averse to risk, if forced to choose among pure strategies, are likely to select a prudent one. Such players may be willing to venture into mixed strategies only if they can expect to do as well as the guarantee of their prudent pure strategy. This motivates the next criterion.

Definition 18.5. A game-playing method satisfies the **security criterion** if all the recommended strategies promise the player an expected payoff at least as good as the player's prudent pure strategy guarantee.

Clearly for both zero-sum and non-zero-sum games, the prudent method satisfies the security criterion. For the equilibrium method, however, the situation is different.

Proposition 18.6. *For matrix games, the equilibrium method satisfies the security criterion. However, for bimatrix games the equilibrium method does not satisfy the security criterion.*

Proof. For a zero-sum game, von Neumann's Min-Max Theorem (Theorem 15.11) shows that the equilibrium method satisfies the security criterion. Now consider the non-zero-sum game of Chicken.

$-10,-10$	$10,-5$
$-5,10$	$0,0$

Recall that row 2 and column 2 are the prudent (pure) strategies for the two players, and each of these strategies has a guarantee of -5. But there are two saddle points in this game: (row 1, column 2) and (row 2, column 1), and thus row 1 and column 1 are both equilibrium (i.e., saddle point) strategies. However, the outcome (row 1, column 1), corresponding to both players choosing these strategies pays each player -10, which is less than -5. □

When one player in Chicken swerves while the other fails to swerve, we have a saddle point outcome, which is a Nash equilibrium. Hence failing to swerve is an equilibrium strategy (as is swerving). But if both players fail to swerve, there will be a collision. The prudent pure strategy in Chicken is always to swerve, which avoids a collision. The fact that equilibrium strategies in non-zero-sum games are not necessarily prudent is part of what makes non-zero-sum game theory difficult to analyze.

The first two criteria are based on what a player can do for himself or herself. The next two criteria are based on the idea that the players form a "society", and we consider how the outcome of the game affects that society as a whole.

Definition 18.7. A method satisfies the **stability criterion** if, whenever all of the players implement the method, the outcome of the game must be a Nash equilibrium.

If an outcome is a Nash equilibrium, it means that the players will not regret their strategy choice when they see which strategy choice their opponent has made. When the outcome is not a Nash equilibrium, the situation is unstable in the sense that at least one of the players would have been better off making a different choice.

Unfortunately, for bimatrix games, it turns out that not many methods satisfy the stability criterion. The example of Chicken shows that the equilibrium method does not satisfy the stability criterion. The same example also shows that the naive method does not satisfy the stability criterion. Moreover, the game of Chase from Chapter 17 shows that even the prudent mixed method need not satisfy the stability criterion. We will see below that the situation is better for zero-sum games. But first we introduce one more criterion.

Definition 18.8. A game playing method satisfies the **Pareto criterion** if, whenever both players implement this method, every possible outcome is Pareto optimal.

Recall that an outcome is not Pareto optimal if there is another outcome that at least one player prefers and that both players agree is at least as good. When the outcome is not Pareto optimal, at least one player will regret the outcome, realizing that he could have had more at no cost to the other player.

Theorem 16.6 shows that all outcomes in a zero-sum game are Pareto optimal, so all zero-sum game methods satisfy the Pareto criterion. The following result shows that the equilibrium method stands out as special among methods for zero-sum games.

Theorem 18.9. *For matrix games, the equilibrium method satisfies the dominance, security, stability, and Pareto criteria. Moreover, any matrix game-playing method that satisfies all four of these criteria must be the equilibrium method, or a method that chooses for each player, a subset of the equilibrium strategies.*

Proof. By the previous discussion, we know that, in the zero-sum and constant-sum cases, the equilibrium method satisfies the Pareto and dominance criteria. We also know that if each player chooses an equilibrium strategy, the outcome will be a Nash equilibrium, so the equilibrium method satisfies the stability criterion. Finally, in a zero-sum game, von Neumann's min-max theorem, Theorem 15.5, shows that a Nash equilibrium always gives the players a payoff of at least their prudent strategy security level. It remains to prove the second statement. For this, we note that if a method satisfies the stability criterion, then it must clearly be a method that chooses a subset of all the equilibrium strategies for each player. □

For zero-sum games, Theorem 16.6 identifies a certain sense in which the equilibrium method is optimal. For non-zero-sum games, however, it seems more difficult to find any methods that satisfy all the criteria. In 1950, Melvin Dresher and Merrill Flood, two mathematicians working for RAND Corporation, devised a game to show that there is no method for non-zero-sum games that is optimal in this sense. That game was the Prisoner's Dilemma, one version of which we considered in Chapter 16.

	Keep quiet	Confess
Keep quiet	$-1, -1$	$-15, 0$
Confess	$0, -15$	$-10, -10$

In the literature on the Prisoner's Dilemma, the strategy of keeping quiet is usually called "Cooperate" because the players — the prisoners — are cooperating with each other. Similarly, the second row or column, which represents the strategy of confessing, is usually called "Defect".

The Prisoner's Dilemma shows that the Pareto criterion is impossible to achieve if we simultaneously demand any one of the other three criteria. This is the impossibility theorem for game theory.

Theorem 18.10. *It is impossible for a bimatrix game-playing method that satisfies either the dominance criterion, the security criterion, or the stability criterion to also satisfy the Pareto criterion.*

Proof. Suppose a certain game-playing method satisfies either the dominance criterion, the security criterion, or the stability criterion. Consider what happens if such a method is used in a game of Prisoner's Dilemma.

	Cooperate	Defect
Cooperate	$-1, -1$	$-15, 0$
Defect	$0, -15$	$-10, -10$

If the method satisfies the dominance criterion, then since defection dominates cooperation, the output the method produces must include the pure strategy of defection. This means that the outcome (defect, defect) is possible. If, on the other hand, the method satisfies the security criterion, then both players must defect, since to cooperate has a smaller guarantee. So again the outcome is (defect, defect). And finally, if the method satisfies the stability criterion, then again, both players must defect, since (defect, defect) is the only Nash equilibrium. However, any method that allows the outcome (defect, defect) violates the Pareto criterion because both players view (cooperate, cooperate) as a superior outcome. □

As social criteria go, Pareto optimality seems to be a compelling requirement. When both players agree that one outcome is worse than another, it certainly seems strange for the players to settle on the worse outcome. The naturalness of the Pareto criterion makes it even more surprising that it clashes with dominance, with prudence, and with stability. In the context of the Prisoner's Dilemma, the Pareto criterion requires cooperation while all of the other criteria require defection. It is this tension that makes non-zero-sum game theory subtle and interesting. On the other hand, as Theorem 18.9 shows, there is no such tension in zero-sum game theory.

18.2 Omnipresence of the Prisoner's Dilemma

The Prisoner's Dilemma is more than just a curious example. Versions of the Prisoner's Dilemma permeate the landscape of political and economic decision-making. It is possible to make an obsession out of finding instances of the Prisoner's Dilemma in the real world. One begins to see the Prisoner's dilemma wherever intractable decision problems loom, and one can use the Prisoner's Dilemma to explain the complex psychology of human behavior in many natural situations.

Let us now look at a number of societal decision-making problems that give rise to the paradox of the Prisoner's Dilemma.

An arms race. Countries around the world must decide how much to invest in their military. Too little invested leaves the country at risk; too much invested leaves a country broke.

Let us build a simple model of this decision problem in the context of two superpowers. We simplify by imagining that there are only two possible choices for each of these players in the game. They must choose between spending a lot on defense (building up) or spending very little on defense (disarming). The primary goal in this decision is to obtain military supremacy over the opponent, but a secondary goal is to avoid spending unnecessarily. Hence the ideal outcome for either player is to build up while the opponent disarms, while the worst outcome is to disarm while the opponent builds up. In the other two cases, where the two countries make the same decision, military parity is achieved, but both countries prefer mutual disarmament to a mutual building up, since the former bolsters both of their economies while the latter drains them.

Here are the four possible outcomes described in words from Row's point of view.

	Disarm	Build up
Disarm	Détente	Defeat
Build up	Victory	Arms race

Let us assign numerical payoffs to each outcome as follows. Military dominance is worth 10 points, while military parity is worth 5 points. Let us assume further that building up costs 2 points. That is, we subtract these two points from what would otherwise be the score. The following chart indicates the points associated with each outcome:

You	Opponent	Outcome	Your score	Opponent's score
Build up	Build up	Arms race	3	3
Build up	Disarm	Victory	8	0
Disarm	Build up	Defeat	0	8
Disarm	Disarm	Détente	5	5

Arranging these data in a bimatrix yields:

	Disarm	Build up
Disarm	$5, 5$	$0, 8$
Build up	$8, 0$	$3, 3$

Each player reasons that the strategy of building up dominates the strategy of disarming. Each player also reasons that building up is the prudent strategy. In other words, both players conclude that they should build up no matter what the opponent does. The result is the outcome (Build up, Build up) in the lower right corner. This is a saddle point, which is the unique Nash equilibrium. With this outcome, each country gets 3 points. It represents an escalating arms race in which military parity is maintained, but each country is also burdened by the expenditure. The paradox is that both countries would prefer to mutually disarm if only they could find their way to that Pareto optimal outcome.

What we have here is an instance of the Prisoner's Dilemma. Disarming is the strategy of "cooperation" while building up is the strategy of "defection". Sometimes treaties can be signed to enforce cooperation on the players and to de-escalate an arms build up. But the incentives to achieve military superiority are strong. While it is in the interest of both parties to negotiate a treaty to encourage the opponent to disarm, it is impossible to eliminate the temptation to defect and to build up, with luck unilaterally. Such thinking leads inevitably to a sustained arms race that is mutually destructive.

The only way out of this trap is for trust to develop between the two players. This may be possible in the context of repeated playing of the game. In fact, decisions on military expenditures are made repeatedly, year after year, and the prospect of retaliation after defection may provide the necessary incentive to cooperate for a player with a long-term vision. But it remains in a player's interest, in any single play of the game, to defect.

In reality, the military expenditure game is played by hundreds of independent countries, each with a wide spectrum of possible strategy choices. The two-person 2-by-2 model we have considered is a vast simplification. Nevertheless, our simple game captures the central tension between harmonious strategies of cooperation, which lead to a better world, and strategies of defection, which always lead to a better outcome for the country that defects but does so at the expense of the other countries.

Pricing gasoline. Two gas stations are located at an intersection in a small town. There are no other gas stations in the area. The two stations must decide how to price their gasoline for the day. Pricing gas too high sends all the customers to the other station. Pricing gas too low leads to vanishing profits. How do the station managers make their decision?

Let us assume that each day there is a market for 10,000 gallons of gas, and all the business goes to the station with the lowest price. However, if the two stations price the same, the business will be evenly split. For simplicity, we assume that the station managers have only two choices: price high (for a profit of 10 cents per gallon) or price low (for a profit of 6 cents per gallon). First we describe the four outcomes in words, taking Row's point of view.

	High	Low
High	Price fixing	No sale
Low	Undercut competition	Price war

Of course the term "price fixing" here should not taken too literally, since the station owners are likely forbidden by law to engage in explicit price fixing. However, it may be in their mutual interest to both post high prices — if they could. Here are the profits associated with the four possible outcomes.

Row	Column	Outcome	Row profit	Column profit
High	High	Price fixing	\$500	\$500
High	Low	No sale	\$0	\$600
Low	High	Undercut competition	\$600	\$0
Low	Low	Price war	\$300	\$300

Arranging these data in a bimatrix yields:

	High	Low
High	$500, 500$	$0, 600$
Low	$600, 0$	$300, 300$

One sees that again this is a Prisoner's Dilemma. Pricing high plays the role of cooperation and pricing low plays the role of defection. Pricing low leads to more profit, no matter how the neighbor prices his product; it is the prudent thing to do. A price war can develop, for the same game-theoretic reason that an arms race can develop. On the other hand, price fixing, either through explicit negotiation or by unspoken trust, leads to reasonably high profits for both players. By cooperating, the two stations can maximize their total income. But the temptation for one station to defect by posting lower prices and cutting out their competitor cannot be eliminated.

Safe and efficient cars. People routinely need to decide what kind of car to buy. Let us restrict our view to the tension between larger, safer cars and smaller, cheaper cars. (For ease of exposition, we ignore fuel costs and environmental concerns.) Although there are millions of cars purchased every year, we simplify by imagining that there are just two players in this game, and that they consider only the cost of the vehicles and the possibility of an accident with the neighbor's vehicle.

Let us also restrict our decision to two options: a sport utility vehicle (SUV) costing \$30,000 and a compact vehicle costing \$25,000. These prices may be considered as retail prices for the vehicles, but they do not take into account matters of safety. In order to be able to compare the outcomes on a linear scale, we convert safety concerns to dollars. A critical element of this game is that the total costs depends on what vehicle the neighbor purchases, owing to safety. Let us make some assumptions about the safety of two vehicles in a crash. When an SUV crashes with a compact car, the driver of the SUV is spared injury while the driver of the compact car may be badly hurt. We assess the cost of this risk at \$13,000 for the driver of the compact. However, in an accident involving two SUVs, the protection afforded by one SUV is precisely canceled by the ferocity of the impact from the other. We assess the cost of this risk at \$7000 per vehicle. We assess the same cost to the risk of a crash involving two compacts.

There are four possible ways that the two neighbors can decide to purchase vehicles. We obtain the following data by summing the retail cost and the safety cost of the vehicles.

Player 1	Player 2	Cost to player 1	Cost to player 2
Compact	Compact	\$32,000	\$32,000
Compact	SUV	\$38,000	\$30,000
SUV	Compact	\$30,000	\$38,000
SUV	SUV	\$37,000	\$37,000

We arrange these data in a bimatrix. We write the scores as negative numbers because they correspond to costs to the players.

	Compact	SUV
Compact	–32,000, –32,000	–38,000,–30,000
SUV	–30,000, –38,000	–37,000,–37,000

Again we have an instance of the Prisoner's Dilemma. The dominant strategy here is to purchase the SUV. When both players play their dominant strategies, the result is the purchase of two SUVs. But the players then realize that they both would have been better off had they both purchased the compact.

There are two critical aspects of this example that are characteristic of all Prisoner's Dilemmas. One is that the safety of one driver is affected by the choice made by the opponent. Without this, there can be no benefit to cooperation. The other is that the price of the vehicles creates an incentive that works to some degree against safety. One can create another version of this "tank vs. tin can" dilemma in which the Prisoner's Dilemma is created by the tension between safety and environmental concerns, or between vehicle cost and environmental concerns.

The American addiction to large cars is as difficult to break as is an arms race. Purchasers reason that they are better off in large cars, no matter what other cars are on the road. On the other hand, Americans as a whole would be mutually better off if they could all simultaneously switch to smaller vehicles. And the very best alternative for any single individual is to persuade all other Americans to drive tin cans while he or she continues to drive a tank for safety. Unfortunately, all other Americans think the same way, and so the stalemate continues.

We explore a number of other instances of the Prisoner's Dilemma game in the exercises.

18.3 Repeated Play

When playing a single instance of the Prisoner's Dilemma, it is difficult to see how a player could be enticed into cooperation. Even if one reaches an agreement in advance to cooperate with an opponent, there is no reason not to defect anyway, since there will be no future involvement with this opponent. Moreover there is no reason for the opponent to adhere to such an agreement either. There is no time to develop trust, and there is no prospect of retaliation.

When the Prisoner's Dilemma is seen as a game that is to be repeated many times, however, there emerges an assortment of ideas about strategies for play over time. Defection may be more profitable in the short term, but defection also dims the hope for cooperation in the long run. It seems as if the personalities of the players become a major ingredient in the decision process. Are they trusting? Greedy? Volatile? Stubborn? Forgiving?

These factors are not easy to model mathematically. The optimal outcome in any repeated Prisoner's Dilemma remains constantly defecting against an opponent who persistently cooperates. But it is hard to imagine why opponents would subject themselves to this kind of abuse. A policy of always defecting is likely to be eventually countered by a decision by the opponent to defect as well.

A strategy in a repeated Prisoner's Dilemma game is a protocol for playing a sequence of repeated instances of the game. Each decision is a mixed strategy. But it can depend on prior choices made by the two players.

One example of a strategy that can be considered is called **Tit-for-tat**, which works as follows. Cooperate in the first round of the game, and thereafter do precisely what the opponent did in the preceding round. The psychology of this approach is that the player starts by reaching out to the other player by cooperating. Then the player responds to cooperation with cooperation but retaliates immediately if the opponent defects. After defection, the player reverts to cooperation in response to the opponent attempting again to cooperate.

Another example of a strategy in a repeated Prisoner's Dilemma game is known as **Pavlov**, which works as follows. Cooperate in the first round of the game. Thereafter, cooperate whenever the preceding round had both players making the same move and defect otherwise. Pavlov attempts to cooperate but punishes defection with a series of counter-defections that only ends when the opponent defects as well. After that, Pavlov immediately attempts to cooperate again.

Let us look at an example of how Tit-for-tat plays against an opponent who tries both cooperation and defection.

Tit-for-tat	C	C	C	D	C	C	C	D	C	D	D	D	C	C
Opponent	C	C	D	C	C	C	D	C	D	D	D	C	C	C

By contrast, here is how Pavlov plays against the same opponent:

Pavlov	C	C	C	D	D	D	D	C	C	D	C	D	D	D
Opponent	C	C	D	C	C	C	D	C	D	D	D	C	C	C

Against either Pavlov or Tit-for-tat, never defecting works best, unless a player knows in advance how many rounds the game will last. If a player knows it is the last round, he is free to defect without facing retaliation. In fact, a series consisting of a fixed number of rounds of the Prisoner's Dilemma turns out to be much the same as a single round. However, when the number of rounds is unknown, choosing an appropriate strategy is more challenging.

Complicated strategies can be constructed that attempt to discern the personality of the opponent and take advantage of weaknesses. Cooperating in every round works well against Tit-for-tat or Pavlov. But that strategy loses badly to the strategy that always defects. In turn, the strategy that always defects does rather poorly against Tit-for-tat. There is no easy way to say what an optimal strategy is unless one knows what kind of opponent one might be facing.

Robert Axelrod of the University of Michigan has run a number of tournaments for computer programs that implement strategies for iterated Prisoner's Dilemma games. Tit-for-tat has emerged as the winner of these tournaments. It seems to encompass the essential personality traits of a good negotiator: optimism, willingness to retaliate, but forgiveness. Still, there is no theorem about the supremacy of Tit-for-tat over all other methods. It will fare poorly against an aggressive and stubborn opponent. However, Axlerod's tournaments show that, in a marketplace of algorithms sporting all kinds of behavioral characteristics, Tit-for-tat tends to do best. This is not a theorem but rather an observation based on what one might call experimental mathematics.

18.4 Irresolvability

Here are the general characteristics of a Prisoner's Dilemma: Among the strategies for each player are a dominated strategy (cooperation) that leads to a win-win situation and a dominant strategy (defection) that leads to a lose-lose situation. The difficulty is that the dominant strategy offers the possibility of a tempting jackpot, better even than the win-win payoff, but it is obtained only if the opponent plays the dominated strategy. When one player wins the jackpot, the opponent gets the sucker's payoff, which is even worse than the lose-lose payoff. The prudent game-playing method and the equilibrium game-playing method both recommend the dominant strategy in a Prisoner's Dilemma, but the result is the lose-lose outcome for both players.

The Prisoner's Dilemma is often taken as an explanation for how self-interest leads naturally to cooperative behavior. At first, self-interest seems to dictate defection in all circumstances. After all, in deciding whether to cooperate or defect, a player always fares better by defecting no matter what the opponent does. Yet when both players reason this way, the lose-lose outcome is obtained. So, paradoxically, self-interest is not in one's own self-interest in the end. Risking the sucker's pay-off, a player may attempt to make the imprudent, non-equilibrium move of cooperation. Against a like-minded opponent, the result is the win-win payoff. Thus cooperative behavior develops among players even as they focus exclusively on their own selfish fortunes. This provides a purely utilitarian rationale for the Golden Rule: "Do unto others as you would wish they did unto you."

Iterating the Prisoner's Dilemma provides a setting in which the opportunity to retaliate and the fear of the opponent's retaliation help promote cooperative behavior. There remains some debate about whether it is rational for a player to cooperate when playing a single instance of the Prisoner's Dilemma. It depends on what one means by "rational", which in turn depends on one's system of beliefs about human behavior. Do other people really reason exactly as you do? If so, unilateral cooperation is appropriate, since your opponent will do the same for the same reason that you do.

The Prisoner's Dilemma often appears when one studies behavioral interactions. Moral philosophers have cited it to explain altruism. Biologists have cited it to explain social development among organisms. Economists have cited it to explain advertising wars. Sociologists and psychologists use it to model an assortment of human behaviors that range from aggressive to cooperative.

One way to resolve the paradox of the Prisoner's Dilemma is to argue that two-player non-zero-sum games are merely zero-sum games for three players with one of the players removed. In other words, we can introduce into the game the banker who funds the payoffs, as a third player who has just one strategy choice. When this is done, it is no longer so clear that the win-win outcome is better than the lose-lose outcome. Think of the innocent public when the two prisoners are released early after cooperating. Think of the gasoline customers when the two service station managers collude to raise prices. Think of the military-industrial complex when the arms race is defused. In all these situations, cooperation helps the two players, but a third player absorbs the loss. Still, it takes a certain cynicism to argue that, when all stakeholders are considered, all games are actually zero-sum. This is a philosophy that argues that wealth is earned only by taking it from others, that happiness is created only by adding to the sadness of others, that the world can't become a better place. In fact, in many circumstances, cooperation can lead to mutual benefit with no collateral damage to others. In such situations, which we might call intrinsically non-zero-sum games, the Prisoner's Dilemma remains a paradox.

The last word on the Prisoner's Dilemma must be that there is an irresolvable tension between the two alternative strategies. The paradox is that both strategies can be defended on the basis of self-interest. The hawk is entirely correct when arguing that we should unilaterally build up our military capability, no matter what choices other countries make. The dove is entirely correct as well when arguing that we would all be better off with no military escalation instead of mutual military escalation. Game theory allows us to put this dilemma into context, but it offers no resolution.

18.5 Exercises and Problems

18.1. Two candidates are running in a close primary election. Each candidate has to make a choice with regard to campaign style. The choice is between running a positive campaign, talking about one's own accomplishments and vision, or running a negative campaign, smearing dirt on one's opponent. Let us make a few natural assumptions:

(a) Negative campaigning is effective. If one candidate campaigns negatively but the opponent chooses a positive campaign (opting not to "fight back"), then the allegations go unchallenged and the negative campaigner wins. The electorate assumes that the negative campaign was honest and justified, and the negative campaigner is not tarnished by his campaign's approach.

(b) Mutual negative campaigning angers the electorate, making it more likely that the winner of the primary will lose to the opponent from the opposing party in the general election.

If you managed one of the primary campaigns, how would you advise your candidate in making this choice? Explain your response in terms of game theory.

18.2. Consider a game with the following payoff bimatrix.

$2, 2$	$1, 3$
$3, 1$	x, x

For what values of x is this the Prisoner's Dilemma? For what values of x is it Chicken?

18.3. Here is the car-dealer game. All issues are settled except the price, which is either "high" or "low". The dealer selects one of the two prices and so does the purchaser. If they agree, then the sale is complete at that price. If dealer bids low and the purchaser bids high, they settle on an average price. If the dealer bids high and the purchaser bids low, then the sale is scuttled — the worst outcome for both parties. Model this by a 2-by-2 bimatrix and analyze it for naive and prudent strategies, dominance, and saddle points. Is this game an instance of the Prisoner's Dilemma? If not, what game is it?

18.4. At a global environmental summit, the Eastern and Western hemispheres each form a coalition to decide whether to sign on to a major air pollution accord. Curiously, owing to prevailing winds, air pollution emitted in one hemisphere has the effect of poisoning the air only in the other hemisphere. Both parties to the potential agreement have environmental concerns as their top priority, but a secondary consideration is that agreeing to the accord will extract an economic cost.

(a) Rank the four possible outcomes (depending on whether the West and East sign or don't sign) from the separate points of view of the West and the East. Use -1 for the best outcome, -2 for the next best, then -3, and then -4 for the worst. The goal is to produce a Prisoner's Dilemma.

West	East	Your cost	Neighbor's cost
Sign	Sign		
Sign	Don't sign		
Don't sign	Sign		
Don't sign	Don't sign		

(b) Put these rankings into a 2-by-2 bimatrix to create a game with the West as Row.

	Sign	Don't
Sign		
Don't		

(c) Is there a saddle point? What is the prudent strategy? What is the reduced form of this game?

18.5. A country suffering from high national debt is in the throes of a two-party presidential election. To address the debt, the next president will have to raise taxes. But the two candidates know that it is political suicide to campaign on a pledge to raise taxes. In fact, the electorate prefers a candidate who vows to *lower* taxes. Unfortunately, any candidate who vows to lower taxes faces an awkward choice if elected: Renege on their promise, alienating the populace, or carry through on their promise, bankrupting the country. Model this scenario as a Prisoner's Dilemma. What assumptions are needed to make certain that the outcomes are ranked in Prisoner's Dilemma fashion?

18.6. In pursuing their foreign relations, countries establish individual personalities. There are those countries whose foreign policy displays resoluteness, and there are countries whose foreign policy displays friendliness. In a summit between two countries, there is a certain gamesmanship. We simplify by assuming that each country must choose one of two strategies: being resolute or being friendly. Friendliness is perceived as catastrophic weakness when facing a resolute opponent, and this leads to lopsided strategic advantage for the opponent. When facing a resolute opponent, on the other band, resoluteness is perceived as belligerence. On the other hand, mutual friendliness leads to peaceful coexistence, though without the benefit of a strategic advantage for either country.

(a) Model this game as a Prisoner's Dilemma by filling in the following bimatrix, with US as the row player, and THEM as the column player. Score the outcomes 4, 3, 2, and 1 (where 4 is best and 1 is worst) for each player.

	Friendly	Resolute
Friendly		
Resolute		

(b) Is there a dominant strategy for US in this game?
(c) What is the prudent strategy for US?
(d) What is the naive strategy for US?
(e) What is the Nash equilibrium strategy for US?

18.7. Imagine a country in which there are two principal political parties, one liberal and one conservative. Each of the two parties must opt to follow either a moderate or an extreme agenda. When one party is extreme while the other is moderate, the extreme party dictates the agenda and wins political control, while the other party is seen as capitulating and loses. When both parties are moderate, a pragmatic bipartisan agenda ensues. When both parties are extreme, a polarized and paralyzed legislature becomes mired in gridlock. Model this by a Prisoner's Dilemma. What does the model predict?

18.8. Two cyclists in a bicycle race have the strength to pull away from all the other cyclists in the pack. They can decide either to pull away unilaterally or to pull away conditionally. If they pull away unilaterally, they compete evenly for the victory in the race. If they pull away conditionally, it means that they keep their eye on the opponent and if the opponent pulls away unilaterally, they follow in the wind stream, riding comfortably in second place. This is known in the racing lingo as drafting. The follower in this position can win the race in the end, because wind resistance tires the racer riding in front, and the racer riding in second place easily passes the leader in the sprint to the finish. If both racers decide to follow conditionally, they remain in the pack, squandering their advantage, and both have in the end only a 10% chance of winning the race. What does game theory suggest?

18.9. Consider the **saddle-equilibrium method** in which a player chooses a saddle point strategy if the game has a saddle point, and otherwise chooses a mixed strategy Nash equilibrium. Consider the four criteria: dominance, security, stability, and Pareto. As a method for zero-sum games, which of the four criteria does the saddle-equilibrium method satisfy? What about as a method for non-zero-sum games?

18.10. We say in a game that row k **strictly dominates** row i if $u_{k,j} > u_{i,j}$ for all j. Similarly, we say that column ℓ **strictly dominates** column j if $v_{i,l} > v_{i,j}$ for all i. Prove that the equilibrium method satisfies the following strong dominance criterion: If one strategy strictly dominates another in a game, then the dominated strategy has probability zero in every Nash equilibrium strategy.

18.11. Prove that for any non-zero-sum game, the guarantee of a player's prudent mixed strategy is greater than or equal to the guarantee of the player's prudent pure strategy. Use this to conclude that the prudent mixed method satisfies the security criterion.

Notes on Part III

While game-theoretic thinking goes back to ancient times, modern game theory began with the 1928 paper [47] published by a 25-year-old mathematician named John von Neumann. In this paper, von Neumann showed that each player in a two-person zero-sum game has an optimal mixed equilibrium strategy (Theorems 15.6 and 15.11). Von Neumann went on to become one of the most influential mathematicians of his time. After immigrating from Europe in 1930, he was a founding member (along with Albert Einstein) of the Institute for Advanced Study in Princeton, New Jersey. He is also known for his work in quantum physics and as one of the inventors of the modern computer. In 1943, von Neumann coauthored the book *The Theory of Games and Economic Behavior*, with economist Oskar Morgenstern [48], which brought game theory to the attention of the world. The scenario in Chapter 15 comes from this book.

Even though von Neumann introduced the equilibrium concept to game theory, the job of extending the idea to non-zero-sum games fell to another mathematician, John Nash. Nash grew up in a small town in West Virginia, and after college, enrolled in the mathematics graduate program at Princeton in 1948, intending to study game theory. Nash tried to discuss his ideas on non-zero-sum games with von Neumann, but von Neumann showed little interest. By this time, von Neumann was interested in cooperative game theory, where players in an n-person game can make bargains. In his 1950 dissertation, Nash showed that every n-person game has at least one mixed strategy equilibrium (Theorem 17.5; our proof is a simplification of the one in [36]). Because of this theorem, equilibria in game theory are now called Nash equilibria. After graduation, Nash accepted a position on the mathematics faculty at MIT, but was forced to resign in 1957 after being hospitalized for schizophrenia. John von Neumann died of cancer in the same year.

Starting in 1950, the federal government sponsored game theory research, hoping to use game theory to advance the US cause in the Cold War. Much of this work was carried out at the RAND Corporation in Santa Monica, California, where both von Neumann and Nash intermittently served as a consultants. It was at the RAND Corporation in 1950 that Melvin Dresher and Merrill Flood (see [19]) invented a counterexample showing that a Nash equilibrium need not be Pareto optimal (see Theorem 18.10). Later, Albert Tucker devised a story about two prisoners to go with this game, and named the game the Prisoner's Dilemma. In 1954, another RAND scientist, J. D. Williams, wrote an entertaining short book, *The Compleat Strategyst* [49], to explain game theory to the public (and especially RAND's military sponsors). The cross-over algorithm comes from that book, as does Problem **15.18**. A famous game theory paper from the Cold War era is O. G. Haywood's [27] game-theoretic analysis of the Battle of the Bismarck Sea, which is the basis for our analysis in Chapter 13. An overview of game theory during this period is given in the William Poundstone's book *The Prisoner's Dilemma* [38]. Game theory is generally not regarded as having lived up to its promise in the Cold War. However, by the 1990s, the impact of game theory on economics was so substantial that the 1994 Nobel Prize in Economics was awarded to John Nash for his earlier contributions to the subject. By that time, Nash was sufficiently recovered from his mental illness to travel to Sweden and receive the prize. The remarkable story of Nash's illness and his recovery is recounted in the book *A Beautiful Mind* by Sylvia Nasar [35]. Nash died in a car crash in 2015 at the age of 86.

The Prisoner's Dilemma game alone is the subject of hundreds of books and papers, having become a metaphor for many social and political problems (see Chapter 18). In his seminal book *The Evolution of Cooperation* [4], political scientist Robert Axelrod discusses the question of how players forced to repeatedly play the Prisoner's Dilemma can learn to cooperate. Game theory has even become an important part of the theory of biological evolution. Evolutionary game theory is discussed in the book *Evolution and The Theory of Games* [34] by the biologist John Maynard Smith, who introduced the idea of an evolutionarily stable strategy. Biologists generally refer to the game of Chicken as the Hawk and Dove game.

There are many good elementary books about game theory. Three of the books that have influenced our treatment of game theory in this book are: *Game Theory and Strategy* by Philip Straffin [43], *Game Theory* by Morton Davis [14], and *Game Theory and Politics* by Steven Brams [11].

Probability theory, the subject of Chapter 14, is not usually considered part of game theory, but at its core it is all about games of chance. The first book on probability was *Liber de Ludo Aleae* ("Book on Games of Chance") written by the 14th century Italian doctor, lawyer, mathematician and gambler Gerolamo Cardano. In it, Cardano describes the method of calculating probabilities by counting equally likely outcomes, as we did in Section 14.1. Starting in 1654, two French mathematicians Blaise Pascal and Pierre de Fermat, began a long correspondence about a gambling problem called the "unfinished game" (Problem **14.14**, see [17]). Out of this correspondence came the idea of expected value. At age 30, Pascal had a mystical experience and gave up mathematics in favor of religion. But still a mathematician at heart, Pascal gave a probabilistic argument in favor of belief in God. The idea of subjective probability, which allows probability to be applied outside gambling, was first proposed by the 18th century Scottish mathematician and Presbyterian minister Thomas Bayes. Bayes was also the first to give a good account of independence.

Part IV

The Electoral College

Introduction to Part IV

Why did our founding fathers see fit to create the Electoral College as a means to decide presidential elections in the United States? One answer is that they did not believe that the populous could be trusted with such an important decision. In the 18th century, the country was sparsely populated and there was no technology to spread information about presidential candidates to voters everywhere. The members of the Electoral College could journey to the nation's capital to become informed about the candidates, and through these electors the people of the nation could have a voice, if indirectly, in the presidential election. Direct election of the president by popular vote was also seen as a threat to the separation of powers between the three branches of government, and some worried that it would accord too much power to the executive branch and lead to tyranny.

But these considerations seem no longer relevant in today's world. It is easy to object to the Electoral College on the grounds that it violates the principle of "one person, one vote". So why do we still retain the Electoral College system? Why not replace the Electoral College with a direct plurality (often called a popular vote) election for president? One answer is that abolishing the current system would necessitate a constitutional amendment, and our Constitution is designed to be difficult to amend. In particular, three quarters of all states must ratify any amendment before it takes effect. There have been numerous attempts to amend the Constitution to abolish the Electoral College, going all the way back to 1825. All have failed. In defense of the Electoral College, some argue that it gives extra weight to small states and that this extra weight is appropriate in order that presidential campaigns pay attention to the constituency of small states. It isn't easy to amend the Constitution when the amendment is deemed harmful to small states. After all, most states are small. To be precise, 34 of the 50 states currently have populations that are below average.

One can adopt two points of view in studying the Electoral College. In the first view, one thinks of the Electoral College as a system in which the states themselves are the voters, and each state is permitted to cast a certain number of votes. We study such weighted voting systems in Chapter 19. In the second view, one thinks of the Electoral College as a system in which individuals are the voters, and the election is conducted via a bloc voting method in which the voting blocs are the states in the union. We adopt this view in Chapter 20, where we address the question of whether the Electoral College really does benefit small states. We draw the surprising conclusion that, in a certain sense, it is the voters in large states who benefit from our current system.

This part of the book draws on material from each of the other three parts. The Electoral College is a voting system (Part I) that depends on the apportionment of representatives in the House (Part II) and whose analysis requires a probabilistic viewpoint (Part III).

19

Weighted Voting

"All animals are equal but some animals are more equal than others."
— George Orwell

19.0 Scenario

Five countries, call them A, B, C, D, and E, form a union (perhaps to provide for common defense or to establish an export cartel). The countries agree on a system of governance by which yes-no decisions are made by voting, assigning each country a number of votes roughly proportional to its population, as follows:

COUNTRY	VOTES
A	9
B	8
C	6
D	3
E	1

Resolutions pass if at least 14 of the 27 votes are in favor.

Is it better to be country D than country E? The question is not merely whether it is at least as good to be country D as country E. The question is whether it is *strictly* better. Is it better to be C than D? B than C? A than B?

Suppose that country D annexes a neighboring island and suddenly increases in population. The union agrees to increase their number of votes to 5, keeping the number of votes for other countries the same, so that the new list of votes is 9, 8, 6, 5, and 1. Resolutions now pass with 15 out of the 29 possible votes. What impact does this increase have on each of the 5 countries?

If you were lobbying this union of countries on some issue, how much effort would you spend trying to persuade the representative from country E to adopt your position? Answer this question for both the voting scheme before the annexation and the one after.

In terms of voting power within the union, which countries benefit from the annexation? Is anything strange about this?

19.1 Weighted Voting Methods

There are times when anonymity is an essential criterion for a voting method, but there are times when it isn't. In certain settings, egalitarianism is a bedrock principle. But in other settings, according voters equal treatment is neither appropriate nor fair. For example, shareholders customarily vote in corporate elections not with one vote per shareholder but rather with as many votes for each shareholder as shares they own. The principle seems to be that a shareholder who has invested twice as much in a company is entitled to twice as much power, twice as much influence, over decisions made at that company. In this chapter, we investigate whether weighted voting really does achieve this goal of distributing power across voters in the anticipated way. We will see that the number of votes assigned to an individual voter is not necessarily directly proportional to the power that the voter has in the election. To make precise sense of this, we will need to explain what we mean by voting power.

The Electoral College is another example of a voting method in which the voting entities do not have equal voices in the process. Here we are modeling the Electoral College as a method with 51 voters — one for each of the states and one for the District of Columbia — in which each voter gets a number of votes equal to the number of members of Congress assigned to them (except for the District of Columbia, which gets 3 votes by agreement). This model is not an exact representation of what actually occurs in presidential elections in the United States, for two reasons. First, it isn't quite correct to say that states vote in the Electoral College. There are actual people who are elected to membership in the Electoral College, and in the end these electors are permitted to vote as they choose, even when this betrays their constituents. Second, while most states have a winner-take-all system, Maine and Nebraska have special rules that allow them to split their votes between candidates. We will ignore these anomalies in this chapter and study an idealized Electoral College in which each of the 51 voters casts one (indivisible) vote that is given a certain prescribed weight.

Those weights are determined by the apportionment of the House of Representatives from the preceding census. The state of Maryland was accorded 8 seats in the House after the 2010 census. With its two senators, the state has 10 members in Congress. So Maryland's vote for president is given a weight of 10 in the Electoral College system.

From now on in this chapter, we will assume, as we did in Chapter 1, that there are just two candidates. We will consider two types of weighted voting methods, each of which comes in two varieties. In all of these methods, we let n be the number of voters, and for each i from 1 to n, we assign a nonnegative number w_i to voter i. We call w_i the **weight** of voter i, but it can also be thought of as the number of votes accorded to voter i. Let $t = w_1 + w_2 + \cdots + w_n$.

In the first type of method, which we call **simple weighted voting**, a candidate wins by getting more than $t/2$ votes. If neither candidate gets more than $t/2$ votes, then the result is a tie. In Chapter 1 we called this a "weighted voting method". One variation on this method is what we call **simple weighted voting with status quo**. To implement it, we designate one of the two candidates as the **status quo** candidate. The other candidate is called the **challenger**. The outcome is the same as with simple weighted voting as long as there is no tie, but if a tie occurs then the status quo candidate is declared the unique winner. The main difference between the version without status quo and the verion with status quo is that the former lacks decisiveness (there can be ties) while the latter lacks neutrality (it treats the candidates differently).

In the second type of method, which we call a **supermajority weighted voting**, we fix a number q called the **quota**. We assume the quota satisfies $t/2 < q \leq t$. A candidate wins under this method by getting q or more votes, and if neither candidate gets q votes, then the result is a tie. We denote this voting method by $V(q; w_1, w_2, \ldots, w_n)$. One can also consider a version of supermajority weighted voting that adds a status quo candidate.

We will require the weights in all of these weighted voting methods to be positive real numbers. Most often they are whole numbers. Weights equal to zero may be permitted, although when $w_i = 0$ it is simpler to remove voter i from the electorate. A voter is called a **dummy** if the results of the election never depend on how the voter's vote is cast. A voter with weight zero is clearly a dummy. Notice that if all the candidates have the same weight (i.e., $w_i = 1$ for all i) then simple weighted voting is the same as the simple majority method from Chapter 1 (with or without a status quo), and supermajority weighted voting is the same as the supermajority method from Chapter 1. Finally, we observe that a simple weighted voting method is a special case of supermajority weighted voting for some q just a little bigger than $t/2$

Example 19.1. In 1958, the European Economic Community, known informally as the Common Market, was formed by the Treaty of Rome as a federation of 6 nations: France, Germany (then West Germany), Italy, Belgium, the Netherlands, and Luxembourg. The Common Market was to be governed by a Council of Ministers, one from each nation, with a fixed method of voting as follows: France, Germany, and Italy would get 4 votes each, Belgium and the Netherlands 2 each, and Luxembourg 1, and 12 votes would be required to pass any resolution. This is the supermajority weighted voting method denoted $V(12; 4, 4, 4, 2, 2, 1)$. If France, Germany, Belgium, and the Netherlands favored a measure in this system, that measure passed with $4 + 4 + 2 + 2 = 12$ votes (or, put another way, with votes of total weight 12). If only France, Germany, Belgium, and Luxembourg favored the measure, it failed to pass, earning only $4 + 4 + 2 + 1 = 11$ votes.

What is notable about this voting method is that Luxembourg, although endowed with a vote, turns out to be a dummy. To see this, note that the weight of every voter other than Luxembourg is an even number and the quota 12 is an even number as well. That means that without Luxembourg, a measure either passes with 12 or more votes in support or fails with 10 or fewer votes in support. In the latter case, Luxembourg's additional vote doesn't help, since that would bring the total weight in favor of the measure to 11 or less. If Luxembourg is in favor of a resolution that passes, then it would have passed even if Luxembourg had voted against it. If Luxembourg is against a resolution that fails, then it would still have failed had Luxembourg voted for it. Hence, Luxembourg's vote never has an impact on the result of any election in this system.

Example 19.2. In 2001, early in the term of the Republican presidential administration of George W. Bush, Senator Jim Jeffords of Vermont announced his resignation from the Republican party to become an Independent. This left behind a Senate of 50 Democrats, 49 Republicans, and Jeffords himself, the lone Independent. Because the Vice President of the United States is President of the Senate and casts tie-breaking votes in the Senate, the body effectively consisted of 50 Democrats, 50 Republicans, and 1 Independent. Assume for the moment that each of these constituencies votes as a bloc. (This assumption is false, but it does model the highly partisan atmosphere for which the modern Senate is notorious.) One can then model the Senate as a body that consists of just three voters, who cast votes of weight 50, 50, and 1, respectively. Since 51 votes are needed to pass legislation in the Senate, Jeffords departure from the Republican party created the simple weighted method $V(51; 50, 50, 1)$.

The wisdom of Jeffords's decision now becomes apparent. In $V(51; 50, 50, 1)$, no single voter has enough weight to make a measure pass, but any two voters taken together do. So one can put into simple terms the way this voting method works: Measures pass if and only if two or more of the three voters support it. There is perfect symmetry between the three voters, and each voter has the same influence over the outcome of any vote as any other voter. While the Democratic party and the Republican party have a vote whose weight is 50 times that of Jeffords, any reasonable definition of voter power will assign these three voters precisely equal power within this system. By leaving the Republican party, Jeffords became as powerful in the Senate as each of the political parties as a whole!

This example illustrates that $V(51; 50, 50, 1)$ operates identically to $V(2; 1, 1, 1)$. To an outside observer interested only in whether a measure passes or fails, these two methods are indistinguishable. In either method, any two voters out of the three create a winning coalition.

19.2 Non-Weighted Voting Methods

How can we tell if a voting method is a weighted voting method? Usually, we know a method is a weighted voting method because someone tells us it is. For example, the current Electoral College is a weighted voting method, with parameters

$$V(270; 55, 38, 29, 29, 20, 20, 18, 16, \ldots, 4, 3, 3, 3, 3, 3, 3, 3, 3).$$

But sometimes a voting method is described not by telling us the weights of the voters. Rather it may be given by describing the rules that define the method in some other sort of way. We might then want to know whether a voting method produces results identical to the results produced by some weighted voting method. If so, we can say that the original method is a weighted voting method.

Recall that a voting method is **monotone** if, whenever a voter changes her vote to a certain candidate, that cannot cause the candidate to go from winning to losing. Recall also that a voting method is **unanimous** if, whenever every voter votes for a certain candidate, that candidate must in fact be the unique winner. In Chapter 1 we saw that every weighted voting method satisfies these two criteria.

For the remainder of the chapter we will study two-candidate voting methods that satisfy the monotonicity and unanimity criteria. We begin with an example of a voting method that does not look like a weighted voting method but turns out actually to be one.

Example 19.3. The United Nations Security Council is a body consisting of 15 members, 5 of whom (China, France, England, Russia, and the United States) are permanent members. A measure before the Council passes if it has 9 members in support of it, provided that no permanent member is against it. (We assume that no nation ever abstains from decision making, so we make no distinction between not supporting a measure and vetoing it.) Is this a weighted voting method? Is it possible to assign weights to the 15 members of the Council and to find a quota q so that measures pass if and only if the weights of the supporting members sum to at least q?

The answer is yes. The UN Security Council voting method turns out to be

$$V(49; 9, 9, 9, 9, 9, 1, 1, 1, 1, 1, 1, 1, 1, 1, 1). \tag{19.1}$$

To see this, note that any measure with the support of 9 members including all 5 permanent members gets at least $9+9+9+9+9+1+1+1+1 = 49$ votes and passes. On the other hand, any measure that fails to win the support of one of the permanent members of the Council can get at most $9+9+9+9+1+1+1+1+1+1+1+1+1+1 = 46$ votes (even if every other member votes in favor), and so such a measure necessarily falls to defeat.

Expression (19.1) is not the only way to represent the United Nations Security Council as a weighted voting method. Certainly doubling all the weights and doubling the quota yields an identical method $V(98; 18, 18, 18, 18, 18, 2, 2, 2, 2, 2, 2, 2, 2, 2, 2)$. Moreover, it is not hard to see that $V(39; 7, 7, 7, 7, 7, 1, 1, 1, 1, 1, 1, 1, 1, 1, 1)$ is also a representation of the Council as a weighted method.

Before we proceed further it will be useful to introduce some terminology. By a **coalition** Y we simply mean a subset of the set of n voters. We imagine the voters who comprise the coalition Y to be like minded in some way. For example, we might take Y to be the set of all voters who support candidate k. One possibility is that *all* the voters support k, in which case Y is the set of all n voters. This is called the grand coalition. At the other extreme, if none of the voters support k, then Y is the empty set ϕ. There are, of course, many coalitions Y in between these two extremes. When there are n candidates, then altogether there are 2^n possible coalitions, including the empty coalition and the grand coalition.

Recall that a voting method is called neutral if it treats both candidates the same. The weighted voting methods with no status quo candidate are all neutral, whereas the weighted voting methods that have a status quo candidates are generally not. One often encounters non-neutral voting methods in legislative bodies whose job it is to approve or reject proposals. We think of these as **yes-no voting methods** — method with candidates "yes" and "no". By design, such methods are usually decisive (cannot result in ties), non-neutral, and "no" is usually the status quo. More generally, when we consider any method that is non-neutral, we will always consider the first candidate (i.e., "yes") to be the challenger and the second candidate (i.e., "no") to be the status quo.

Definition 19.4. Suppose a voting method is neutral. A coalition Y is called a **winning coalition** if a candidate wins whenever every voter in Y votes for that candidate. In a non-neutral voting method, a coalition Y is a winning coalition if the challenger wins whenever every voter in Y votes for the challenger. In both situations, we include the case where a candidate and the opponent tie as wins.

Now we are ready to look at an example of a voting system that is not a weighted voting system.

Example 19.5. Proposed legislation in the United States becomes law if it has the support of a majority of the Senate, a majority of the House of Representatives, and the president. If the president does not support a bill (and we assume she vetoes it), then the bill can still become law if it has the support of 2/3 of the Senate and 2/3 of the House of Representatives (in which case the president's veto is overridden). The vice president also has a role: He votes to break ties in the Senate. We can think of this entire process as a voting method involving 537 voters: 100 senators, 435 representatives, one president, and one vice president. Is it a weighted method?

It is instructive to try to find weights for these 537 voters and a quota that represents the manner in which legislation is passed in the United States. After trying for a while, one begins to suspect that it is impossible to do so. It is in fact impossible, as the following proposition attests.

Proposition 19.6. *The United States system for passing legislation is not weighted.*

Proof. Suppose that the United States system is, in fact, a weighted method, with quota q and weights for each of the 537 voters. We aim to derive a contradiction from this assumption. To help visualize the proof, imagine assigning the colors red, white, and blue to the members of the House of Representatives and the Senate. For example, red could be assigned to the most conservative members, white to those in the middle, and blue to the most liberal. Let us suppose that 3/7 of the members are colored red, 2/7 of the members are colored white, and 2/7 of the members are colored blue. It really doesn't matter which members get which colors, nor does it matter that the number 435 isn't precisely divisible by 7. It is sufficient if the fractions are merely approximate.

Let the symbols HR, HW, HB, SR, SW, and SB stand for the total weight (the sum of the weights of the votes), respectively, of the red representatives, white representatives, blue representatives, red senators, white senators, and blue senators.

Consider the coalition of all representatives and senators who are either red or white. Since this coalition consists of 5/7 of each body, and since our Constitution gives such a coalition the power to override a veto, it can, without the support of the president, vice president, or a single blue members of the House or Senate, force a bill to pass. This means that the total weight of this coalition must be at least q. We must therefore have

$$HR + HW + SR + SW \geq q. \tag{19.2}$$

Next consider the coalition consisting of all representatives and senators who are either red or blue. Similar reasoning leads to

$$HR + HB + SR + SB \geq q. \tag{19.3}$$

Now consider the coalition consisting of all the senators but only the red representatives. This coalition cannot force a bill to pass without further support, because they are short of the majority needed in the House to pass the bill. It follows that they weigh less than the quota, which implies that

$$HR + SR + SW + SB < q. \tag{19.4}$$

Similarly, the coalition consisting of all the representatives but only the red senators is not a winning coalition, so

$$HR + HW + HB + SR < q. \tag{19.5}$$

Summing the two inequalities (19.2) and (19.3) yields

$$2HR + HW + HB + 2SR + SW + SB \geq 2q. \tag{19.6}$$

On the other hand, summing the two inequalities (19.4) and (19.5) yields

$$2HR + HW + HB + 2SR + SW + SB < 2q. \tag{19.7}$$

But inequalities (19.6) and (19.7) are contradictory. Hence the method is not weighted. □

19.3 Voting Power

In a weighted voting method, the justification for assigning different weights to votes is usually that different voters are somehow entitled to different influence over the outcome of the election. But what does it mean to have influence over the outcome of an election? The only situation in which a particular voter influences an election is if the candidate she voted for won but would have lost had it not been for her vote. Absent this situation, either the candidate loses, and there is nothing the voter can do about it, or the candidate wins, but the voter's support was not needed. If it were not for the possibility of altering the outcome of an election, most voters would just as soon stay home.

This is the argument that attorney John Banzhaf (currently a law professor at The George Washington University) used in 1964 to argue a case in the New York State Supreme Court on behalf of a group of Nassau County (NY) voters who believed their district was underrepresented in the County Board of Supervisors (see Problem 19.6). Banzhaf went on to apply a similar analysis to the Electoral College, a version of which we will present in Chapter 20. But first, in this chapter, we study this idea in a more general context.

To estimate how much influence a voter has in an election, we will model the election as a random process. Suppose voter i, who is one of the n voters in the election, supports a particular candidate (who we assume to be the challenger in the case that the voting method is non-neutral). We assume the other $n-1$ voters, besides voter i, cast their ballots at random, independently, and choosing one of the two candidates, each with a probability of 1/2. (The basic ideas of probability theory are discussed in Chapter 14.) This model of an election is not meant to be a realistic portrayal of voter behavior. Rather, it is meant to reflect a complete lack of information about how the voters other than i will behave. As a result of these assumptions, all the possible profiles in which voter i votes for the given candidate are considered to be equally likely. In particular, this number of profiles is equal to 2^{n-1}, the number of coalitions to which voter i belongs. Since all outcomes are equally likely, probabilities can computed by merely counting profiles.

Definition 19.7. Let Y be a winning coalition that has voter i as a member. Voter i is called **critical** in coalition Y if the coalition obtained from Y by removing voter i is no longer a winning coalition.

Definition 19.8. The **Banzhaf power** of a voter k is equal to the number of winning coalitions in which voter k is critical.

It is easy to see that if a weighted voting method gives all the voters the same weights, then it endows every voter with equal Banzhaf power. This includes the simple majority method. Anonymity does indeed impose egalitarianism. But our interest here is primarily on methods that violate anonymity, like weighted voting methods, that award different degrees of influence on decision-making to different voters. In this setting, we expect voters with greater weight to have greater power.

Example 19.9. Consider the voting method $V(11; 6, 5, 4, 3, 2)$. Call the 5 voters A,B,C,D,E in decreasing order of weight. Thus the coalition consisting of the voters A and B, which we denote simply AB, has weight 11 and is therefore a winning coalition. Voter A is critical in this winning coalition, because without voter A, coalition AB becomes the coalition consisting of voter B by himself, a coalition of weight only 5, which is below the quota. To compute the Banzhaf power of voter A, we must count all the winning coalitions that have A as a critical member.

There are $2^5 = 32$ possible coalitions, and we list them here, according to their weight:

ABCDE (20); ABCD (18); ABCE (17); ABDE (16); ACDE, ABC (15);
BCDE, ABD (14); ABE, ACD (13); ACE, BCD (12);
ADE, BCE, AB (11); BDE, AC (10); CDE, AD, BC (9); AE, BD (8);
BE, CD (7); CE, A (6); DE, B (5); C (4); D (3); E (2); ϕ (0).

This is a neutral method, and 15 of these are winning coalitions by virtue of having at least 11 votes.

Now consider just the coalitions to which A belongs. There are 16 of them:

ABCDE (20); ABCD (18); ABCE (17); ABDE (16); ACDE, ABC (15);
ABD (14); ABE, ACD (13); ACE (12); ADE, AB (11);
AC (10); AD (9); AE (8); A (6).

Note that A belongs to $2^4 = 16$ coalitions, which is exactly half of the $2^5 = 32$ coalitions altogether.

Of the coalitions to which A belongs, 12 are winning, with a weight of at least 11 (the quota):

ABCDE (20); ABCD (18); ABCE (17); ABDE (16); ACDE, ABC (15);
ABD (14); ABE, ACD (13); ACE (12); ADE, AB (11).

Among these, 9 have voter A as a critical member:

ABDE (16); ACDE, ABC (15); ABD (14); ABE, ACD (13);
ACE (12); ADE, AB (11).

This is because the coalition that remains after A leaves any of these coalitions is no longer winning. The Banzhaf power of voter A is therefore 9. This reflects the fact that the probability that the vote of voter A will alter the outcome of the election is 9/16.

We note in passing that A is not critical in the three winning coalitions

ABCDE (20), ABCD (18), ABCE (17),

because if A leaves any of these, the remaining coalitions,

BCDE (14), BCD (12), BCE (11),

are all still winning. In these cases, the vote of A is not necessary for a victory.

We are interested not so much in the Banzhaf power of any one voter but in comparing the Banzhaf power of two different voters. In elections with many voters, the probability that a single voter's contribution will be critical is certain to be minuscule. Our interest is not so much to complain about the smallness of this number and therefore about the futility of voting but rather to compare the relative sizes of two such numbers to discern whether voting power is really reflected well by the number of votes accorded each voter.

Continuing the analysis of the example, voter B is critical in precisely 7 winning coalitions:

ABC, BCDE, ABD, ABE, BCD, BCE, AB;

voter C in 5:

BCDE, ACD, ACE, BCD, BCE;

voter D in 3:

ACD, BCD, ADE;

and voter E in 3:

ACE, ADE, BCE.

Therefore the Banzhaf powers of the 5 voters are 9, 7, 5, 3, and 3.

These numbers reveal some surprising information. First, it is notable that voters D and E, with 3 and 2 votes respectively, turn out to have precisely equal power in this voting method. So if it had been the intention that voter D should have a voice that is 50% stronger than voter E, the effort has failed. Moreover voter A, who has precisely twice as many votes as voter D, turns out to have precisely three times the power. This illustrates that doling out votes as a way to apportion influence over decision-making does not work exactly as one might expect.

Let us revisit Example 19.1, from which even stranger results emerge. Recall that in this voting scheme, there were 6 voters representing countries with initials F, G, I, B, N, and L, with votes of weight 4, 4, 4, 2, 2, and 1, respectively. We could try to write down all the coalitions and check to see which are winning ones, but with $2^6 = 64$ coalitions, this would be a large undertaking. Instead, because the quota is set to 12 in this example, we find that there are only 14 of the 64 coalitions that are winning. We list them, along with their weights and their critical members, in Table 19.1.

Table 19.1 European Economic Community of 1958.

Coalition	Weight	Critical members
FGIBNL	17	None
FGIBN	16	None
FGIBL	15	F, G, I
FGINL	15	F, G, I
FGIB	14	F, G, I
FGIN	14	F, G, I
FGIL	13	F, G, I
FGBNL	13	F, G, B, N
FIBNL	13	F, I, B, N
GIBNL	13	G, I, B, N
FGI	12	F, G, I
FGBN	12	F, G, B, N
FIBN	12	F, I, B, N
GIBN	12	G, I, B, N

We determine the Banzhaf power of each voter here by counting the number of times a voter appears in the third column. This is 10 for France, Germany, and Italy, 6 for Belgium and the Netherlands, and 0 for Luxembourg. This calculation reveals what we observed earlier and what must have become soon apparent to Luxembourg back in 1958. Luxembourg was entirely powerless in this voting method, because the probability that their vote mattered was exactly 0. There is no possible circumstance in which changing Luxembourg's vote would have altered the outcome of any election. Also the ratio of the number of votes for France and the number of votes for Belgium was 4-to-2, or 2-to-1. But the ratio of their power was 10-to-6, or 5-to-3. France could well have argued that their constituency was in effect less well represented by their vote in the Council, by comparison with Belgium, than the vote of weight 4 would suggest.

We also revisit Example 19.2, in which we modeled the defection of Jim Jeffords from the Republican party. We modeled this by assuming there were three voters. These voters use the weighted voting method $V(51; 50, 50, 1)$, and we now abbreviate them naturally as D, R, and I. Here the winning coalitions and critical members are summarized in the Table 19.2.

Table 19.2 The party of Jim Jeffords.

Coalition	Weight	Critical members
DRI	101	None
DR	100	D, R
DI	51	D, I
RI	51	R, I

The Banzhaf power is 2 for each voter. Although voter D has 50 times as many votes as voter I, we learn that voter I has a precisely equal voice in the method as does voter D.

19.4 Power of the States

If the primary target of our study is the Electoral College, we should endeavor to compute the Banzhaf power of each of the states in order to check whether there are the kinds of curious anomalies that one finds with $V(12; 4, 4, 4, 2, 2, 1)$ or with $V(51; 50, 50, 1)$. But this task is easier said than done. The Electoral College voting method has 51 voters, and so the number of coalitions is $2^{51} = 2{,}251{,}799{,}813{,}685{,}248$, which is a large number. As a result, even the fastest computers cannot compute the exact Banzhaf power of a state in a reasonable time.

So we adopt the following approach to approximating the power of the states in the Electoral College system. The actual numerical Banzhaf power of each state will be a number in the trillions. What interests us is not the precise magnitude of this number but the ratio of two of these numbers. For example, we may wish to focus on the states of California, with 55 electoral votes, and South Dakota, with 3 electoral votes. The ratio between these numbers is $55/3 \approx 18.33$. We'd like to estimate the ratio between their Banzhaf powers. If this number should turn out to be close to 18.33, then we will conclude that the Electoral College accomplishes what it seems to. If this number should turn out to be far from 18.33, then we have an indication that the Electoral College does not apportion influence as we expect it to.

The ratio of the Banzhaf power of California to the Banzhaf power of South Dakota is equal to the ratio of two probabilities, the probability that a randomly chosen coalition will have California as a critical member divided by the probability that a randomly chosen coalition will have South Dakota as a critical member. Unfortunately, we cannot compute these theoretical probabilities exactly, owing to limitations of computers. Instead, we estimate these theoretical probabilities with empirical probabilities, by repeatedly sampling from randomly chosen coalitions and counting how often each state is a critical member. This approach is known as a **Monte Carlo simulation**, because it introduces randomness, like the machines in the casinos in Monte Carlo, to estimate a quantity whose exact value is inaccessible owing to computer limitations.

We present the results of this simulation in Table 19.3. There we list the 51 voting jurisdictions in the Electoral College, with the number of electoral votes for each jurisdiction and the probability that each jurisdiction is a critical member of a randomly selected coalition. This probability was estimated by a computer program that selected 1,000,000,000 coalitions at random and counted how many times each state was a critical member. The probability in Table 19.3 is this number divided by 1,000,000,000, rounded to the nearest hundredth.

For example, in our simulation, California was a critical member of 235,472,993 of the 1,000,000,000 randomly selected coalitions. We therefore put 23.55% as our estimate of the probability that California is a critical member. By contrast, South Dakota was a critical member of just 11,310,640 of those 1,000,000,000 coalitions. Hence we estimate the probability that South Dakota is a critical member to be 1.13%. As expected, and as fairness demands, California has more power in the Electoral College system than does South Dakota. And the Banzhaf analysis measures the advantage of California over South Dakota numerically: California is $23.55/1.13 \approx 20.84$ times as powerful as South Dakota in the Electoral College.

The data reveal no surprising anomalies. There are no dummies. Whenever a state has more electoral votes than another, it has more power as well. Maryland has twice as many electoral votes as West Virginia, and it has almost precisely twice the power as well. Indeed, it can be seen that the probability P here depends on the number of electoral votes E approximately according to the formula $P = E/250$. That this

Table 19.3 The number of electoral votes of each of the 51 voting jurisdictions and the estimated probability of being a critical member of a winning coalition.

STATE	EVs	Prob
California	55	23.55%
Texas	38	14.94%
New York	29	11.20%
Florida	29	11.20%
Illinois	20	7.62%
Pennsylvania	20	7.62%
Ohio	18	6.84%
Michigan	16	6.07%
Georgia	16	6.07%
North Carolina	15	5.69%
New Jersey	14	5.30%
Virginia	13	4.92%
Washington	12	4.54%
Massachusetts	11	4.16%
Indiana	11	4.16%
Arizona	11	4.16%
Tennessee	11	4.16%
Missouri	10	3.78%
Maryland	10	3.78%
Wisconsin	10	3.78%
Minnesota	10	3.78%
Colorado	9	3.40%
Alabama	9	3.40%
South Carolina	9	3.40%
Louisiana	8	3.02%

STATE	EVs	Prob
Kentucky	8	3.02%
Oregon	7	2.64%
Oklahoma	7	2.64%
Connecticut	7	2.64%
Iowa	6	2.26%
Mississippi	6	2.26%
Arkansas	6	2.26%
Kansas	6	2.26%
Utah	6	2.26%
Nevada	6	2.26%
New Mexico	5	1.89%
West Virginia	5	1.89%
Nebraska	5	1.89%
Idaho	4	1.51%
Hawaii	4	1.51%
Maine	4	1.51%
New Hampshire	4	1.51%
Rhode Island	4	1.51%
Montana	3	1.13%
Delaware	3	1.13%
South Dakota	3	1.13%
Alaska	3	1.13%
North Dakota	3	1.13%
Vermont	3	1.13%
DC	3	1.13%
Wyoming	3	1.13%

is a direct proportion is evidence from this point of view that there is no great mischief with the Electoral College.

Yet the relationship between P and E is not in fact precisely a direct proportion. The comparison of the data for California and South Dakota demonstrates this. If it were a direct proportion then the ratio 55/3 of electoral votes for these states would equal precisely the ratio 23.55/1.13 of their probability of being a critical member. In fact, California, with 18.33 times as many electoral votes as South Dakota, has 20.84 times the power. This Banzhaf analysis reveals a slight extra advantage for large states beyond what the electoral numbers suggest.

Certainly California deserves more votes in the Electoral College system than does South Dakota. But exactly how much more is hard to say. The ratio of the 2010 populations of California to South Dakota is 37,341,989/819,761 $\approx$ 45.55. The ratio of their apportionment numbers is $53/1 = 53$. The ratio of their weight in the Electoral College is $55/3 \approx 18.33$, demonstrating the advantage to small states of adding 2 to the apportionment numbers to obtain the number of electoral votes. Finally, the ratio of the Banzhaf powers of the two states is estimated by Monte Carlo simulation to be 20.84. Which of these ratios to focus on is a question of political philosophy; there is no precise definition of fairness that mathematics can dictate here.

The Banzhaf power analysis of weighted voting, which reveals astonishing anomalies with certain small examples with 3 or 6 voters, is not a source of significant distortion or unfairness in the Electoral College system. The addition of two electoral votes to the apportionment number for each state substantially influences the balance between large states and small, to the benefit of small states. The analysis of Banzhaf power seems to exert a small correction factor back in favor of large states, although this effect is relatively small. This supports the widely held belief that the Electoral College system is a boon for small states, who would be foolish to support its abolition.

But there is after all a significant distortion and unfairness in the Electoral College system, and we analyze it in the next chapter. The Banzhaf power emerges as a red herring in this analysis. And even the addition of two seats for senators turns out not to be the dominant factor in this analysis. The important realization, also due to Banzhaf, is that states don't actually vote in elections; rather, people do. And when you ask not about the states of California and South Dakota themselves but instead whether *individuals* in California or *individuals* in South Dakota are in a stronger position with regard to influence in our presidential voting method, the change in perspective leads to a different computation and a surprising answer.

19.5 Exercises and Problems

19.1. Compute the Banzhaf power of each voter in the method $V(21; 20, 19, 1)$.

19.2. Compute the Banzhaf power of each voter in the method $V(10; 10, 1, 1, 1, 1, 1, 1, 1, 1, 1)$.

19.3. Compute the Banzhaf power of each voter in the method $V(15; 9, 7, 5, 3, 1)$.

19.4. Compute the Banzhaf power of each voter in the method $V(13; 9, 7, 5, 3, 1)$.

19.5. A certain voting method with 10 voters allows measures to pass if they are favored by voter A and any 3 other voters, voters B and C and any 3 other voters, or any 6 voters. Show that this is a weighted voting method.

19.6. Give an example of a weighted voting method with 10 voters in which no two voters have the same weight and yet the method is anonymous.

19.7. Suppose that we amend a weighted voting method by giving one voter, call her voter A, "veto power" in the following sense: A measure under the new method passes if and only if it passed under the old method *and* voter A is in favor of it. Show that the new method is a weighted voting method.

19.8. The Nassau County (New York) Board of Supervisors in 1964 consisted of members representing six districts known as Hempstead 1, Hempstead 2, Oyster Bay, North Hempstead, Long Beach, and Glen Cove. Respecting the different populations of the districts, the county rules assigned varying numbers of votes to the six supervisors: 31 for Hempstead 1, 31 also for Hempstead 2, 28 for Oyster Bay, 21 for North Hempstead, 2 for Long Beach, and 2 also for Glen Cove. A simple majority of 115 total votes was needed to pass resolutions. Determine the Banzhaf power of each of the six districts. (This was the example that prompted John Banzhaf's original work.)

19.9. Consider a voting method with 26 voters, one for each letter of the alphabet. Suppose that a proposal passes if it has the support of 4 of the 5 vowels or if it has the support of 3 of the 5 vowels and at least 10 of the consonants. Show that this is a weighted voting method.

19.10. Two voting systems are called equivalent if they always have the same winning coalitions.

(a) Show that $V(5; 4, 3, 2, 1)$ is equivalent to $V(9; 8, 7, 2, 1)$.
(b) Show that $V(5; 4, 3, 2, 1)$ is not equivalent to $V(9; 8, 7, 3, 2)$.
(c) For what values of x is $V(5; 4, 3, 2, 1)$ equivalent to $V(9; 7, 6, 5, x)$?

19.11. Suppose that a mayor and city council of 6 members operate under a city charter in which legislation passes if it is supported either by the mayor and 3 council members or by any 5 council members.

(a) Is this a weighted voting system?

(b) What is the ratio of the Banzhaf power of the mayor to the Banzhaf power of a council member?

19.12. Suppose that a mayor and city council of 4 members operate under a city charter in which legislation passes if it is supported either by the mayor and 2 council members or by all 4 council members.
(a) Is this a weighted voting system?
(b) What is the ratio of the Banzhaf power of the mayor to the Banzhaf power of a council member?

19.13. Assume in some weighted voting method that $w_i > w_j$ (which is to say that voter i has more weight than voter j). Examples show that voter i and voter j may have the same Banzhaf power or that voter i may have greater Banzhaf power than voter j. Explain why voter i cannot have smaller Banzhaf power than voter j.

19.14. Assume in some weighted voting method that $w_i = w_j$ (which is to say that voters i and j have the same weight). Explain why voters i and j have the same Banzhaf power.

19.15. Canada is a federation of 10 provinces whose populations according to the 2011 census are given in the following table:

PROVINCE	POPULATION
Alberta	3,645,257
British Columbia	4,400,057
Manitoba	1,208,268
New Brunswick	751,171
Newfoundland	514,536
Nova Scotia	921,727
Ontario	12,851,821
Prince Edward Island	140,204
Quebec	7,903,001
Saskatchewan	1,033,381

The Canadian Constitution is amendable by what is known as the "7-50 formula": In order to pass, a proposed amendment must have the support of 7 of the 10 provinces representing at least 50% of the Canadian population. Show that the Canadian system is not a weighted voting method.

19.16. Suppose that we have a voting scheme that can be weighted in two different ways. One way has quota q_1 and assigns weight a_1 to voter A and weight b_1 to voter B, while the other way has quota q_2 and assigns weight a_2 to voter A and weight b_2 to voter B. Suppose that $a_1 > b_1$ and that $a_2 < b_2$. Show that there is a third way of assigning weights to the voters that has quota q_3, assigns weight a_3 to voter A and weight b_3 to voter B, and has $a_3 = b_3$.

19.17. Consider a voting method with four voters A, B, C, and D and with winning coalitions AB, CD, ABC, ABD, ACD, BCD, and ABCD. Show that this is not a weighted voting method.

19.18. Suppose that in some voting method, coalitions ABCDE and FGH are both winning but coalitions ABCFG and DEH are both losing. Show that this is not a weighted voting method.

19.19. Suppose that we have two weighted voting methods, call them method 1 and method 2, involving the same set of voters, and suppose that we create from them a new voting method by calling a coalition winning if it is winning in either method 1 or method 2. Show that this new method need not be a weighted voting method.

20

Whose Advantage?

"Before God we are all equally wise ... and equally foolish." — Albert Einstein

20.0 Scenario

Let S be the set of voters who vote for candidate A in a presidential election in the United States, and suppose that, owing to these votes, candidate A wins the Electoral College tally and becomes president. What is the smallest size that S can have? It is known that S can be of size less than 50% of the size of the electorate. In fact, George W. Bush won the 2000 election for president with only about 48% of the popular vote. Is it possible for someone to win a presidential race with only 40% of the (popular) vote? What about 30%? What about 20%? What about 1%?

20.1 Violations of Criteria

It is not states but people who vote in presidential elections. In the previous chapter, we addressed the question of whether the Electoral College system is fair to the states. In this chapter, we change our perspective and address the question of whether the Electoral College system is fair to the individual people who vote. The principle of "one person, one vote", so fundamental to our democracy that it has become a cliché, seems to demand that the voting system that we use in national elections be anonymous. Each vote should count the same as every other. The principle of "majority rules" (yet another cliché) seems to demand that our voting method should respect a view supported by a majority of the electorate. Yet the Electoral College system violates both of these principles.

Proposition 20.1. *The Electoral College system violates the anonymity criterion.*

Proof. We need to provide an example of an Electoral College election in which the outcome would have differed had various voters traded ballots. Such a trade would not affect the popular vote tally or even a tabulated profile of the electorate. But it could actually alter the outcome of an Electoral College election. To see why, we provide an example. In the year 2004, George W. Bush defeated John Kerry in the Electoral College by a margin of 286 to 251. Ohio, with its 21 electoral votes, was a critical member of the coalition of states that supported Bush. But if 200,000 Bush supporters in Ohio had traded ballots with 200,000 Kerry supporter in California, Kerry would have easily carried both states. This is because Kerry carried California by a margin of approximately 1,236,000 votes, while Bush's margin in Ohio was approximately 119,000 votes. This shows that the Electoral College system violates anonymity. □

Proposition 20.2. *The Electoral College system violates the majority criterion.*

Proof. What is demanded here is an example of an election in which one candidate has more than 50% of the popular vote but nonetheless loses in the Electoral College. History provides one such example. In the presidential election of 1876, Samuel J. Tilden received 51.0% of the popular vote but lost the presidential race in the Electoral College, 185 to 184, to Rutherford B. Hayes. □

In fact, it is possible for a minority of about 22% of the electorate in the United States to elect a president, even if the remaining 78% vote for the opponent. (See Exercise **20.8**.) This demonstrates that it is not necessary to have a lot of voters supporting a candidate for president, provided that the right voters support her. Moreover, to win a presidential election, it is not sufficient to have a majority of the electorate. It isn't sufficient even to have an overwhelming majority of the electorate.

Actually, there is a scenario in which a candidate could conceivably ascend to the presidency in the Electoral College system with a mere 12 voters. All that is needed is a single voter in the 12 largest states, which control a total of 283 electoral votes, a majority of the total number of electoral votes. If no other voters in those states cast ballots — an assumption that, while straining credibility, does not defy possibility — then those 12 voters carry the entire election. Even if 50,000,000 voters from the remaining 39 states and the District of Columbia all vote for the opponent, the Electoral College vote will still be 283 to 255, a narrow victory for a candidate who has only 12 supporters out of 50,000,012 voters all told. This shows that it is possible in theory for a candidate to win in the Electoral College with only $12/50{,}000{,}012 \approx 0.000024\%$ of the popular vote.

An election of the president by the plurality method (in this context often called "direct popular vote") would satisfy the anonymity and majority criteria as well as giving precisely equal power to every voter. Movements to abolish the Electoral College system in favor of a direct popular vote have been a regular part of our history. And yet there has been little momentum behind such movements in recent years. Amendments to alter election procedures face particularly daunting political obstacles, because every such change is seen as favoring some and harming others. No matter who has the impression that they will be harmed, and no matter whether there is any truth to this impression, they are likely to oppose the change out of perceived self-interest. And the procedure for amending our constitution requires the support of an overwhelming majority of states. So the Electoral College system persists.

20.2 People Power

In much of this chapter, we consider the Electoral College to be a voting method that is restricted to the case of two candidates. This restriction is primarily for the convenience of exposition. The conclusions that we draw apply equally well to the Electoral College in the real world, where the restriction to two candidates is not imposed. We also ignore the small concern that the Electoral College can end in a tie, with 269 electoral votes for each candidate. (The Constitution spells out rules about how such ties are to be resolved by Congress.)

Because the Electoral College system is not anonymous, some voters are more influential than others in determining the outcome of a presidential election. Let us attempt to quantify the notion of influence so that we can compare voters in different states. We adopt a similar paradigm to the one from Chapter 19. In particular, we assume that the influence of a voter is proportional to the probability that her vote alters the outcome of the election. As before, in order to speak of probabilities, we introduce randomness. In particular, we consider what happens if each voter independently casts a ballot at random giving equal probability to each of the two candidates. One could imagine that each voter flips a fair coin to determine how to vote. With this model, what is the probability that a single voter will cast a critical deciding vote in such a situation?

This probability is certainly minuscule. In order to cast the deciding vote in the Electoral College system, one must cast a vote that breaks a tie among the voters in one's own state. This requires that the other voters in the state come to an exact tie between the two candidates, leaving one's own ballot to be the deciding, tie-breaking vote. That in itself is unlikely, since in every state the number of voters is very large. But even more must occur if a single vote is to change the outcome of the presidential election. Not only must that vote swing the state to the opposite result, but then the state must go on to swing the Electoral College election to the opposite result. That too is unlikely.

Although these probabilities are minuscule, the degree that they differ from zero is critical. People go to the polls to vote because they know that there is some chance, however remote, that their vote will change the outcome of the election. Although voters know that there is almost no chance that their vote matters, the "almost" there is significant. They know that there is *some* positive probability that their vote will matter, and therein lies our democracy. We do not go to the polls just to express ourselves or to partake in a ritual of democracy. We go because there is a positive probability that our vote will make a difference.

In fact, it won't be necessary to calculate these probabilities. Our interest is only to compare two of these minuscule numbers by computing the ratio between two of these probabilities for voters in different states, to see if the ratio is near 1. In other words, we are interested in the relative sizes of two of these numbers, not in the absolute size of either of them. A ratio near 1 would be an indication that the system is nearly fair.

Suppose that a voter is in a state that has population p and that has a members in the House of Representatives. That state is allotted $a + 2$ electoral votes. What is the probability that the voter casts a critical vote in the presidential election? We would like to know how this number depends on p and a. Of course a itself depends on p via a roughly direct proportion, or such is the goal of the apportionment process.

Our first insight is that casting a critical vote can be thought of as a compound event, comprised of, first, the voter casting a critical vote in the statewide election and, second, the state casting a critical vote in the Electoral College. Moreover, the intrastate and interstate processes are independent. This independence comes from the assumption that voters toss their coins independently. The result of the intrastate process (the state popular vote) depends only on the coins tossed by voters within the state. The result of the interstate process (the Electoral College) depends only on the coins tossed by voters *not* within the state. Let R be the probability that the voter casts a deciding vote within the state, and let S be the probability that the state casts a deciding votes in the Electoral College. Then the probability that an individual will cast a deciding vote in the presidential election is the product $R \times S$, owing to our assumption of independence. Our interest in the end is to ascertain how this depends on p and a. One's intuition is that R gets smaller as p increases while S gets larger as p (and therefore a) increases. Our calculations will confirm this intuition.

We begin with a computation of R. For convenience, let us assume that the population p of the state in question is an odd number. In this case, we can write p in the form $2n+1$ for a positive integer n. It may be instructive for the reader to imagine that the state consists of the reader and $2n$ other voters. The probability that the reader casts a deciding vote in the plurality election within the state is equal to the probability that the remaining $2n$ residents split their votes exactly, n for one candidate and n for the other. There are exactly 2^{2n} ways that the $2n$ residents can vote, since there are 2 choices for each voter. These form a sample space of equally likely outcomes. To determine the probability of any set of outcomes in this sample space, we need only count the number of outcomes in the set and divide by 2^{2n}.

We are interested in the set of outcomes in which exactly half of the $2n$ voters vote for each of the two alternatives. So the question becomes: How many of the 2^{2n} profiles have exactly n votes for each alternative?

For example, in a state of population $p = 11$, we have $n = 5$, and we want to know in how many ways 10 voters can tie precisely. The profile

A	A	B	A	B	B	B	A	A	B

is an example of such a tie among $p - 1 = 2n = 10$ voters. We identify this particular profile with the subset $\{1, 2, 4, 8, 9\}$ of elements of $\{1, 2, 3, 4, 5, 6, 7, 8, 9, 10\}$ associated with voters who vote for candidate A. So we can translate the question to: How many subsets of $\{1, 2, 3, 4, 5, 6, 7, 8, 9, 10\}$ have exactly 5 elements? We denote the answer to this question by $C(10, 5)$.

More generally, for $0 < k < m$, we let $C(m, k)$ denote the number of subsets of the set $\{1, 2, 3, \ldots, m\}$ that have exactly k elements. Out of a total of 2^{2n} possible profiles with $2n$ voters, exactly $C(2n, n)$ of them represent an exact tie. So the probability of a voter casting the deciding vote in the state is $C(2n, n)/2^{2n}$.

We now argue that $C(m, k) = m!/(k!\,(m-k)!)$, where we write $r!$ to mean $r \times (r-1) \times \cdots \times 3 \times 2 \times 1$. (The expression $r!$ is known as "r factorial".) To see this, note that there are $m \times (m-1) \times \cdots \times (m-k+1) =$

$m!/(m-k)!$ ways to line up k distinct numbers from $\{1,2,3\ldots,m\}$, but each subset of k such numbers can be arranged into a line in $k!$ ways. There are therefore

$$C(m,k) = \frac{m!}{k!\,(m-k)!}$$

subsets of $\{1,2,3\ldots,m\}$ with exactly k elements. Substituting $2n$ for m and n for k, we obtain

$$C\,(2n,n) = \frac{(2n)!}{n!\,n!}\,. \tag{20.1}$$

This is an exact expression, but it does not convey much information about the rate of growth of $C\,(2n,n)$ as a function of n. In order to get such information, we use an estimation for $r!$ known as Stirling's formula, which tells us that

$$r! \approx r^r e^{-r}\sqrt{2\pi r}.$$

Here $\pi \approx 3.14159$ is the well known constant from the formula for the area of a circle, and $e \approx 2.71828$ is another well known constant, familiar as the base of the natural logarithm. Using Stirling's formula on (20.1), we obtain

$$\begin{aligned} C\,(2n,n) &= \frac{(2n)!}{n!\,n!} \\ &\approx \frac{(2n)^{2n}e^{-2n}\sqrt{2\pi(2n)}}{\left(n^n e^{-n}\sqrt{2\pi n}\right)\left(n^n e^{-n}\sqrt{2\pi n}\right)} \\ &= \frac{(2n)^{2n}e^{-2n}\sqrt{4\pi n}}{n^{2n}e^{-2n}2\pi n} \\ &= \frac{2^{2n}}{\sqrt{\pi n}}\,. \end{aligned}$$

While this is only an approximation, it does a better job than (20.1) in giving a sense of the magnitude of $C\,(2n,n)$. In particular, our goal was not merely to count the ways in which $2n$ voters could tie; rather, it was to estimate the probability of that occurrence. To obtain a probability, divide this last expression by 2^{2n} and use $p = 2n+1$ to obtain the simple expression

$$R = \frac{C(2n,n)}{2^{2n}} \approx \frac{1}{\sqrt{\pi n}} = \sqrt{\frac{2}{\pi}} \cdot \frac{1}{\sqrt{p-1}}. \tag{20.2}$$

As expected, this gets smaller as p gets larger. This is consistent with intuition: The larger the state, the less likely a resident is to cast a decisive vote in the statewide popular vote election.

Now we move on to calculate the probability S that a state contributes the critical votes in the Electoral College. This is nothing more than the Banzhaf power of the state in question divided by 2^{51}. (There are 51 "states" from the point of view of the Electoral College.) We don't have any means to compute this precisely, but we can use the Monte Carlo simulation from Chapter 19 to estimate these probabilities. (See Table 19.3.) One can summarize the data there as follows: With an error of at most 20% or so, the Banzhaf power of a state is proportional to the number $a + 2$ of electoral votes that the state is accorded. In fact, the data in Table 19.3 suggests that the probability S is approximately equal to $(a + 2)/250$. The denominator 250 in this expression is not derived from theory but rather is obtained by fitting to experimental data. As expected, S gets larger as a gets larger. The larger the state, the more likely that state is to cast decisive electoral votes.

Let us express S in terms of p instead of a. The standard divisor in the United States is roughly 700,000. An apportionment method endeavors therefore to make a equal to roughly $p/700{,}000$, the standard quota. So for a rough approximation, we substitute this for a and obtain the estimate

$$S \approx \frac{(p/700{,}000) + 2}{250} = \frac{p + 1{,}400{,}000}{175{,}000{,}000}. \tag{20.3}$$

This approximation applies for p in the range from about 500,000 to 30,000,000 and with an error of roughly ±20%. But US states do indeed have populations roughly in that range.

The probability that an individual in a state with population p casts the deciding vote in a presidential election is the product of R and S. Using (20.2) and (20.3), we obtain as an estimate for this probability,

$$R \cdot S \approx \frac{\sqrt{2/\pi}}{\sqrt{p-1}} \cdot \frac{p + 1{,}400{,}000}{175{,}000{,}000} \approx K \cdot \frac{p + 1{,}400{,}000}{\sqrt{p}}, \tag{20.4}$$

where we have written K for the constant

$$K = \sqrt{\frac{2}{\pi}} \cdot \frac{1}{175{,}000{,}000} \approx 0.00000000456.$$

Notice that we have replaced $p - 1$ by p without introducing any appreciable error. Expression (20.4) is our final formula for the probability we were seeking, in a form that shows us how this probability is affected by the population p of the state.

20.3 Interpretation

Formula (20.4) is only approximate, for several reasons. First, we have assumed that the apportionment a is given by $p/700{,}000$. While this is the right order of magnitude, it ignores the rounding involved with apportionment. In particular, for the state of Montana, this represents an error of roughly 40%, since their $a = 1$ member of the House must represent a full $p = 994{,}416$ constituents. Second, we have assumed that there is a direct proportion between the number of votes cast in the Electoral College by a state and the Banzhaf power of the state. But we have already seen that this introduces another error of up to 20% or so. Moreover, we employed Stirling's approximation, which is another reason that (20.4) is not exact. But Stirling's approximation the way we have used it introduces errors of far less than 1%, so it is not a major source of error here.

Despite these approximations, there is still useful information to be culled from (20.4). For California, with $p \approx 37{,}300{,}000$, (20.4) tells us that the probability of a single voter casting a critical vote in a presidential election is

$$K\frac{37{,}300{,}000 + 1{,}400{,}000}{\sqrt{37{,}300{,}000}} \approx K\frac{38{,}700{,}000}{6107} \approx 6337K.$$

For South Dakota, with $p \approx 800{,}000$, (20.4) tells us that the probability of a single voter casting a critical vote in the presidential election is

$$K\frac{800,000 + 1,400,000}{\sqrt{800,000}} \approx K\frac{2,200,000}{894} \approx 2460K.$$

The surprise here is that the voter in California carries substantially more influence, substantially more power, than does the voter in South Dakota, by a factor of more than 2.5. This stands in opposition to the conclusion we drew in Chapter 19, where we observed that South Dakota, by virtue of the extra two electoral votes associated with senatorial representation, seemed to have an advantage in the Electoral College. But when one asks if individual South Dakotans have an advantage, the answer is no. Even with the extra two electoral votes for each state, voters in South Dakota have a voice that is less powerful than the voice of Californians, by a factor of more than 2.5.

A careful look at (20.4) explains what is going on. The constant K in (20.4) is immaterial in this analysis, since we are always comparing two states by computing the ratio of (20.4) for the two appropriate values of p. The constant K cancels from the numerator and denominator of such ratios. The p in the numerator of (20.4) reflects the fact that larger states have proportionally larger influence on the Electoral College. The 1,400,000 in the numerator of (20.4) comes from the two electoral votes added to the apportionment number for each state. The effect is to increase (20.4), but the increase is relatively small for large states and relatively large for small states. This is of clear benefit to the small states. Finally, the $\sqrt{p}$ in the denominator of (20.4) reflects the fact that voters in large states are less likely to cast the deciding vote in their statewide elections. This is the factor that our intuition tells us should balance out the extra electoral votes in the large states. But the mathematics shows us that it does not in fact balance it out. The square root diminishes the impact of the population on the denominator in (20.4), making the population in the numerator the dominant force in this formula. There is therefore an advantage to living in California as compared to South Dakota: You are more than 2.5 times as likely to have the deciding voice in the presidential election.

If the approximations involved with the use of (20.4) raise concerns, then there is an alternative that produces a much more accurate assessment of these probabilities. That alternative is to use the data from Table 19.3. This avoids the two estimates that introduce the major errors associated with (20.4), although it introduces a new kind of error, the statistical kind, owing to the randomness of the Monte Carlo simulation. In Table 20.1, we compute the value of $K \times B_i/\sqrt{p_i}$, where B_i is the Banzhaf power of state i, where p_i is the population of state i, and where K is a constant, chosen for our convenience so that the value of this expression is equal to 1 when the chosen state is Nevada (the state that gives the lowest value). We use estimated 2014 populations, extrapolated from 2010 census data by the United States Census Bureau, and we rank the states in decreasing order of voter power. The numbers in Table 20.1, therefore, represent by what ratio voters in each state are more powerful than Nevada voters. It turns out that California voters are a full 3 times more powerful than South Dakota voters.

Why then do we have an Electoral College? It was established by our founding fathers in the Constitution for a variety of reasons. In the 18th century, the nation was sparsely populated, with no simple means of communication or transportation that could connect the citizenry. There was doubt therefore that the citizenry could be in a position to make an informed choice. Moreover, campaigning for public office was seen at this time as a distasteful practice, and there was concern that a popular vote mandate for a president could lead to an imbalance of power of the executive branch over the other two branches of government.

These arguments do not seem persuasive in the modern era. So why haven't any of the efforts to abolish the Electoral College been successful? It is because the Electoral College is seen as a boon to smaller states,

Table 20.1 Relative strengths of voters in the 51 voting jurisdictions, in decreasing order of power.

STATE	EVs	Index
California	55	3.38
Texas	38	2.57
New York	29	2.25
Florida	29	2.25
Pennsylvania	20	1.91
Illinois	20	1.90
Ohio	18	1.80
Michigan	16	1.73
Georgia	16	1.71
North Carolina	15	1.61
New Jersey	14	1.59
Virginia	13	1.53
Washington	12	1.53
Tennessee	11	1.45
Indiana	11	1.45
Minnesota	10	1.45
Arizona	11	1.43
Massachusetts	11	1.43
Wisconsin	10	1.41
South Carolina	9	1.38
Maryland	10	1.38
Alabama	9	1.38
Missouri	10	1.37
Wyoming	3	1.32
Colorado	9	1.31
Rhode Island	4	1.31

STATE	EVs	Index
Kentucky	8	1.29
Vermont	3	1.28
Louisiana	8	1.25
Connecticut	7	1.25
Dist. of Columbia	3	1.25
West Virginia	5	1.24
Nebraska	5	1.23
Oklahoma	7	1.20
Nevada	6	1.20
Kansas	6	1.19
Oregon	7	1.19
Utah	6	1.18
Arkansas	6	1.18
Alaska	3	1.18
North Dakota	3	1.18
New Hampshire	4	1.17
Mississippi	6	1.17
Maine	4	1.17
New Mexico	5	1.17
Iowa	6	1.15
Hawaii	4	1.13
South Dakota	3	1.10
Idaho	4	1.06
Delaware	3	1.05
Montana	3	1.00

who are naturally unwilling to support such a change and relinquish their advantage. The irony is that if these small states acted on behalf of their residents, they would realize that these residents are significantly harmed by the Electoral College system. It is the voters in small states who would be empowered by a change to a direct popular vote.

It is not the case that the Electoral College system survives today because it is seen as fair. It survives because, although it is seen as unfair, it is seen as unfair in a way that is politically desirable or politically unpalatable to correct. From the point of view of states, it gives an advantage to small states. From the point of view of voters, as one can see from looking at the general trend in Table 20.1, it gives an advantage to voters in large states. Either way, it does not give equal treatment to all voters, as does, among various possible methods discussed in Part 1, the plurality method. But it is never easy to amend the Constitution to redistribute power, influence, or representation. The parties involved have too much at stake, and for every party that imagines that it would benefit from such a change, there is another party that believes that it would be harmed.

And so the quirky Electoral College system remains, despite its disadvantages, and is likely to be with us for some time. At least we can say this for it: It is a voting method, and it is both neutral and monotone. The rules are clear and known to everyone in advance. We are certainly better off with these rules than with a system that is secret, dictatorial, or inconsistent. So it survives, for better or for worse, like Hill's method for computing apportionment and the plurality method for conducting gubernatorial elections, as a part of the American political heritage.

20.4 Exercises and Problems

20.1. Compute $C(4,2)$, and explicitly list each of the $C(4,2)$ profiles with four voters in which two voters vote for candidate A and two voters vote for candidate B.

20.2. Compute $C(6,3)$, and explicitly list each of the $C(6,3)$ profiles with six voters in which three voters vote for candidate A and three voters vote for candidate B.

20.3. What is the probability that a family of four children consists of two boys and two girls?

20.4. Compare the value of $m!$ with Stirling's approximation for $m!$ for $m = 1, 2, 3, \ldots, 10$. For each of these values of m, compute by what percentage the approximation differs from the exact value.

20.5. What value does Equation (20.4) assign to the probability that a voter in Wyoming will cast a deciding vote in a presidential election? Express the answer as a fraction with numerator equal to 1.

20.6. Generally, the states at the top of the list in Table 20.1 are the largest states in the union. What is the factor that causes New York voters to be slightly stronger than Texas voters, even though Texas is a larger state?

20.7. The seven smallest states and the District of Columbia are accorded three members in the Electoral College. These eight jurisdictions appear in Table 20.1 in reverse order of their populations, with the largest state (Montana) at the bottom. Why is this?

20.8. Construct an example of a coalition of US voters satisfying the following two properties: (1) The coalition contains less than 23% of the registered voters nationwide; and (2) if every member of the voting coalition supports one particular candidate for president, that candidate will win in the Electoral College even if every registered voter not in the coalition votes for an opponent.

20.9. Consider a union of three states that conducts presidential elections by a system that works like the US Electoral College. The three states have just 3, 5, and 7 voters, respectively, and each state gets as many votes in the Electoral College as it has voters. This is an example of what we called a bloc voting system.

(a) Voters in which state are most powerful?

(b) What is the exact probability that a voter in each of the three states casts a deciding presidential vote?

(c) How would these probabilities change if the number of voters and the number of Electoral votes in the three states changes from 3, 5, and 7 to 3, 5, and 9?

(d) How would these probabilities change if the number of voters and the number of Electoral votes in the three states changes from 3, 5, and 7 to 1, 5, and 7?

20.10. Consider a union of five states that conducts presidential elections by a system that works like the US Electoral College. The five states have just 1, 3, 5, 7, and 9 voters, respectively, and each state gets as many votes in the Electoral College as it has voters. What is the probability that a voter in the state of size 5 influences the outcome of the national election? How does this compare that the lone voter in the first state influences the election?

20.11. Equation (20.2) gives both an exact value of R and an approximate value of R. How do these differ when $n = 2$ and $p = 5$? What about when $n = 7$ and $p = 15$. What about when $n = 1000$ and $p = 2001$?

20.12. Suppose that the Electoral College system were altered so that each state was permitted a vote of weight $n+2$, where n, instead of being the apportionment number for the state, was set equal to the standard quota for the state. For example, Montana, whose standard quota is 1.40, would get a vote of weight 3.40 instead of just 3. How would such a change impact the disproportion noted in Table 20.1?

20.13. What positive value of p yields the smallest probability in equation 20.4? (Answer this question by calculus, if you can, or by drawing a graph, perhaps with a calculator.) This is the worst population for a state to have in terms of the presidential election power of its citizens, although still ignoring the rounding associated with apportionment. Which US state has a population closest to this value?

20.14. Consider the case of the US Senate after Jim Jeffords's departure from the Republican party (see Example 19.2). Analyze this example from the point of view of individual senators rather than political parties. Assume that all senators express a view for or against a proposal on the basis of the random flip of a coin. Assume also that all senators agree to cast their votes in favor of whichever position is expressed by the majority of senators from their party. For the sake of convenience, assume that a party that splits evenly on a proposal, with exactly 25 in favor and 25 against, flips another coin and casts all 50 votes in favor or against, depending on the result of the coin. What is the probability that a single senator from the Democratic party, say, tips the Senate result by casting a vote? What is the corresponding probability for Senator Jeffords?

Notes on Part IV

The idea of computing a power index in a weighted voting system originated with Lionel Penrose in 1946 [37]. Lloyd Shapley and Martin Shubik proposed a competing index in 1954 [41]. John Banzhaf [7] invented his index to analyze the distribution of power on the county board of Nassau County, New York in the early 1960s. The index is named after Banzhaf, although it is the same as the index proposed earlier by Penrose. Other indices to measure power were introduced in [16] and [30]. In [22], Dan Felsenthal and Moshe Machover review these indices and make a case for the Shapley-Shubik index as the one that best avoids paradoxes.

In our treatment, we side with the Banzhaf power as the one that best captures and measures the notion of the value of a vote. It answers the question: How likely is a vote to change the outcome of an election? In our presentation of the Banzhaf power, we avoid the common practice of rescaling the Banzhaf power so that the powers of the voters sum to 1 (or 100%). Although the Banzhaf power of a voter can be interpreted as a probability, these probabilities for different voters live in different sample spaces, and it makes little sense to add them. Indeed, it seems as if measures of the *percentage* of power held by each voter in a voting system are not meaningful. In fact, Taylor and Pacelli [46] attempt to compute the percentage of the power in the US federal system that is held by the president; they obtain 77% when using one index but 0.4% when using another. It isn't clear what these numbers mean or how they might inform policy.

Banzhaf applied his index to the problem of the Electoral College in 1968 [8]. The Banzhaf power index has become a common topic in textbooks for general education courses in mathematics. One can find an elementary presentation in [13] or [44]. The simulations leading to Table 19.3 and Table 20.1 took about 12 hours on a desktop computer using MATLAB®. The simplicity of the Banzhaf model makes it attractive, but many have argued (e.g., [42], [32], [25]) that the assumption that voters behave like random coin tosses is unrealistic and therefore that the model is not applicable. Indeed, the Banzhaf assumption can be likened to the false assumption that all states are swing states. In fact, some states are strongly partisan and there is a negligible chance that any vote is critical there. The empirical frequency of close elections is much smaller than would be predicted under the random-voter hypothesis. Still, such criticisms miss the point of the Banzhaf power. It is not meant to measure the empirical probability that a voter will change the election result, but instead to determine how responsive the voting system itself is to a vote, independent of any empirical assumptions about voters.

One way to think about the Electoral College is as an agreement by voters to pool their votes as in a bloc voting system. There are reasons to be concerned about such agreements. In particular, if California were to grow until it had more than half of the electoral votes, then the Electoral College system would degenerate into a plurality election in which only voters in California are counted. Put another way, if the 11 largest states were permitted to come to an agreement to cast all of their electoral votes unanimously, then they would reduce all the remaining states to irrelevancy. In fact, such an idea is the basis of the current plan known as the National Popular Vote Interstate Compact, a plan by which states agree to cast their electoral votes for whichever presidential candidate has the plurality of votes nationwide. This is seen as a way of abolishing the Electoral College de facto without amending the Constitution. As soon as a collection of states that controls a majority of the electoral votes signs on to such an interstate compact, the Electoral College election is, by agreement, handed to the winner of the plurality vote nationwide. On the other hand, allowing the Electoral College to be hijacked by such a maneuver would also allow it to be hijacked for more sinister and inappropriate purposes. For example, an interstate compact to cast electoral votes for whichever presidential candidate has the plurality of votes *within the participating states* would be an excellent tool for the disenfranchisement of non-participants.

Problem **1.15** illustrates the danger of allowing bloc voting in the Supreme Court. Consider what would occur if 5 of the justices agreed to a compact to vote as a bloc. Clearly this disenfranchises the remaining 4 justices. Consider further what would occur if, among the 5 justices in the bloc, 3 of them agree to vote as a sub-bloc before they meet with their bloc of 5. Consider further yet what would occur if, among the 3 justices in the sub-bloc, 2 of them agree to vote as a sub-sub-bloc before they meet with their sub-bloc of 3. The result would be a Supreme Court that is controlled entirely by 2 justices, with the remaining 7 justices reduced to dummies. This demonstrates the dangers inherent in bloc voting compacts. One way to defuse these dangers is to forbid such compacts by insisting that justices cast votes that indicate their honest views, even if they are minority views. Another protection against this kind of smothering of the Supreme Court is to make certain that no 2 or 3 or 5 justices are in lockstep agreement on all issues. Justices must be independent and free to vote their conscience.

The most recent serious attempt to repeal the Electoral College was in 1969, following the narrow presidential victory of Richard Nixon. An amendment to the Constitution passed in the House of Representatives by an overwhelming bipartisan majority. But the amendment was opposed in the Senate, mostly by senators from small states, who perceived that their constituents benefited from the Electoral College (despite Banzhaf's analysis). The effort was ended by a filibuster in the Senate.

Arguments in favor of the Electoral College system run along several lines. Some argue that it enhances the federal nature of our republic. In particular, the Electoral College system allows individual states the autonomy to conduct elections in whatever manner they see fit. Others argue that it enhances minority rights, especially regional minorities. Some believe that it discourages third-party candidates and thereby offers stability to the two-party system. It is sometimes argued that, in recent years, the Electoral College system has benefited the Republican party, which is more rural and more concentrated in small states than is the Democratic party.

Still, most arguments in favor of the Electoral College seem either anachronistic, such as the view that direct election of a president would lead to tyranny, or dubious, such as the view that voters in small states benefit. Yet the Electoral College remains, owing to rules that make our Constitution difficult to amend, as a quaint, if quirky, element of the American political system. And it seems likely that it will be with us for some time to come.

Solutions to Odd-Numbered Exercises and Problems

Chapter 1

1.1. (a) Candidate B wins with 11 votes, which is a majority of the 17 total votes. (b) It is a tie, because neither candidate has 2/3 or more of the votes. The quota q for this method with $t = 17$ voters would be $q = \lceil pt \rceil = \lceil (2/3)17 \rceil = \lceil 34/3 \rceil = 12$. Since neither candidate has 12 votes, neither candidate wins. (c) Neither candidate has 12 votes, but candidate A, as the status quo, is declared to be the winner. (d) Candidate A gets 16 votes, while candidate B gets only 15 votes. Candidate A, therefore, has a majority of the (weighted) votes and is the winner.

1.3. (a), (b), and (c) Yes. One example is the method that chooses whichever option is preferred by fewer than half the voters (the "simple minority method"). (d) Yes. Consider the method that selects your preferred option if at least 5% of the electorate favors that position and otherwise declares the election to be a tie. This method is not neutral, but it is monotone. (e) No. The generalization of May's Theorem (Theorem 1.24) asserts that the only such methods are simple majority, a super-majority, or the all-ties method. None of these methods allow for a victory for any position with only 10% support.

1.5. (a) Since votes are simply counted in this method, it does not matter which voters contribute which votes. (b) The rules for selecting one candidate are exactly the same as the rules for selecting the other. (c) A candidate with a majority of the votes who then changes the mind of some voters in his favor will still have a majority of the votes.

1.7. (a) This method is anonymous, because it makes no special reference to any particular voters, and neutral, because the rules are the same for the two candidates. But the method is not monotone, because A wins if A gets 5 votes and B gets 12 votes, but B wins if one of the 12 voters who prefers B switches preferences, so that A gets 6 votes and B gets 11. (b) This method is neutral, because the two candidates get equal treatment, and is monotone, because a vote for one candidate can never harm that candidate. But, under the assumption that the electorate contains both male and female voters, the method is not anonymous, because the voters do not get equal treatment. Male voters are treated one way, females another. (c) This method is anonymous, because all voters get equal treatment (all voters are ignored), and monotone, because a vote for one candidate never harms that candidate (no votes have any effect at all). But the method is not neutral, because the two candidates are clearly not on equal footing.

1.9. (a) Assuming that there are voters of widely varying weight, the answer is no. Heavier voters have a greater opportunity to affect the outcome of the election than do lighter voters. (b) No. Candidate B has an advantage over candidate A with these rules, since candidate B has the advantage of being the status quo. (c) Yes. If candidate A wins (garnering at least 70% of the total weight) and then some voters change their preference from B to A, then the total weight preferring candidate A can only increase. Therefore candidate A continues to win. On the other hand, if candidate B wins (garnering more than 30% of the total weight) and then some voters change their preference from A to B, then the total weight preferring candidate B can only increase. Therefore candidate B continues to win. (d) Yes, because the method specifies that candidate B wins whenever candidate A does not win. (e) Yes, because the method is in fact decisive.

1.11. The rules appear to specify special rules for the vice president. Nevertheless, the method is anonymous. One can imagine that the vice president submits her preference in all cases but that her preference is revealed publicly only when the senators vote exactly 50:50. A measure passes if at least 51 senators favor it or if exactly 50 senators and the vice president favor it. In other words, a measure passes if at least 51 of the 101 voters favor it. This is the simple majority method, an anonymous method, for 101 equally weighted voters.

1.13. The simple minority method is anonymous, because the result depends only on how many votes the candidates receive and not further on precisely which voters voted for which candidates. The method is neutral because the condition for getting elected is the same for one candidate as it is for the other. The method is not monotone, however, because obtaining more votes can in fact be harmful to a candidate in this method. For example, with 9 voters, a candidate wins with 4 votes but loses with 5.

The method is almost decisive, because there are only three possibilities for the number of votes that candidate A receives: Either it is larger than the number of votes that candidate B receives (candidate B wins), it is smaller than the number of votes that candidate B receives (candidate A wins), or it is equal to the number of votes that candidate B receives (it is a tie). Thus, ties occur only when the two candidates receive an identical number of votes.

1.15. (a) No, it is not anonymous, because 5 justices matter and 4 don't. (b) Yes, it is neutral, because the rules treat the two alternatives equally. (c) Yes, it is monotone, because "yes" votes never do harm to the "yes" cause and "no" votes never help the "yes" cause. (d) Because it makes no difference how you vote; your vote is immaterial.

1.17. (a) Yes. Either candidate wins if he or she receives 50 or more votes. This is the simple majority method. (b) No. A candidate with 89 votes wins but would lose if she obtained one more vote. (c) Yes. Because there are exactly 99 voters, precisely one of the two candidates must receive an odd number of votes. So the method applied to this particular electorate with 99 voters is decisive. (d) Yes. All voters are treated (i.e., ignored) equally. Swapping ballots among voters never changes the outcome of this election.

Chapter 2

2.1. (a) Candidate A has 4 first-place votes, more than any other candidate. Therefore candidate A wins the plurality election. (b) The total number of Borda count points for candidate A is 19, for candidate B is 27, for candidate C is 24, for candidate D is 25, and for candidate E is 15. Therefore candidate B wins the Borda count election. (c) Candidate C wins a head-to-head election against any other candidate. To see this, note that 6 of the 11 voters prefer C to A, 6 of the 11 voters prefer C to B, 6 of the 11 voters prefer C to D, and 6 of the 11 voters prefer C to E. Hence C tallies a perfect score by the Copeland method and wins the Copeland election. (d) The candidate with the most last-place votes is E, with 6 last-place votes, and so we eliminate E. With E gone, candidate A has 6 last-place votes (among the remaining candidates) and so is eliminated. With A and E gone, C has 5 last-place votes and is the next eliminated. Finally, the race is between B and D, and candidate D is favored by 6 of the 11 voters. Hence D wins the Coombs election. (e) One voter ranks candidate E first. If this voter is declared to be the dictator, then candidate E wins.

2.3. (a) Candidate B wins with 13 votes. (b) The Borda count tallies for the candidates are as follows: A, 55; B, 59; C, 35; D, 31. Candidate B wins again. (c) First eliminated is candidate D with zero first-place votes. Next eliminated is candidate C, with 7 first-place votes. This leaves candidates A and B for the final round: Candidate B defeats A, 16 to 14. (d) Candidate B defeats every other candidate in a head-to-head match-up, tallying 3 points in the Copeland scoring system. Since no other candidate can obtain more than 2 points, candidate B is the winner. (e) Candidate C is eliminated first with 16 last-place votes. After that, candidate D is next eliminated with 15 last-place votes. This leaves candidates A and B again, and A is eliminated with 16 last-place votes to candidate B's 14. Candidate B wins.

2.5. (a) The profile represented here has candidate A winning the Hare method immediately but also being eliminated by the Coombs method in the first round. (Candidate B wins the Coombs election.)

4	3	2
B	A	A
C	B	C
A	C	B

(b) The profile here is the "dual" of the profile in part (a). A is knocked out of the Coombs election in the first round, with 5 last-place votes, but A wins the plurality election with 4 votes, more than either B or C.

4	3	2
A	C	B
C	B	C
B	A	A

(c) In the profile represented here:

2	2	3	2
A	A	B	C
B	C	C	B
C	B	A	A

candidate A wins the plurality election, but obtains only 8 points in the Borda count, fewer than the 9 points that candidate C earns and the 10 points that candidate B earns. (d) The profile pictured below, with just 3 voters, has candidate B winning the Borda count with 6 points, more than the 4 points earned by each of the other candidates.

A	D	C
B	B	B
C	A	D
D	C	A

2.7. The following profile does the job.

3	4	6
C	C	B
B	A	A
A	B	C

Candidate A wins the antiplurality election with 3 last-place votes. Candidate B wins Borda count with 15 points. Candidate C is the Condorcet candidate and so wins the Copeland election.

2.9. Some profiles might have no candidate with 25% of the first-place votes. (b) One cannot determine how a voter will respond to this question from the voter's preference order. Our social choice functions require preference ballots, not approval ballots. (c) Drawing straws is improper, because the output of a social choice function can depend only on the profile. (d) Not every profile has such a candidate. For example, in the profile

A	C	B
B	A	C
C	B	A

candidate A defeats candidate B, candidate B defeats candidate C, and candidate C defeats candidate A.

2.11. Candidate B's score will be the same as candidate D's score, except that B has one more first-place vote and D has one more last-place vote. Since first-place votes are worth at least as much as last-place votes, B's score is at least as high as D's in any positional method. So D can't be the unique winner.

2.13. The profile

A	E	B	C	D
B	A	E	D	C
C	D	A	E	B
D	B	C	A	E
E	C	D	B	A

is such an example. Every positional method gives a 5-way tie, because every candidate has exactly one first-place, one second-place, one third-place, one fourth-place, and one last-place vote. Hence no positional method can distinguish between the 5 candidates. Copeland's method, however, gives 3 points to candidate A, because candidate A defeats candidates B, C, and D in head-to-head match-ups, while candidates B, C, and D each get 2 Copeland points and candidate E gets just 1 Copeland point. Hence candidate A wins the Copeland election.

2.15. In the following profile

A	C	B
B	A	C
C	B	A

A defeats B and then C defeats A in the second round. If candidates A and C are switched on each voter's ballot, then we arrive at the profile

C	A	B
B	C	A
A	B	C

in which A is eliminated in the first round by B, and then C defeats B in the second round. Thus C, not A, wins.

2.17. (a) No, because plurality winners must receive at least one first-place vote. (b) Yes, because in the following profile

B	C	D
A	A	A
D	B	C
C	D	B

candidate A wins the Borda count election while candidates B, C, and D are the winners of the nomination method. (c) No, because a Hare winner must have at least one first-place vote to survive even the first round of elimination. (d) Yes, because the example of part (b) has candidate A also as the Coombs winner.

2.19. Any profile in which one candidate has a majority of the first-place votes but also a plurality of the last-place votes can serve as the required example. The profile that solves exercise **2.7** is an example.

2.21. Information at www.archives.gov indicates that the popular vote was distributed roughly as follows: Gore, 48.4%; Bush, 47.9%; Nader, 2.7%; Buchanan, 0.4%; Other, 0.6%. For the purposes of argument, let us ignore the effect of all candidates except Gore, Bush, and Nader. Also, let us assume that 90% of Nader voters had preference lists N>G>B (the remainder N>B>G), that 75% of Gore voters had G>N>B (the remainder G>B>N), and that 75% of Bush (and Buchanan) voters had B>N>G (the remainder B>G>N). Finally we distribute the "other" voters equally among the six possible preference orders. The resulting profile is:

12.1	36.0	12.2	36.4	0.4	2.5
B	B	G	G	N	N
G	N	B	N	B	G
N	G	N	B	G	B

With these assumptions, Gore wins plurality, but only by a small margin. Gore is also the Condorcet winner, and therefore wins by the Copeland method. Gore also wins Borda count and Hare elections, where in the latter case, Nader is the first eliminated. Interestingly, in the Coombs system, Bush is the first eliminated, and Gore beats Nader in the run-off. Nader wins the antiplurality election — which with just three candidates is the same as the vote-for-two election — since he is the least disliked candidate.

Chapter 3

3.1. Candidate B is the Condorcet candidate, defeating each of the other four candidates in head-to-head competition by a score of 3-to-2.

3.3. The following profile has D as the anti-Condorcet candidate but has no Condorcet candidate, since A defeats B, B defeats C, and C defeats A head-to-head.

A	C	B
B	A	C
C	B	A
D	D	D

3.5. Suppose a profile that has candidate A with every first-place vote is presented to a Pareto method. Since candidate A is preferred by every voter to candidate B, candidate B must lose. This assertion about candidate B is true for every candidate who is not candidate A, so every candidate other than candidate A must lose. Hence A must be the unique winner.

3.7. (a) Candidate D must win, because he is the Condorcet candidate. (b) Candidate C must lose. The only difference between the profiles is that A and C have switched places. Since the method is neutral and since A lost before, C must lose after. (c) Candidate D must lose. Every voter prefers B to D. (d) No conclusion can be drawn. Candidate A may or may not win after. For example, if the method is a dictatorship with the first voter as dictator, then A still wins. But if the method is that the top two candidates for voter three win (a monotone method!), then A loses after. (Examples demonstrate that no conclusion can be drawn about candidates B or C either.) (e) No conclusion can be drawn. The first voter changes the relative standing of candidates B and C, so independence has nothing to say here. If the first voter is dictator, then C wins and B loses after. If the second voter is dictator, then B wins and C loses after. If the method is best-of-B-or-C (an independent method), then A and D lose after. But if the method is A, B, and D always win (another independent method), then A and D win after. So anything is possible for the four candidates.

3.9. (a) Profile Q may be seen as obtained from profile P by having one voter (the last one) move candidate A up from second-place to first-place. The result of this change, we are told, is that candidate A falls to defeat. This violates monotonicity, because in the presence of monotonicity, candidates are not harmed by being moved up on the preference lists of voters. (b) Although one preference order is changed, no voter has changed his mind about whether A is preferable to C or C is preferable to A. That the winner in profile P is A but the winner in profile Q is C therefore serves as witness to the fact that the method in use violates the independence criterion.

3.11. The Pareto criterion reduces to the unanimity criterion when restricted to methods for two candidates. This is because the only way for every voter to put one candidate ahead of another is for every voter (unanimously) to favor the one candidate.

3.13. *All* social choice functions for two candidates are independent. Two profiles in which every voter gives the same relative preference toward candidates A and B must in fact be identical profiles. Therefore they give the same result, no matter what the social choice function is.

3.15. Any candidate who loses a head-to-head election with any other candidate cannot be a winner of a stable social choice function. In other words, only Condorcet candidates can be chosen to be winners of stable social choice functions. But then, facing the Condorcet paradox — a profile with no Condorcet candidate — a social choice function must select some winner, and that violates stability.

3.17. (a) Consider the before-and-after profiles below. A wins “before”, moves up on one voter’s list, yet loses “after”. (b) The “after” profile for the solution to part (a) shows that the Pareto criterion fails. There

C	B	A
A	A	B
B	C	C

before

C	A	A
A	B	B
B	C	C

after

candidate A is ahead of candidate B on every voter's list, but B is selected as the winner. (c) The method is anonymous because the description of the method makes no distinction between voters. The method is neutral because the rules for winning apply equally to all candidates.

Chapter 4

4.1. (a) Vote-for-two does not satisfy the Condorcet criterion, as the following profile illustrates. Candidate A is the Condorcet candidate.

C	C	A	B	D
E	F	G	H	I
A	A	B	A	A
B	B	D	D	B
D	D	E	E	E
F	E	F	F	F
G	G	H	G	G
H	H	I	I	H
I	I	C	C	C

(b) Vote-for-two does not satisfy the anti-Condorcet criterion either, since an anti-Condorcet candidate can in fact have more first-place and second-place votes than any other candidate. In the profile for part (a), candidate C is the anti-Condorcet candidate but wins the vote-for-two election. (c) The method does not satisfy the Pareto property either, because if every voter puts candidate A in first place and candidate B in second place, candidate B ties candidate A and both are winners. (d) Vote-for-two is not independent, since, in the following profile candidate A wins, but if candidate C moves to the top of the list for the last two voters, candidate B would win instead.

3	1	2
A	C	B
B	A	A
C	B	C

4.3. (a) The French method violates the Condorcet criterion, because a Condorcet candidate may not have any first-place votes and would therefore be eliminated in the first round. (b) The French method satisfies the anti-Condorcet criterion, however, because an anti-Condorcet candidate, even if he does make it to the run-off (which is possible!), will lose the run-off. (c) The French method is not monotone, because there is an example involving three candidates without ties that shows that the Hare method is not monotone. That same example has to work here, since the French method operates exactly like the Hare method, provided there are just three candidates and no ties.

4.5. (a) Yes. A Condorcet candidate wins every head-to-head match-up, and such a candidate is never eliminated in the SET system. (b) No. If the voter profile is

A	A	D	C	B
B	B	A	D	C
C	C	B	A	D
D	D	C	B	A

then A defeats B, C defeats D, and A goes on to defeat C in the finals round. However if candidates A and C switch places in the profile, leading to

C	C	D	A	B
B	B	C	D	A
A	A	B	C	D
D	D	A	B	C

then B defeats A, D defeats C, and B goes on to defeat D in the final round. This violates neutrality, since switching C into the position of winner A should cause C to become the winner.

4.7. (a) Yes, because the rules pertain equally to all voters. (b) Yes, because the rules pertain equally to all candidates. (There is a potential confusion here: The recall election by itself is not neutral, since the incumbent is at a disadvantage. But the instant recall method is neutral, since it combines a plurality election with immediate recalls, and the rules for the compound method treat all candidates equally.) (c) No, because a Condorcet candidate may have a plurality of the first-place votes and as such get eliminated (recalled) in the first round. (d) We answer equivocally. If one adheres to the requirement to ignore ties, then the candidates are eliminated sequentially. In this case, the answer is yes. A candidate that manages to win the instant recall must advance to a majority status by the sequential removal of other candidates but while some other candidate remains under consideration. The instant recall winner is preferred to this remaining candidate by a majority of voters. Hence the instant recall winner is not the anti-Condorcet candidate. However, if one does permit ties — assuming that all plurality winners are eliminated at each stage, then it is easy to imagine a profile in which three candidates tie for plurality and are eliminated, allowing a fourth candidate to win . . . even if the fourth candidate is last on every voter's preference list! (e) No. A candidate can win the instant recall election after another candidate is eliminated. Such an instant recall winner was not the plurality winner originally and hence could have been eliminated had some voters moved her up on their preference lists. (f) No. In the profile below, every voter prefers candidate A to candidate D, yet D wins the instant recall election, following the elimination of A and then B.

1	2	3	3
A	C	B	A
B	A	A	D
D	B	D	B
C	D	C	C

(g) Yes, because a candidate with a majority of first-place votes wins an instant recall election by definition. (h) No. The profile from part (f) above has candidate D winning. If the first voter promotes candidate B from second place to first place, then candidate A becomes the instant recall winner. This violates independence. (i) No, as the example of part (f) illustrates.

4.9. (a) Yes. The new candidate would not affect any voter's first choice candidate, so the plurality tally will end up exactly as it was before the new candidate was added. (b) No. After the new candidate is added, the antiplurality election declares all candidates except the new one to be a winner, independent of how the election may have turned out prior to the addition of the new candidate. (c) Yes. The new candidate earns no Borda points at all. The Borda count score of the remaining candidates is indeed affected by the addition of the new candidate, but in a uniform way. Each candidate increases his or her Borda count score by exactly n points, where n is the number of voters in the electorate. Adding n to each number in a list does not alter the numerical rank of those numbers, so whichever candidate or candidates won the Borda count election before the new candidate arrived will also win it afterwards.

4.11. (a) Yes, because a candidate without any top votes can't have the most top votes. (b) No. The antiplurality winner need not have any first-place votes, as in the following profile, where candidate B wins the antiplurality election.

A	C
B	B
C	A

(c) No. The Borda count winner need not have any first-place votes, as is illustrated in the following profile, where candidate B wins the Borda count election.

A	C
B	B
D	D
C	A

(d) Yes, because a candidate without any top votes will necessarily be eliminated in the first round of the Hare method. (e) No, as the profile above for part (b) illustrates. (f) No, since a Condorcet candidate will always win the Copeland method but need not have any top votes. The following profile shows candidate A as a Condorcet candidate (and therefore a Copeland winner) but without top votes.

B	C	D
A	A	A
D	D	B
C	B	C

(g) Yes, because a candidate without any top votes can't have the dictator's top vote, so can't win a dictatorship election.

4.13. Consider the before-and-after tabulated profiles below. In the "be-

5	4	2	1
A	B	B	C
C	A	C	B
B	C	A	A

before

5	4	2	1
A	B	B	C
C	A	A	B
B	C	C	A

after

fore" profile, the Coombs method eliminates candidate B in the first round, leaving candidate A to defeat candidate C by a score of 9 to 3 in the second round. Now two voters move candidate A up on their preference lists, and the "after" profile is obtained. Facing this profile, the Coombs method eliminates candidate C in the first round, and candidate B defeats candidate A by a score of 7 to 5 in the second round. Candidate A moved up on some preference lists, yet this change results in his defeat. This violates monotonicity.

4.15. Bloc voting is neutral because the rules for electing one candidate are the same as the rules for electing any other. It satisfies the unanimity criterion because if every voter puts candidate A in first place, every bloc votes for candidate A, and it is a landslide. It satisfies monotonicity because moving candidate A up on a voter's preference list cannot remove candidate A from a first-place ranking from any voter, nor can it cause any other candidate to obtain a new first-place ranking from any voter. Finally, it satisfies Pareto because if every voter prefers candidate A to candidate B, candidate B gets no first-place votes and therefore no consideration at all from the bloc voting method.

To see why the bloc voting method violates anonymity, consider a profile in which the first-place votes are distributed as follows:

- the bloc of 9 voters: 4 votes for A and 5 votes for B;
- the bloc of 7 voters: 4 votes for A, 1 vote for B, and 2 votes for C;
- the bloc of 5 voters: 4 votes for A and 1 votes for C;
- the bloc of 3 voters: 1 vote for A and 2 votes for B; and
- the bloc of 1 voters: 1 vote for B;

In this profile, candidate B wins the bloc voting election by winning the blocs of size 9, 3, and 1, and therefore garnering 13 of the blocked votes. However, if one vote for B in the bloc of size 9 is traded for one vote for A in the bloc of size 7, the result is a win for A, who now wins the blocks of size 9, 7, and 5, gathering 84% of the total blocked votes.

The bloc voting method can certainly end in a tie, for example if candidate A wins the blocs of size 9 and 1, candidate B wins the blocs of size 7 and 3, and candidate C wins the bloc of size 5. It is therefore indecisive. The method violates the majority criterion, as the above blocked profile shows. Candidate A has a majority (13 out of 25) of the individual votes there, but loses the bloc voting election. Since the method violates the majority criterion, it must also violate the Condorcet criterion. It also violates the anti-Condorcet criterion, because candidate B in the profile above wins the bloc voting election, but it is not difficult to complete the information above to a full profile in such a way that candidate B is also the anti-Condorcet candidate. Finally, bloc voting does not satisfy the independence criterion. To see this, imagine that, in the profile above with candidate B as winner, two of the voters in the first bloc who voted for candidate B promote candidate C to a first-place ranking. This promotes candidate A to victory in the bloc voting election, violating independence.

Chapter 5

5.1. The plurality method is anonymous and Pareto. Dictatorships are independent and Pareto. It is somewhat more challenging to find an example of a method that is anonymous and independent, but an example can be found in Problem **5.5** below.

5.3. (a) Yes, since all voters are treated equally by the rules. (b) No. Consider the before-and-after profiles pictured below. In the "before"

A	A
B	C
C	B

before

A	C
B	A
C	B

after

profile, candidate A wins. In the "after" profile, all candidates win. This violates independence, since candidate B loses "before" but is a winner "after". (c) No. This fact is witnessed by the "after" profile above, in which candidate A is preferred to candidate B by every voter and yet candidate B manages to be a winner.

5.5. (a) Suppose candidate X wins and candidate Y loses in a best-of-two election when facing profile P, and suppose profile Q has every voter putting X and Y in the same relative position as they do in profile P. Clearly candidate X must be one of the princes, since they are the only two candidates eligible to win the best-of-two election. Without loss of generality, say that X is Prince A. Candidate Y might be Prince B or might not. If he is, then Prince A defeats him head-to-head in profile Q, because the head-to-head election between Prince A and Prince B is identical in Q and in P. If he is not, then Y loses in profile Q because he loses in all profiles. Either way, candidate Y loses in profile Q, which confirms that this method is independent. (b) The method is not Pareto, since even if every voter prefers a particular commoner to both princes, one of the princes wins anyway. Another argument is that, were the method Pareto, we would have a counterexample to Corollary 5.2.

5.7. (a) The influential coalitions are those that consist of more than half the voters. To see this, simply observe that if a majority of voters put a candidate in first place, that candidate is guaranteed to win the plurality election. No coalition with at most half of the voters has this property. (b) The influential coalitions are those that contain the dictator. If every voter in such a coalition puts candidate A in first place, then in particular the dictator puts candidate A in first place and candidate A wins. A coalition without the dictator clearly is not influential, since they have no influence of any kind over the election. (c) The influential coalitions are those consisting of at least 8 voters. If every voter in such a coalition puts candidate A in first place, then candidate A will earn at least 24 Borda count points, and the maximum Borda count score for any other candidate would be 22, obtained by getting two first-place votes and all the remaining second-place votes. So candidate A would win. On the other hand, a coalition of 7 voters is not influential, as the profile below demonstrates.

7	3
A	B
B	C
D	D
C	A

Candidate B wins the Borda count election with 23 points, beating candidate A, who earns just 21 points.

5.9. In the profile below, 9 out of the 10 voters prefer candidate A to candidate B, 9 out of the 10 voters prefer candidate B to candidate C, and so forth, and 9 out of the 10 voters prefer candidate I to candidate J. Then, to complete the loop, 9 out of the 10 voters prefer candidate J to candidate A. A nearly Pareto method could not choose any winner when facing this profile, violating the universal domain requirement.

A	J	I	H	G	F	E	D	C	B
B	A	J	I	H	G	F	E	D	C
C	B	A	J	I	H	G	F	E	D
D	C	B	A	J	I	H	G	F	E
E	D	C	B	A	J	I	H	G	F
F	E	D	C	B	A	J	I	H	G
G	F	E	D	C	B	A	J	I	H
H	G	F	E	D	C	B	A	J	I
I	H	G	F	E	D	C	B	A	J
J	I	H	G	F	E	D	C	B	A

5.11. Consider a profile with an odd number of voters that does not have a Condorcet candidate. Call a sequence of candidates **strange** if each candidate in the sequence is defeated by the next candidate in the sequence in a head-to-head match-up and the last candidate in the sequence is defeated by the first candidate in the sequence. The goal of this problem is to show that there is a strange sequence of length 3. We show first that there is a strange sequence of some length.

Let A_1 be any candidate. Since A_1 is not a Condorcet candidate, there is some candidate A_2 who defeats A_1 head-to-head. Since A_2 is not a Condorcet candidate, there is some candidate A_3 who defeats A_2 head-to-head. Continue in this way to produce a sequence of candidates each of whom is defeated by the next. Since there are no Condorcet candidates, the process can be repeated indefinitely. But there are only finitely many candidates, so eventually the process must repeat a candidate. At that moment, a strange sequence of candidates is identified.

We now show that there is a strange sequence of exactly three candidates. Suppose that $A_1, A_2, \ldots, A_k$ is a strange sequence with $k > 3$ candidates. Imagine a head-to-head match-up between candidates A_1 and A_3. If A_1 defeats A_3, then A_1, A_2, A_3 is a strange sequence of length 3, and we are finished. Otherwise, A_3 defeats A_1, in which case $A_1, A_3, \ldots, A_k$ is a strange sequence with $k - 1$ candidates (candidate A_2 has been excised). In other words, if there is a strange sequence with k candidates, and $k > 3$, then there is necessarily a strange sequence with fewer than k candidates. Applying this principle repeatedly starting with any strange sequence at all yields a strange sequence of length 3. Naming the candidates in this sequence C, B, and A, respectively, yields the Condorcet paradox of the problem statement.

(A similar argument shows that any profile that does not have an anti-Condorcet candidate must also have a strange sequence consisting of three candidates.)

Chapter 6

6.1. (a) One can still use a positional voting scheme that gives 2 points for every first-place vote and 1 point for every second-place vote. More generally, one can consider a scheme that gives a points for every first-place vote and $b \leq a$ points for every second-place vote. (b) One can certainly eliminate the candidate or candidates with the fewest first-place votes. One can then continue as in the Hare method, but with one proviso: Once two candidates are eliminated, it is possible that a voter's ballot expresses no view at all about the candidates that remain. In this case, the voter is in effect eliminated from the electorate. Nevertheless, one can repeatedly eliminate the candidate or candidates with the fewest first-place votes until all remaining candidates tie for the fewest first-place votes, at which point those candidates are named winners. (c) A voter indicates his preference between two candidates only if at least one of those candidates appears on the voter's ballot as a first-place or second-place choice. One can still imagine each pair of candidates facing each other in a head-to-head match-up, but one should be prepared to ignore voters who have ranked neither candidate in first-place or second-place. Nonetheless, one can declare a head-to-head victory for one candidate over another if the number of voters who prefer the one to the other exceeds the number of voters who prefer the other to the one. (d) The two ballots give identical information in the case when there are only three candidates, because once two candidates are identified as a first choice and a second choice, the third candidate is naturally left to be the third and last choice.

6.3. The ballot could look like:

B	B	A	A	A
A	C	B	B	C
C	A	C	C	B

Candidate A has the majority of first-place votes, but candidate B is approved by 4 of the 5 voters and is therefore the approval voting winner.

6.5. Suppose that a social choice function satisfies the Pareto criterion and that every voter prefers candidate A to candidate B. Then candidate B cannot win. Nor can candidate B win any elections taking place with candidates eliminated from the slate as long as candidate A remains in the slate. Therefore the iterated social preference function can never rank candidate B until candidate A is ranked and eliminated. Therefore the iterated social preference function will satisfy the Pareto criterion.

Conversely, imagine that a social choice function violates the Pareto criterion. Then for some profile and for some pair of candidates A and B, every voter places A higher than B, yet B wins. Facing this profile, the iterated social preference function ranks B in first place (possibly tied with candidate A or other candidates). This violates the Pareto criterion for social preference functions, which requires that A be preferred to B in the social ranking.

6.7. Consider the profile of range ballots given by the following table:

3 voters	2 voters	1 voter
A: 0.6	A: 0	A: 0
B: 0.4	B: 0.6	B: 0.2
C: 0.3	C: 0.4	C: 1.0

Since the first 3 voters approve of candidate A and disapprove of the other two candidates, candidate A wins the approval election. The candidate with the second most approvals is candidate B with 2 approvals. Hence approval voting ranks the candidates in the order A>B>C. However, with range voting, the total scores of candidates A, B, and C, respectively, are 1.8, 2.6, and 2.7, so range voting ranks the candidates in the reverse order C>B>A.

6.9. Each of the n candidates can be paired with any of the other $n-1$ candidates. This makes it seem as if there are $n(n-1)$ pairs of candidates, but since we consider the pair A-B to be the same as the pair B-A, our count of $n(n-1)$ actually counts every pair twice. The correct number of (unordered) pairs is therefore $n(n-1)/2$.

6.11. The Borda count score for a candidate A is equal to the number of pairs (v, C) where v is a voter, C is a candidate other than A, and voter v places candidate A ahead of candidate C. For any particular candidate C, the number of voters v who place A ahead of C is just the entry of the Condorcet matrix in row A and column C. Therefore, the sum of the entries in row A gives the Borda count score.

6.13. In the profile

A	B	C	C
B	A	D	D
D	D	A	B
C	C	B	A

candidate C is the Hare winner, after candidate D and then candidates A and B are eliminated. Switching C and D in this profile yields a profile in which candidate D is the winner. Yet the Condorcet matrix for both these profiles is the matrix all of whose off-diagonal entries are 2.

6.15. A Condorcet method is called **monotone** if, whenever a voter switches from A>B to B>A, it cannot happen that B wins before the switch but not after the switch. If such a change is made, the effect on the Kemeny score of every ranking of candidates is to increase by 1 the Kemeny score if B is ahead of A or to decrease by 1 the Kemeny score if A is ahead of B. If B was at the top of the ranking with the highest Kemeny score before the switch, it remains at the top of this ranking after the switch and this ranking increases its Kemeny score by 1. No other ranking can increase in Kemeny score by more than 1, so the same ranking that was selected by the Kemeny method before the switch is selected by the Kemeny method after the switch. Hence candidate B remains the Kemeny winner after the switch.

Chapter 7

7.1. The standard quota for each state is $1/17$ of the House size 60, which is $60/17 \approx 3.5294$. The upper quota is 4, and the lower quota is 3.

7.3. (a) This is not an apportionment method, since the sum of the lower quotas is generally smaller than h. (b) This is an apportionment method, although not a reasonable one. (c) This is an apportionment method as well. (d) This is not an apportionment method, since the sum of the seats need not be h. For example, with 3 states of equal population and $h = 10$, the standard quota for each state is $3.3\overline{3}$, so this method would assign 3 seats to each state. But $3 + 3 + 3 \neq 10$.

7.5. (a) The standard quotas for $h = 10$ are 0.54, 2.43, and 7.03, so the Hamilton apportionment is 1, 2, and 7. The standard quotas for $h = 11$ are approximately 0.59, 2.67, and 7.73, so the Hamilton apportionment is 0, 3, and 8. Note the appearance of the Alabama paradox: The first state's apportionment decreases when the house size increases. (b) The new standard quotas for $h = 11$ are 0.53, 2.50, and 7.97, so the Hamilton apportionment is 1, 2, and 8. Note that the smallest state decreased in population while the second state increased, and yet the first state acquired a seat from the second state. This is a population paradox.

7.7. The role of population is played here by the ridership. Hence $p = 130,000$ is the total average daily ridership on the six routes, $h = 130$ is the total number of buses available, and $s = p/h = 1000$ is the average number of daily riders per bus. The standard quotas are the ideal number of buses per route. The lower quota and upper quotas are what we would get if we rounded down or up, respectively, the standard quotas. If we use the lower quotas, then we do not use all the available buses, wasting resources. If we use the upper quotas, we do not have enough buses, requiring additional expenditures. An apportionment represents a plan for doling out exactly the $h = 130$ available buses among the routes in a fair way, relative to the ridership. The standard quotas in this problem are:

A	B	C	D	E	F
45.30	31.07	20.49	14.16	10.26	8.72

which leaves room to round four numbers down and two numbers up. Hamilton's method therefore results in the apportionment:

A	B	C	D	E	F
45	31	21	14	10	9

7.9. (a) The standard divisor for this problem, representing the cost of each piece of candy, is \$1.10/27 $\approx$ 4.074 cents. The standard quotas are obtained by dividing 4.074 into 10, 25, and 25, obtaining approximately 2.45, 12.27, and 12.27 for Bobby, Jessie and Sandy, respectively. These round to 3, 12, and 12, respectively, since the fractional part of Bobby's quota is larger than that of Jessie or Sandy. (b) This time the standard divisor is \$1.10/28 $\approx$ 3.929 cents. The standard quotas are approximately 2.55, 12.73, and 12.73 for Bobby, Jessie, and Sandy, respectively. These round to 2, 13, and 13, respectively, since the fractional part of Bobby's quota is now smaller than that of Jessie or Sandy. Notice that Bobby loses a candy when the extra piece is discovered.

7.11. Suppose that the Hamilton apportionment of a state is given by the whole number m. This means that the standard quota of the state must be greater than $m - 1$. The effect of increasing h is to increase the standard quota of every state. So the standard quota of the state after h is increased remains greater than $m - 1$. Hence the Hamilton apportionment of the state after h is increased is at least $m - 1$.

7.13. Let $n = 11$, and let the population of the first 10 states be $1,000,000$ and the population of the 11th state be $1,000,001$. Suppose that $h = 12$. Then the standard quota of each state is roughly 1.09, with the 11th state having a just slightly higher standard quota than the other 10 states. Hamilton's method provisionally assigns one seat to each state, and then it gives the single leftover seat to the 11th state. So that state gets 2 seats with a standard quota of less than 1.1.

7.15. Let $n = 2$, $h = 10$, $p_1 = 19,000$, and $p_2 = 81,000$. Then the standard quotas are 1.9 and 8.1 for the two states, respectively. Hamilton's method awards upper quota to the smaller state. The Gross method awards upper quota to the larger state.

7.17. The following census provides an example.

state	population	standard quota	lower quota	Hamilton apportionment
1	$p_1 = 1{,}200{,}000$	1.2	1	1
2	$p_2 = 2{,}400{,}000$	2.4	2	2 or 3
3	$p_3 = 4{,}400{,}000$	4.4	4	5 or 4
	$p = 8{,}000{,}000$			$h = 8$

Chapter 8

8.1. (a) The standard divisor is $40/5 = 8$. (b) This makes the quota for state 1 equal to $13/8 = 1.625$. (c) The Hamilton apportionment is $a_1 = 2$ and $a_2 = 3$. (d) and (e) The modified quota of $d = 6.5$ yields modified quotas of $13/6.75 \approx 1.926$ and $27/6.75 = 4$, which lead to a Jefferson apportionment of $a_1 = 1$ and $a_2 = 4$. (f) and (g) The modified divisor of $d = 9$ yields modified quotas of $13/9 \approx 1.444$ and $27/9 = 3$. These give an Adams apportionment of $a_1 = 2$ and $a_2 = 3$. Moreover no smaller divisor can work, since a divisor smaller than 9 will give a quota for state 2 that is larger than 3.

8.3. (a) The sum of the four populations given is $p = 8{,}000{,}000$. (b) The standard divisor is $p/h = 8{,}000{,}000/160 = 50{,}000$. (c) The table gives the data.

	A	B	C	D
Population	3,310,000	2,670,000	1,330,000	690,000
Standard quota	66.2	53.4	26.6	13.8
Lower quota	66	53	26	13
Upper quota	67	54	27	14

(d) The table gives the results.

	A	B	C	D	total	divisor
Hamilton	66	53	27	14	160	50,000
Jefferson	67	54	26	13	160	49,300
Adams	66	53	27	14	160	50,500
Webster	66	53	27	14	160	50,000

8.5. (a) Since the total population is $p = 1{,}000$, the standard divisor is

$p/h = 1{,}000/10 = 100$. (b) The standard quota for state 4 is $p_4/100 = 590/100 = 5.9$. The upper quota is therefore 6. (c) With standard quotas of 1.2, 1.4, 1.5, and 5.9, states 3 and 4 get rounded up to yield an apportionment of $a_1 = 1$, $a_2 = 1$, $a_3 = 2$, and $a_4 = 6$. (d) and (e) The divisor $d = 80$ works, yielding modified quotas of 1.5, 1.75, 1.875, and 7.375, which round down to $a_1 = 1$, $a_2 = 1$, $a_3 = 1$, and $a_4 = 7$. (f) Yes. The apportionment $a_4 = 7$ exceeds the upper quota 6 for state 4. (g) The critical divisor $d = 590/7$ is the largest possible Jefferson divisor. If d is larger than this, state 4 will receive at most 6 seats. (h) and (i) The divisor $d = 130$ yields modified quotas of 120/130, 140/130, 150/130, and 590/130, which round up to $a_1 = 1$, $a_2 = 2$, $a_3 = 2$, and $a_4 = 5$, the Adams apportionment.

8.7. We have a total population of 100 and 5 classes to be apportioned. (Think of $h = 5$ as the house size.) The standard divisor (average class size) is 20, and the standard quotas are 2.60, 1.65, and 0.75. The Hamilton method first assigns 2, 1, and 0 classes to calculus, set theory, and topology, respectively, and then chooses set theory and topology to receive an extra class, because 0.75 and 1.65 present a stronger claim for rounding up than does 2.60. So the Hamilton apportionment is 2 classes of calculus, 2 classes of set theory, and 1 class of topology. The modified divisor 16 gives modified quotas of 52/16, 33/16, and 15/16, which are easily seen to be 3.something, 2.something, and 0.something. Hence the Jefferson apportionment is 3 classes of calculus, 2 classes of set theory, and no classes of topology.

8.9. (a) The divisor 460 gives modified quotas of 2.72, 7.93, and 11.09. Hence the Jefferson apportionment with 20 pearls is 2, 7, and 11. The divisor 530 gives modified quotas of 2.36, 6.89, and 9.62. Hence the Adams apportionment with 20 pearls is 3, 7, and 10. (b) The divisor 450 gives modified quotas of 2.78, 8.11, and 11.33, so the Jefferson apportionment with 21 pearls is 2, 8, and 11. The extra pearl goes to Barbara. The divisor 520 gives modified quotas of 2.40, 7.02, and 9.81, so the Adams apportionment with 21 pearls is 3, 8, and 10. Again the extra pearl goes to Barbara. The Jefferson and Adams methods are divisor methods and are therefore house monotone. So they do not suffer from the Alabama paradox, as the Hamilton method does. Compare this problem with Problem **7.4**.

8.11. The problem asks for the value of x such that if State A has exactly population x, then the method is unable to decide — because it's an exact tie — whether to give 3 or 4 seats to State A and unable to decide whether to give 6 or 7 seats to State B. For Webster's (and Hamilton's) methods, this occurs precisely when State A has modified quota 3.5 and State B has modified quota 6.5. This occurs when State A has population precisely 3,500,000. For Adams's method, this occurs precisely when State A has modified quota 3 and State B has modified quota 6. This occurs when State A has precisely 1/3 of the population, using a modified divisor that is simultaneously one-third of State A's population and one-sixth of State B's. Hence $x = 10{,}000{,}000/3 \approx 3{,}333{,}333$. For Jefferson's method, this occurs precisely when State A has modified quota 4 and State B has modified quota 7. This occurs when State A has precisely $4/(4{+}7) = 4/11$ of the population, using a modified divisor that is simultaneously one-fourth of State A's population and one-seventh of State B's. Hence $x = (4/11) \cdot 10{,}000{,}000 \approx 3{,}636{,}363$. For Dean's method, this occurs when the divisor d satisfies $x/d = 24/7$ and $(10{,}000{,}000 - x)/d = 84/13$. Eliminating d from these equations and solving for x yields $x = 10{,}000{,}000/\,(1 + (84/13)(7/24)) \approx 3{,}466{,}666$. For Hill's method, this occurs when the divisor d satisfies $x/d = \sqrt{12}$ and $(10{,}000{,}000 - x)/d = \sqrt{42}$. Eliminating d from these equations and solving for x yields $x = 10{,}000{,}000/\left(1 + \sqrt{42}/\sqrt{12}\right) \approx 3{,}483{,}315$.

8.13. Let $p_1 = 3400$ and $p_2 = 6600$. With $h = 10$, the standard quotas are 3.4 and 6.6 for the two states, respectively, so, since $0.6 > 0.4$, the Hamilton apportionment is $a_1 = 3$ and $a_2 = 7$. With divisor $d = 1110$, however, the modified quotas are roughly 3.06 and 5.95, so Adams's method rounds these up to $a_1 = 4$ and $a_2 = 6$.

8.15. The example below illustrates a lower quota violation for the Adams method with $h = 10$ and with divisor $d = 1{,}400{,}000$.

State	Population	Standard Quota	Lower Quota	Modified Quota	Adams
1	1,700,000	1.7	1	1.21	2
2	1,600,000	1.6	1	1.14	2
3	1,500,000	1.5	1	1.07	2
4	5,200,000	5.2	5	3.71	4

The quota violation is with state 4, whose standard quota is 5.2 but whose Adams apportionment is just 4.

8.17. (a) The identric mean of 1 and 2 is $4/e \approx 1.472$, while the identric means of 2 and 3 is $27/(4e) \approx 2.483$. Hence with the standard divisor of $d = 100$, the standard quotas of 1.48, 2.48, and 6.04 are rounded to 2, 2, and 6, respectively. (b) The example of part (a) works. Webster's apportionment is obtained with a divisor of $d = 99$, obtaining modified quotas of approximately 1.495, 2.505, and 6.101. These round to 1, 3, and 6 using arithmetic rounding, which yields Webster's apportionment.

8.19. First imagine dividing the standard divisor into the two populations. One obtains the standard quotas for the two states. If these are both whole numbers, then the rounding function leaves them both alone, and we have our apportionment. If one of these numbers is rounded down while the other is rounded up, then we also have our apportionment, with one state getting its lower quota and one state getting its upper quota. There is no quota violation in this case.

But what if the rounding function rounds both numbers down or both numbers up? In the first case, we have a provisional apportionment assigning both states their lower quota and leaving just one seat still to be allocated. By modifying (lowering) the divisor, we eventually reach a point where the rounded modified quotas allocate that final seat. Whichever way that seat goes, one state obtains its lower quota and the other obtains its upper quota. Hence there is no quota violation in this case.

A similar argument shows that there can be no quota violation in the case that both standard quotas are rounded up.

Chapter 9

9.1. (a) Yes, because the method discriminates between states only on the basis of the population of the states. (b) Yes, because the largest state remains the largest state if all states increase or decrease by a fixed percentage. (c) Yes, because the only way for a state to lose representation is for it to be the largest state, but then for another state to overtake it as the largest state. This can never occur when the largest state gains in population but the other state loses population. (d) No. For example, in the US census from the year 2010, California's standard quota is approximately 52.5, but this method would assign California 435 seats. That is a dramatic quota violation.

9.3. This problem is most easily solved by computation with a spreadsheet. The resulting apportionments for State 20 are 81 by Hill's method, 81 by Jefferson's method, 62 by Adams's method, and 81 by Webster's method. Since the standard quota for State 20 is approximately 75.1, these represent dramatic quota violations.

9.5. (a) Clearly yes, since if a quota violation occurs with the Jefferson method, then Hamilton's method takes over, and Hamilton's method prohibits quota violations. (b) Yes. In the end the hybrid method gives either Jefferson's or Hamilton's apportionment. Both of these methods satisfy the order-preserving property, therefore so does the hybrid. In other words, if $p_i > p_j$, then both Jefferson's and Hamilton's methods guarantee that $a_i \geq a_j$. Hence the hybrid method guarantees this as well. (c) No, because the impossibility theorem says that no method that satisfies the quota rule can satisfy population monotonicity.

9.7. (a) No. With $h = 10$ and $n = 3$, the most populous state gets 4 seats, even if its fair share is 9.5. Similarly, the least populous state gets 3 seats, even if its fair share is 0.0001. These illustrate quota violations. (b) Yes. If h is increased, then one extra seat is dealt out to the next state in line. Hence seats, once distributed, are never recalled with an increase in h. (c) Yes. Seats are dealt to larger states before smaller ones, so a smaller state can never obtain a seat before a larger one. (d) Every state gets exactly 2 seats no matter what its population. This is the apportionment method (if you can call it that) used by the United States Senate.

9.9. (a) Yes, because the apportionment is not affected by a proportional change in the populations of the states. (b) No. To illustrate a quota violation, one can contrive examples. Alternatively, one can compute the Jefferson-plus-one apportionment for the US census from the year 2010. California gets only 50 seats by this method, a lower quota violation. (c) Yes, because an increase in h increases $h - n$, which in turn can only increase the Jefferson apportionment of the $h - n$ seats (because Jefferson's method is house monotone). (d) Yes, because if the apportionment a_i of state i decreases and the apportionment a_j of state j increases, then the Jefferson apportionment of the $h - n$ seats also decreases for state i and increases for state j. Since Jefferson's method is population monotone, this requires either that the population p_i decreases or the population p_j increases. (e) The Adams modified divisor for h works as the Jefferson modified divisor for $h - n$ and gives an extra seat to each state.

9.11. Consider the standard divisor for Adams's method. The standard divisor yields the standard quota, which when rounded up gives each state its upper quota. This always allocates too many seats, so we must increase the divisor making all the quotas smaller than the standard quotas. This shows that the modified divisor for Adams's method is always larger than the standard divisor. Therefore the modified quotas are all smaller than the standard quotas, and hence the Adams apportionment can never exceed the upper quota.

To show that Adams's method violates the lower quota rule, we need to find a profile such that one state gets less than its lower quota. Here is one possibility, with House size $h = 18$, total population $p = 130,000$, standard divisor $s \approx 7222.2$, and modified divisor $d = 9000$.

	A	B	C	D
Population	100,000	10,000	10,000	10,000
Standard quota	13.8642	1.3846	1.3846	1.3846
Lower quota	13	1	1	1
Modified quota	11.1111	1.1111	1.1111	1.1111
Adams apportionment	12	2	2	2

Notice that State A gets fewer seats than its lower quota.

9.13. An example with $h = 15$ and $n = 3$ has populations $p_1 = 1120$, $p_2 = 4420$, and $p_3 = 9460$. The chart tallies the apportionment by the three methods.

State	1	2	3
Standard quota	1.12	4.42	9.46
Hamilton apportionment	1	4	10
Lowndes apportionment	2	4	9
$x/\sqrt{m}$ method	1	5	9

Since this method satisfies the quota rule and is neutral, the impossibility theorem guarantees that it violates population monotonicity.

9.15. (a) and (b) No. With $h = 3$, $n = 2$, and two states of approximately equal population, it is the smaller state that gets 2 seats by the Klein method, while the larger state gets only 1 seat. This violates the order-preserving property and therefore population monotonicity. (c) No. An example illustrates that the Klein method is subject to the Alabama paradox. Suppose that $n = 4$ and $p_1 = 1000$, $p_2 = 3000$, $p_3 = 48,000$, and $p_4 = 48,000$. With $h = 10$, the standard quotas are 0.1, 0.3, 4.8, and 4.8, yielding lower quotas of 0, 0, 4, and 4. The Klein method gives the remaining 2 seats to the 2 smaller states, yielding an apportionment of 1, 1, 4, and 4. With $h = 11$, the standard quotas are 0.11, 0.33, 5.28, and 5.28, yielding lower quotas of 0, 0, 5, and 5. Only one seat remains, and the Klein method assigns it to the smallest state, yielding an apportionment of 1, 0, 5, and 5. The second state loses a seat as h increases from 10 to 11. (d) Yes. Every state receives either its lower quota or its upper quota.

Chapter 10

10.1. The Jefferson apportionment for $h = 15$ would be $a_1 = 1$, $a_2 = 1$, and $a_3 = 13$, using $d = 6{,}000{,}000$ as a divisor. This represents a quota violation for the third state, since its standard quota would be $15 \times 0.79 = 11.85$. Therefore this cannot be the Balinski-Young apportionment. With $h = 14$, the Jefferson apportionment would be $a_1 = 1$, $a_2 = 1$, and $a_3 = 12$, using $d = 6{,}500{,}000$ as a divisor. This does not display any quota violations (since the standard quota for the third state would be $14 \times 0.79 = 11.06$), and so this is the Balinski-Young apportionment for $h = 14$. To obtain the Balinski-Young apportionment for $h = 15$, note that the Jefferson critical divisors for the second seat for states 1 and 2 are 5,000,000 and 5,500,000, respectively, while the third state is ineligible. Therefore, the second state gets its second seat when h increases to 15, and the final Balinski-Young apportionment for $h = 15$ is $a_1 = 1$, $a_2 = 2$, and $a_3 = 12$.

10.3. Suppose that $p_1 = 17{,}000$, $p_2 = 18{,}000$, and $p_3 = 65{,}000$. The Jefferson apportionment for $h = 3$ is $a_1 = 0$, $a_2 = 0$, and $a_3 = 3$, witnessed by the modified divisor $d = 20{,}000$. This is a quota violation for state 3, however, since the standard quota for state 3 with $h = 3$ is $3 \times 0.65 = 1.95$.

10.5. Consider a union of 20 states of approximately equal population. In particular, assume that each of the states has a population that is between 4% and 6% of the total population of the union. According to the new standard for eligibility, every state is ineligible for the very first seat, since the standard quota for any state with $h = 1$ is at most 0.06.

10.7. Imagine that a census undergoes a proportional increase, with every state growing precisely by the same percentage, but that h does not change. We must show that this sort of proportional increase does not alter the Balinski-Young apportionment. The effect of such an increase is (1) to leave all of the standard quotas unchanged but (2) to increase all of the Jefferson critical divisors by the same fixed percentage. Owing to (1), the eligibility standard is unchanged. Owing to (2), the priority for receiving seats is unchanged. Therefore the Balinski-Young apportionment will be unchanged for any h.

10.9. The Balinski-Young method doles out the seats in the following order: California, Texas, New York, Florida, Illinois (California is ineligible), Pennsylvania, Ohio (Texas is ineligible, as California continues to be), Michigan, California (now eligible for its second seat), Georgia (California and New York ineligible), North Carolina, New Jersey, Texas (now eligible for its second seat), Virginia, Washington, New York (now eligible for its second seat), California (now eligible for its third seat). By contrast, the Jefferson method gives an extra seat to California (4) and Florida (2), at the expense of Virginia (0) and Washington (0).

10.11. The Jefferson apportionment is $a_1 = 4$, $a_2 = 3$, and $a_i = 0$ for $i \geq 3$, realized with the divisior of $d = 10$. The Balinski-Young apportionment method doles out the seats in the following order: state 1, state 2, state 1, state 2, state 1, state 3 (states 1 and 2 are ineligible), and then state 1, yielding $a_1 = 4$, $a_2 = 2$, $a_3 = 1$, and $a_i = 0$ for $i \geq 4$. Note that the Jefferson apportionment witnesses no upper quota violations, and yet the two apportionments do not agree.

Chapter 11

11.1. (a) State A would have average district size $5{,}000{,}000/8 = 625{,}000$, while State B would have average district size $6{,}770{,}000/12 \approx 564{,}167$. Since smaller districts are better than larger districts, this apportionment favors State B. (b) State A would have average district size $5{,}000{,}000/9 \approx 555{,}556$, while State B would have average district size $6{,}770{,}000/11 \approx 615{,}455$. This time State A has the advantage. (c) The difference in average district sizes in part (a) is approximately $625{,}000 - 564{,}167 = 60{,}883$, while the difference in part (b) is approximately $615{,}455 - 555{,}556 = 59{,}899$. The latter difference is smaller, so the apportionment in part (b) is preferable from this point of view. (d) Degree of representation is the reciprocal of average district size. In part (a), the degrees of representation of States A and B are, respectively, $1/625{,}000 = .000001600$ and $1/564{,}167 \approx .000001773$, for a difference of approximately .0000000173. In part (b), the degrees of representation of States A and B are, respectively, $1/615{,}455 = .000001625$ and $1/555{,}556 \approx .0000018$, for a difference of approximately .000000175. The former is smaller, so the apportionment in part (a) is preferable from this point of view.

11.3. If $h = 4$, $p_1 = 4000$, and $p_2 = 7000$, then the divisor $d = 2950$ yields modified quotas of roughly 1.36 and 2.37, respectively. Using harmonic rounding yields $a_1 = 2$ and $a_2 = 2$ for the Dean apportionment. On the other hand, with divisor $d = 2850$, the modified quotas are roughly 1.40 and 2.46, respectively, which when rounded geometrically yield $a_1 = 1$ and $a_2 = 3$ for the Hill apportionment.

11.5. (a) The Dean critical divisors take the form $p_i(2m+1)/(2m(m+1))$

for $i = 1, 2, 3$ and for small nonzero integers m. (When $m = 0$, the critical divisor is infinite.) There are exactly ten of these numbers that exceed 100,000 — counting the three infinite ones — and they correspond to pairs (i, m) that are $(1, 0)$, $(1, 1)$, $(2, 0)$, $(2, 1)$, $(2, 2)$, $(3, 0)$, $(3, 1)$, $(3, 2)$, $(3, 3)$, and $(3, 4)$. Hence the Dean apportionment is $a_1 = 2$, $a_2 = 3$, and $a_3 = 5$. Note that 100,000 is the standard divisor and also serves as a Dean modified divisor. (b) The Hamilton apportionment is $a_1 = 1$, $a_2 = 4$, and $a_3 = 5$.

11.7. Since $p < q$, it follows by multiplying both sides by p that $p^2 < pq$. Taking the square root of both sides of this inequality yields $p < \sqrt{pq}$. Similarly, multiplying both sides of $p < q$ by q yields $pq < q^2$. Taking the square root of both sides of this inequality yields $\sqrt{pq} < q$. (b) The square of any nonzero real number is always positive. Hence $(p-q)^2 > 0$. Expanding the square yields $p^2 - 2pq + q^2 > 0$. Adding $4pq$ to both sides of this inequality produces $p^2 + 2pq + q^2 > 4pq$ or $(p+q)^2 > 4pq$. Taking the square root of both sides of this inequality yields $p + q > 2\sqrt{pq}$, and dividing by 2 gives the desired result.

11.9. Put $p_1 = p_2 = 149,000$, $p_3 = 702,000$, and $h = 10$. Then a modified divisor for Dean's method can be $d = 110,000$, yielding modified quotas 1.35 for states 1 and 2 and modified quota 6.38 for state 3. These gives $a_1 = a_2 = 2$ and $a_3 = 6$, a quota violation for state 3.

11.11. Let the standard quotas for the three states be s_1, s_2, and s_3, and let a_1, a_2, and a_3 be the Webster apportionment for house size h. Then $a_1 + a_2 + a_3 = h = s_1 + s_2 + s_3$. Suppose that the Webster modified divisor d is less than or equal to the standard divisor. Then the modified quotas q_1, q_2, q_3 are no smaller than s_1, s_2, s_3, respectively. We now show that there is no quota violation for state 1. First note that $s_1 - a_1 \leq q_1 - a_1 \leq 1/2$, where the last inequality comes from the Webster method. It follows that the Webster apportionment for state 1 is no less than the standard quota minus 1/2. Also note that $a_1 - s_1 = s_2 - a_2 + s_3 - a_3 \leq q_2 - a_2 + q_3 - a_3 \leq 1/2 + 1/2 = 1$. It follows that the Webster apportionment for state 1 is no more than the standard quota plus 1. Taken together, this shows that the Webster apportionment for state 1 is within one unit of the standard quota. A similar argument works when d is greater than the standard divisor.

11.13. A number of the form p_k/m for a nonnegative integer m is an Adams critical divisor for state k. Adams's method chooses the largest h of these Adams critical divisors, counting first the ones with $m = 0$ (they may be thought of as infinite critical divisors). If p_i/a_i is the largest district size for the Adams apportionment with house size h, then there are h Adams critical divisors larger than this number. Any other apportionment for house size h will assign some state fewer seats than Adams's method does. But then that state will have larger district size than p_i/a_i.

Chapter 12

12.1. The answers appear in Table A.1.

Table A.1 Apportionments for 1790 by various methods.

State	Pop.	Ham/Web	Jeff	Adams/Dean/Hill
Virginia	630,560	18	19	18
Massachusetts	475,327	14	14	14
Pennsylvania	432,879	13	13	12
North Carolina	353,523	10	10	10
New York	331,589	10	10	10
Maryland	278,514	8	8	8
Connecticut	236,841	7	7	7
South Carolina	206,236	6	6	6
New Jersey	179,570	5	5	5
New Hampshire	141,822	4	4	4
Vermont	85,533	2	2	3
Georgia	70,835	2	2	2
Kentucky	68,705	2	2	2
Rhode Island	68,446	2	2	2
Delaware	55,540	2	1	2

12.3. The states with the smallest district sizes, in order, are Rhode Island, Wyoming, Nebraska, West Virginia, and Vermont. The states with the largest district sizes, in order, are Montana, Delaware, South Dakota, Idaho, and Oregon.

12.5. In 2010, it was North Carolina, whose apportionment would have increased from 13 to 14. (After North Carolina, subsequent extra seats would go in order to Missouri, New York, New Jersey, and Montana.) In 2000, it was Utah, whose apportionment would have increased from 3 to 4. (Utah got 4 seats in the 2010 apportionment.)

12.7. The cut-off between 11 and 12 for Hill's method is $\sqrt{11 \cdot 12} \approx 11.489$, while the cut-off between 12 and 13 is $\sqrt{12 \cdot 13} \approx 12.490$. Hence the population of Maryland must be between $900{,}000 \cdot 11.489 \approx 10{,}340{,}000$ and $900{,}000 \cdot 12.490 \approx 11{,}240{,}000$.

12.9. Montana's apportionment population in the 2010 census was 994,416. Had that number increased by a mere 8000 residents, their population would have been over 102,000, sufficient for them to have gotten two seats.

Chapter 13

13.1. Here are the flow diagrams:

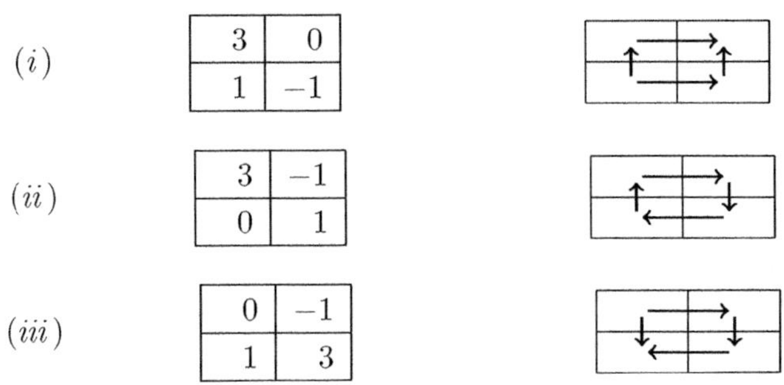

Game (*i*) has a saddle point at (row 1, column 2). Game (*ii*) has no saddle points. Game (*iii*) has a saddle point at (row 2, column 1).

13.3. (a) The naive strategies for both players in all three games are shown below by arrows.

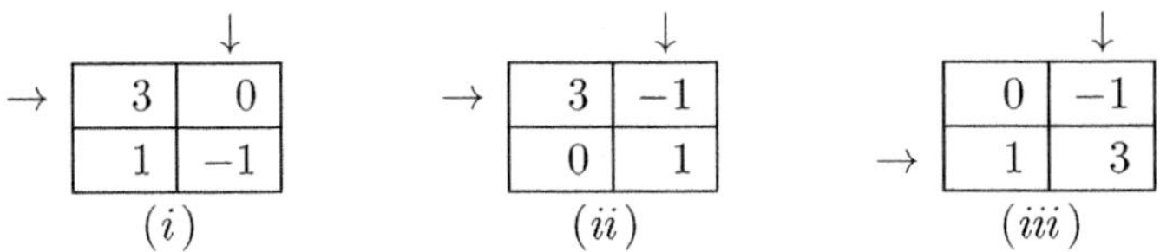

(b) In game (*i*), the payoff for the doubly-naive outcome is 0, so neither player comes out ahead. Notice that it is the same as the saddle point in this game. In game (*ii*), the payoff for the doubly-naive outcome is -1, so the outcome favors Column. In game (*iii*), the payoff for the doubly-naive outcome is 3, so the outcome favors Row.
(c) The counter-naive strategies for both players in all three games are indicated below by double arrows.

(i) Row's arrow ⇒ points to row 1; Column's arrow ⇓ points to column 2.

3	0
1	−1

(ii) Row's arrow ⇒ points to row 2; Column's arrow ⇓ points to column 2.

3	−1
0	1

(iii) Row's arrow ⇒ points to row 2; Column's arrow ⇓ points to column 1.

0	−1
1	3

(d) In game (*i*), the naive and counter-naive strategies are the same for both players (they are actually saddle point strategies), so the result is still a tie. In game (*ii*), Row plays row 2 instead of row 1, and her payoff improves from −1 to 1. In game (*iii*), Row's counter-naive strategy is the same as her naive strategy, so the result does not change. (Note, however, that Column's counter-naive strategy is different from his naive strategy.)

13.5. (a) In game (*i*), row 1 dominates row 2, and column 2 dominates column 1. In game (*ii*), there are no dominated strategies. In game (*iii*), row 2 dominates row 1, but there are no dominated columns. (b) Game (*i*) completely reduces in a single step to the 1-by-1 game with entry 0. Game (*ii*) is already completely reduced. In game (*iii*), we delete the dominated row 1 to obtain the 1-by-2 game | 1 | 3 |. In this game, column 1 dominates column 2. After deleting the dominated column, we obtain a 1-by-1 game with entry 1.

13.7.

The last column gives the row minima and the bottom row gives the column maxima; the arrows (←, ↑) mark row 4 and column 2.

3	0	4	1	1	0
0	0	3	1	−2	−2
−1	2	0	3	−1	−1
3	1	1	1	3	1 ←
3	2	4	3	3	
	↑				

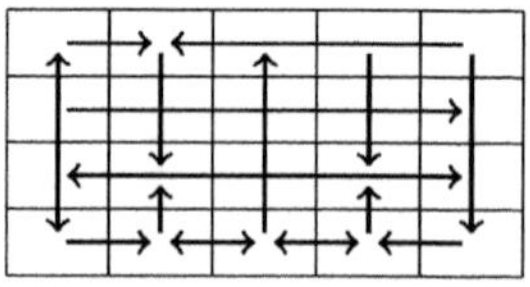

The min-max algorithm shows $r = 1$ and $c = 2$, and since $r \neq c$, there are no saddle points. This is verified by the flow diagram.

13.9.

3	4	5
2	0	6
1	6	0

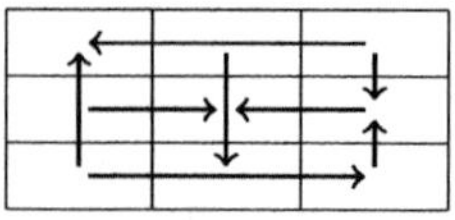

There is a saddle point at (row 1, column 1). But if the outcome were to start at (row 2, column 2) and follow the flow diagram, it would cycle through (row 3, column 2), (row 3, column 3), (row 2, column 3) and (row 2, column 2) indefinitely. In particular, the saddle point would never be reached.

13.11. We set up this game with the terrorists as Row and the FBI as Column. The off-diagonal entries (where the terrorists and FBI go to different locations) are given scores of 100% and 80%. If both the terrorists and the FBI go to A, the terrorists will be stopped, which we score as 0%. If they both go to B, the chance that the terrorists will succeed is 40%. Given the 80% value the terrorists put on this target, we give this outcome a score of $40\% \times 80\% = 32\%$. This results in the following matrix.

	A	B
A	0	100
B	80	32

There is no saddle point as can be seen from either the flow diagram or the min-max algorithm.

The terrorist's naive strategy is to go for their favorite target, A. The FBI naive strategy is also A, hoping to catch the terrorists attacking their favorite target. In this scenario the FBI prevails, catching the terrorists.

The terrorist's prudent strategy is to attack B, because even in the worst case, they still have a 32% chance of success (or really 40% chance of 80% success). The FBI prudent strategy is to guard A because not guarding A affords the terrorists a chance for a successful attack on their primary target, A. In this scenario, the terrorists are guaranteed to succeed in attacking target B, although they would prefer target A.

13.13. In the game below, Row 1 is naive, Row 2 is prudent and Row 3 has average 7/3 which is greater than the averages of the first two rows.

5	−1	−2
0	1	3
−1	4	4

13.15. No, the game below has a saddle point in the lower right corner, but no row or column dominates any other.

4	−5	−3
−5	4	−1
3	2	0

There is, however, no 2-by-2 example: any 2-by-2 game with a saddle point completely reduces to a 1-by-1 game.

Chapter 14

14.1. (a) $\{hhhh, hhht, hhth, hhtt, hthh, htht, htth, httt, thhh, thht, thth, thtt, tthh, ttht, ttth, tttt\}$.
(b) The probability of four heads is $1/16$. The probability of four heads or four tails is $1/8$. The probability of not all the same is $7/8 = 1 - 1/8$.
(c) There are six ways to get two heads and two tails, so the probability is $6/16 = 3/8$. But there are only four ways to get three heads and one tail, so the probability is $4/16 = 1/4$. In other words, tossing three heads is less likely.
(d) The probability of three of one kind and one of the other is $1/4 + 1/4 = 1/2$. So this is more likely than two of each kind, which by (c) has probability $3/8$.

14.3. The table below has the four outcomes on the top line, and their probabilities below. In the boxes are the payoffs.

hh	tt	th	ht
\$6	\$3	−\$5	−\$5
1/4	1/4	1/4	1/4

We calculate

$$E = (1/4)6 + (1/4)3 + (1/4)(-5) + (1/4)(-5) = -0.25,$$

so the bet favors the huckster, but only by a small margin.

14.5. Let x denote the amount you are willing to pay for this bet.

Rain	Shine
\$5	$-x$
0.3	0.7

To break even, you need $E = 0$, but $E = 5(0.3) + (-x)(0.7) = 1.5 - 0.7x$. We solve $1.5 - 0.7x = 0$ to obtain $x = (1.5)/(0.7) \approx 2.14$. If you pay any less than this, then it is a good bet. If 30% is replaced by p, then the break-even point becomes $5p/(1-p)$.

14.7. First

$$E(P, Q_1) = (4/5)1 + (1/5)3 = 7/5,$$

and

$$E(P, Q_2) = (4/5)2 + (1/5)(-1) = 7/5.$$

Then

$$E(P, Q) = (1/2)E(P, Q_1) + (1/2)E(P, Q_2) = 7/5.$$

It would not matter if Column changed his mixed strategy, because $E(P, Q)$ is always a weighted average of two numbers that are the same.

14.9. Below, on the left, is the game with the mixed strategy probabilities written along the sides. In the matrix on the right are the probabilities of the six outcomes. Since we assume the mixed strategies are independent, these are the products of the probabilities written to the right and below the matrices.

1	−2	7	2/5
−5	3	−6	3/5
1/7	4/7	2/7	

2/35	8/35	4/35	2/5
3/35	12/35	6/35	3/5
1/7	4/7	2/7	

Then

$$E(P, Q) = \frac{2}{35}1 + \frac{8}{35}(-2) + \frac{4}{35}7 + \frac{3}{35}(-5) + \frac{12}{35}3 + \frac{6}{35}(-6) = -\frac{1}{35}.$$

14.11. (a) 1/101, (b) 8 to 1, (c) You should pay her \$4, because then $E = (4/5)1 + (1/5)(-4) = 0$.

14.13. We set it up like this:

\$10,000	-\$50,000	0
.17	.03	.80

Thus $E = .17(\$10{,}000) + .03(-\$50{,}000) + .80(0) = \$200$, which is a gain on average, so you should probably vote for the bill.

14.15. We compute

$$\begin{aligned} p_1q_1 + p_1q_2 + \cdots + p_mq_n &= p_1(q_1 + q_2 + \cdots + q_n) + \cdots \\ &\quad + p_m(q_1 + q_2 + \cdots + q_n) \\ &= p_1 + p_2 + \cdots + p_m = 1, \end{aligned}$$

since $q_1 + q_2 + \cdots + q_n = 1$ and $p_1 + p_2 + \cdots + p_m = 1$.

14.17. In Powerball, you pick five white ball numbers from 1 to 69, and one red ball (the "power ball") number from 1 to 26. Five white balls and one red ball are drawn. Here is the table of prizes for a \$2 ticket and their probabilities, as published on the Powerball website. The Grand Prize or Jackpot increases each time there is no winner, but starts at \$40,000,000, which is the value we will use. The first column shows the number of correct white + number of correct red balls that a player needs to have in order to win.

Balls	Prize	Probablity
$5+1$	Grand Prize	1/292,201,338.00
$5+0$	\$1,000,000	1/11,688,053.52
$4+1$	\$50,000	1/913,129.18
$4+0$	\$100	1/36,525.17
$3+1$	\$100	1/14,494.11
$3+0$	\$7	1/579.76
$2+1$	\$7	1/701.33
$1+1$	\$4	1/91.98
$0+1$	\$4	1/38.32

Note that the Powerball website refers to "odds" even though the numbers given are really probabilities. To compute the expected payoff we multiply the value of each prize by its probability

$$E = (1/292{,}201{,}338.00)(\$40{,}000{,}000) + \cdots + (1/55.41)(\$4) \approx \$0.46.$$

Considering a ticket costs \$2.00 it is not a good bet.

Sometimes the Grand Prize (jackpot) is high enough that a Powerball ticket is a good bet from the point of view of this expected value analysis. If J denotes the jackpot in dollars, then the expected value of the game exceeds the cost of the ticket when

$$E = (1/292{,}201{,}338)J + .3199 > 2.$$

This occurs whenever the jackpot J is more than \$490,933,824, which happens occasionally. Note, however, that this analysis ignores taxes, present value of future payments, decreasing marginal utility of money, and the possibility of having to share the jackpot.

Chapter 15

15.1. One can check (see Exercise **13.1**) that games (*i*) and (*iii*) have saddle points, so these are the solution. For (*i*) this is Row 1, Column 2, $v = 0$, and for (*iii*) it is Row 2, Column 1, $v = 1$. Game (*ii*), however, has no saddle point so we apply the cross-over algorithm, starting with the cross-over matrix

$$\begin{array}{|c|c|}\hline 3 & -1 \\ \hline 0 & 1 \\ \hline \end{array} \quad \longrightarrow \quad \begin{array}{|c|c|}\hline 1 & 0 \\ \hline 1 & 3 \\ \hline \end{array}\,.$$

Adding the row and column entries

1	0	1
1	3	4
2	3	

we have that $N = 1 + 4 = 2 + 3 = 5$. Dividing the numbers outside the matrix by N gives the mixed strategy equilibrium probabilities.

1	0	1/5
1	3	4/5
2/5	3/5	

Finally, $\Delta = (3)(1) - (-1)(0) = 3$, and $v = \Delta/N = 3/5$. In summary, the solution is Row $P = (1/5, 4/5)$, Column $Q = (2/5, 3/5)$, and $v = 3/5$.

15.3. It is easy to see that (Row 1, Column 1) is a saddle point with value $v = 3$. Note, however, that if we were to start at (Row 2, Column 2) and follow the flow diagram, we would never reach the saddle point. This suggests that we look at the "sub-game" that we get by deleting Row 1 and Column 1.

0	6
6	0

This new 2-by-2 is easily seen to have no saddle point. Using the cross-over algorithm, we see that $P = Q = (1/2, 1/2)$ is an equilibrium for the new game with $v = 3$. We claim that $P = Q = (0, 1/2, 1/2)$ is an equilibrium for the original game. In particular, P and Q are neutralizing strategies. We show this for P by playing it against each basic strategy Q_1, Q_2, and Q_3. The result is $E(P, Q_1) = (0)3 + (1/2)3 + (1/2)3 = 3$, $E(P, Q_2) = (0)3 + (1/2)0 + (1/2)6 = 3$, and $E(P, Q_3) = (0)3 + (1/2)6 + (1/2)0 = 3$.

15.5. The game in Problem **13.11** has the matrix

	A	B
A	0	100
B	80	32

Applying the cross-over algorithm we have

32	−80	48/148
−100	0	100/148
68/148	80/148	

with $N = -148$ (the row and column sums are all negative). Then $\Delta = -8000$ and $v = 8000/148 \approx 54\%$. This number is an average of several things: the values the terrorists place on the two targets, their chances of success in an attack on each, and the probabilities that the mixed strategies recommend for attacking targets. Let us call it the terrorists' success rate. The equilibrium strategy says the terrorists should attack landmark A with a probability of about 1/3, and if they do, their success rate is around 54%. This improves on their prudent strategy, which promises a 32% success rate. But also, the FBI can hold the terrorists to the 54% success rate by using the mixed strategy $Q = (68/148, 80/148)$. It recommends guarding the more valuable A 45% of the time and guarding B the rest of the time.

15.7. The third column dominates the first, so Column's best strategy will never selection the first column. After reducing the game by elimination of the first column, we obtain a 2-by-2 game with no saddle point. We obtain the optimal strategy for the two players and the value of the 2-by-2 completed reduced game by the cross-over method (i.e., Corollary 15.13), obtaining $(17/27, 10/27)$ as the optimal Row strategy, $(8/27, 19/27)$ as the optimal Column strategy, and $-1/27$ as the value of the game. The optimal strategy for Column in the 3-by-2 game is $(0, 8/27, 19/27)$, while the Row strategy and the value remain the same as for the 2-by-2 game. (This game favors the Column player, though just slightly.)

15.9. Consider the votes Rose gets. If the candidates visit different states, Rose gets all the electoral votes from the state he visits (29 from Florida or 18 from Ohio), but none from the other state. If both candidates visit Florida, then they split the Florida votes equally (14.5 votes, on average), and Rose gets 25% of Ohio (4.5 votes). Similarly, if both candidates visit Ohio, Rose gets 75% of Florida (21.75 votes) and they split Ohio (9 votes). Rose's votes are summarized in the table below.

	Florida	Ohio
Florida	14.5+4.5	29+0
Ohio	0+18	21.75+9

Since the total number of electoral votes between the two states is fixed at $27 + 20 = 47$, this game is constant sum, and we express it as a matrix game. Then we draw a flow diagram.

19	29
18	30.75

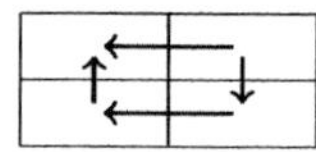

From the flow diagram, we see that there is a saddle point: Both Rose and Collins go to Florida. In this way, Rose gets 19 votes, but Collins gets 28, despite Rose's seeming advantage. This is counter-intuitive.

15.11. The value of a game with no saddle point is $v = \Delta/N$ (where N cannot be 0). If $\Delta = 0$, then $v = 0$, and the game is fair. However a game with $\Delta = 0$ can have a saddle point with a nonzero value, and thus not be fair. For example:

1	2
2	4

15.13. (a) In going from the old game to the new, all the arrows in the flow diagrams reverse direction, so if the original game has no saddle point, the new game does not have one either. Since neither game has a saddle point, we can solve both with the cross-over algorithm. Notice that the determinant Δ does not change, but the sign of N does. Since $v = \Delta/N$, v in the old game becomes $-v$ in the new. (b) If the original game has a saddle point, then since the arrows all change direction, the new game also has a saddle point. (c) Any game with all entries positive becomes a game with all entries negative, so in this case the sign of v changes (whether or not there is a saddle point). However Row is favored in both the new and old games with

2	1
−1	−2

15.15. The first sub-game (obtained by deleting Row 3)

1	0
2	−3

has a saddle point at (Row 1, Column 2) with value $v = 0$. We play Column $Q = Q_1 = (0, 1)$ against the deleted Row 3 and get $E = 0(-1)+1(4) = 4$ which is greater than $v = 0$.

The second sub-game (obtained by deleting Row 2) has no saddle point so is solved by the cross-over algorithm

1	0
−1	4

$\longrightarrow$

4	1	5/6
0	1	1/6
4/6	2/6	

with $v = \Delta/N = 4/6 = 2/3$. We play $Q = (4/6, 2/6) = (2/3, 1/3)$ against the deleted Row 2 to get $E = (2/3)2 + (1/3)(-3) = 1/3$, which is less than $v = 2/3$. Thus, we can stop here without looking at the third sub-game. The solution to the 3-by-2 game has $P = (5/6, 0, 1/6)$, $Q = (2/3, 1/3)$, and $v = 2/3$.

15.17. Since Row 2 dominates Row 1, we delete Row 1 to get the 2-by-3 game:

2	−1	−2
−1	0	3

We employ the technique of Exercise **15.15** on this 2-by-3 game, adjusting appropriately to reflect that the third option is now Column's rather than Row's. The 2-by-2 game obtained by deleting Column 3 is:

2	−1
−1	0

Applying the cross-over algorithm, we see that the solution is $P = (1/4, 3/4)$, $Q = (1/4, 3/4)$ and $v = -1/4$. We play this P against the deleted Column 3 to get $E = (1/4)(-2) + (3/4)(3) = 7/4$. This is greater than $v = -1/4$, so we need not look at the other two 2-by-2 sub-games. In the original 3-by-3 game, we make $P = (0, 1/4, 3/4)$ (Row plays the dominated strategy Row 1 with probability 0) and $Q = (1/4, 3/4, 0)$. The value of the game is $v = -1/4$.

Chapter 16

16.1. On the left, we show (a) the guarantees and prudent strategies, and, on the right, (b) the flow diagram.

1, −3	3, −1	1	←
−2, 4	−4, 2	−4	
−3	−1		
	↑		

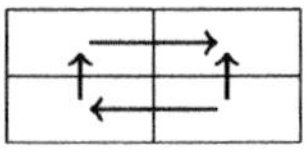

The game has a saddle point at (Row 1, Column 2). (c) Here is the payoff polygon.

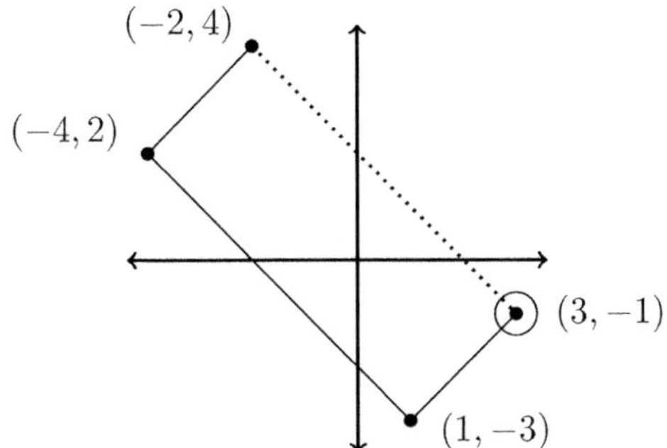

(d) Row 1 dominates Row 2. After deleting Row 2, Column 2 dominates Column 1, so the game reduces to the 1-by-1 game consisting of the saddle point from the original game. The saddle point is Pareto optimal, so it is a good choice for both players.

16.3. We show (a) the guarantees and prudent strategies, and (b) the flow diagram.

3, 3	9, 4	6, 6	3	
4, 5	8, 8	5, 9	4	←
2, 1	7, 7	1, 2	1	
1	4	2		
	↑			

The game has a saddle point at (Row A, Column C). (c) Row B dominates Row C. After deleting Row C, Column C dominates Columns A and B. Removing these, Row A dominates Row B, so the resulting 1-by-1 game has Row A and Column C. This corresponds to the saddle in the 3-by-3 game.

In this game, the doubly prudent outcome is also doubly naive, and it offers each player a good payoff of 8. Unfortunately, it is not a saddle point, so both players have an incentive to move away from it, and the guarantee is only 4. The saddle point offers each player 6, only a little less than the doubly prudent outcome, but its guarantee is lower (3 for Row and 2 for Column). Good advice is difficult to give here. Rows A and B and Columns B and C form a Prisoner's Dilemma.

16.5. First we fill in a matrix with the outcomes "meet" and "don't" meet.

	North	South
North	meet	don't
South	don't	meet

(a) Here we suppose both players want to find each other. We put 1 for a meeting and 0 for a miss.

	North	South
North	1, 1	0, 0
South	0, 0	1, 1

(b) In this part, the change is that the child does not want to be found, so only his payoffs change.

	North	South
North	1, 0	0, 1
South	0, 1	1, 0

(c) The child's motives become increasingly complex — the parent still just wants to find the child.

	North	South
North	1, 2	0, 1
South	0, 0	1, 3

(Some disagreement is possible here about the relative placement of 1 and 2 for Column.) Notice that in this last version of the game, (North, North) and (South, South) are both saddle points. However, the child will probably opt for South since it his naive strategy. Since the parent knows the child, she will probably go there too.
(c) Game (a) is a coordination game, Game (b) is strictly competitive, and Game (c) is mixed motive.

16.7. Make one village Row and the other Column. The strategies are "One" boat or "Two". The chart below (which is not yet the game) shows the pounds of fish per boat, depending on the total number of boats.

	One	Two
One	50	40
Two	40	15

We multiply these numbers by the number of boats each village has to obtain the game bimatrix.

	One	Two
One	50, 50	40, 80
Two	80, 40	30, 30

The prudent strategy for each village is to stay with one boat, but the naive strategy is to build a second. There are two Pareto optimal saddle points. Both occur when one village has two boats and the other has just one. While such a saddle point would probably seem unacceptable to the village with just one boat, building a second boat will not help (although it will punish the other village). This game is essentially Chicken. A negotiator would probably try to get the villages to share a second boat. However, the three boat total catch of 120 pounds is only a little more than the two boat total catch of 100 pounds. Considering it means 50% more work for each village, it may not be worth it.

16.9. We set this up with the government as Row and the worker as Column, using somewhat arbitrary payoff numbers that reflect the players' preference orders as given in the problem.

	Look	Stop
Pay	$10, 5$	$-10, 10$
Withhold	$-5, 1$	$0, 0$

The government's naive strategy is to pay unemployment benefits, hoping it will encourage job seeking. But the worker's naive strategy is to stop looking. The temptation to loaf is why some oppose welfare. The government's prudent strategy is to withhold benefits, and the worker's prudent strategy is to look for work. While this outcome will probably reduce unemployment, it may seem heartless. The outcome (Pay, Look) is Pareto optimal. This outcome is probably what the architects of unemployment benefits had in mind. However, since this game has a pursuit-type flow diagram, it has no saddle points.

16.11. (a) In Chicken the two outcomes where one driver swerves and the other does not are maximum sum. In Battle of the Sexes, it is the two outcomes where the couple goes out together. In both of these cases the outcomes are saddle points. In the Prisoner's Dilemma, however, the outcome where the two prisoners keep quiet is maximum sum. It is not a saddle point. (b) In a zero-sum game, all outcomes are maximum sum, but the maximum is zero. (c) Suppose an outcome is Pareto inferior to another outcome. Moving to the other outcome would not decrease either player's payoff but would increase the payoff of at least one. Thus the sum would go up. (d) The examples in (a) work for this.

16.13. (a) Here are the zero-sum parts of these games

	Hockey	Ballet
Ballet	0	2.5
Hockey	−2.5	0

	Don't	Swerve
Don't	0	7.5
Swerve	−7.5	0

	Confess	Quiet
Confess	0	7.5
Quiet	−7.5	0

	North	South
North	2	2
South	1	3

(b) Battle of the Sexes, Chicken, and the Prisoner's Dilemma all have essentially the same zero-sum part. Each has a saddle point in the upper left. This outcome corresponds to the doubly prudent outcome in the non-zero-sum game. Since Battle of the Bismarck Sea is already zero-sum, it is its own zero-sum part. (c) Sometimes a player in a non-zero-sum game attempts to get as far ahead of the opponent as possible, rather than concentrating on his or her own payoff. Such a player is essentially playing the zero-sum part.

Chapter 17

17.1. (a) This is a coordination type flow diagram with saddle points in the upper right and lower left corners.

0, −1	3, 4
2, 1	−1, −2

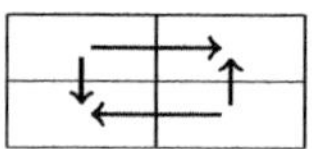

(b) Because of the flow diagram type, there is also a mixed strategy Nash equilibrium. To find it, we have Column apply the cross-over algorithm to Row's game

0	3
2	−1

$\longrightarrow$

−1	−2
−3	0
−4	−2
4/6	2/6

to get $Q = (4/6, 2/6) = (2/3, 1/3)$, and since $N = -6$ and $\Delta = -6$, $E = \Delta/N = 1$. We also have Row apply the cross-over algorithm to Column's game

−1	4
1	−2

$\longrightarrow$

−2	−1	−3	3/8
−4	−1	−5	5/8

to get $P = (3/8, 5/8)$, with $N = -8$, $\Delta = -2$ and $F = \Delta/N = (-2)/(-8) = 1/4$. So (P, Q) is a Nash equilibrium with payoff pair $(E, F) = (1, 1/4)$. (c) Here is the payoff polygon

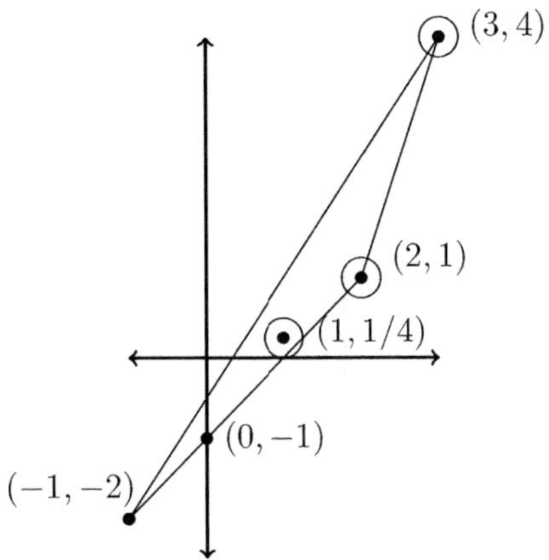

(d) The only Pareto optimal Nash equilibrium is the saddle point (Row 1, Column 2).

17.3. (a) To find prudent pure strategies, we use the min-max algorithm

0, −1	3, 4	0	←
2, 1	−1, −2	−1	
−1	−2		
↑			

So Row 1 is prudent with $r = 0$ and Column 1 is prudent with $c = -1$.
(b) To find the prudent mixed strategies, we first check that neither Row's nor Column's game has a saddle point. Thus Row and Column can each solve their own games using the cross-over algorithm. For Row, this yields

0	3
2	−1

⟶

−1	−2	−3	1/2
−3	0	−3	1/2

so $P = (1/2, 1/2)$ and $\overline{r} = v = \Delta/N = 1$. For Column,

−1	4
1	−2

⟶

−2	−1
−4	−1
−6	−2
3/4	1/4

so $Q = (3/4, 1/4)$ and $\overline{c} = v = \Delta/N = 1/4$.

17.5. The flow diagram for the Prisoner's Dilemma is of doubly dominant type.

17.7. In our solution to Problem **16.9** we set this game up as follows

	Look	Stop
Pay	10, 5	−10, 10
Withhold	−5, 1	0, 0

This game has a pursuit-type flow diagram, so it has no saddle points, but it does have a mixed strategy Nash equilibrium. To find it, Column applies the cross-over algorithm to Row's game

10	−10
−5	0

$\longrightarrow$

0	5
10	10
10	15
10/25	15/25

to get $Q = (10/25, 15/25) = (2/5, 3/5)$. Since $\Delta = -50$ and since $N = 25$, $E = \Delta/N = -2$. Column has made it so this is Row's score no matter what Row does.

Also, Row applies the cross-over algorithm to Column's game

5	10
1	0

$\longrightarrow$

0	−1	−1	1/6
−10	5	−5	5/6

to get $P = (1/6, 5/6)$. Here $N = -6$, $\Delta = -10$ so $F = \Delta/N = (-10)/(-6) \approx 1.67$.

Here is a possible interpretation. The worker looks for work about 40% of the time and the government usually pays unemployment. But the government cuts off benefits $1/6 \approx 17\%$ of the time. In this behavior pattern, the worker is a little more happy than with (Withhold, Stop) and the government is a little less happy. But neither has an incentive to change unilaterally.

17.9. If $f(x) = x$, then we obtain $x = x/2$, which implies that $x = 0$. This number is not in I. Here I is not compact, since it does not contain its endpoints 0 and 1.

17.11. If $f(x) = x$ then either $x + 1/2 = x$ or $x - 1/2 = x$. In both cases, there is no such number. Here, the function f jumps from 1 to 0 at $x = 1/2$, so it is not continuous.

17.13. First we compute the following: $E(P_1, Q) = (4/7)2 + (3/7)8 = 32/7$, $E(P_2, Q) = (4/7)4 + (3/7)5 = 31/7$, $E(P, Q) = (1/5)(32/7) + (4/5)(31/7) = 156/35$, and $F(P, Q_1) = (1/5)6 + (4/5)1 = 10/5 = 2$, $F(P, Q_2) = (1/5)3 + (4/5)7 = 31/5$, $F(P, Q) = (4/7)(2) + (3/7)(31/5) = 19/5$. Now, $a_1 = \max(0, 32/7 - 156/35) = 4/35$ and $a_2 = \max(0, 31/7 - 156/35) = 1/35$, so $1 + a_1 + a_2 = 40/35 = 8/7$. Thus $p_1' = (4/35 + 1/5)/(8/7) = 11/40$ and $p_2' = 29/40$. Similarly, $b_1 = \max(0, 2 - 19/5) = 0$ and $b_2 = \max(0, 31/5 - 19/5) = 12/5$, so $1 + b_1 + b_2 = 17/5$. Thus $q_1' =$

$(4/7+0)/(17/5) = 20/119$ and $q_2' = 99/119$. It follows that $N(P,Q) = (P', Q')$ where $P' = (11/40, 29/40)$ and $Q' = (20/119, 99/119)$.

17.15. We write this as a bimatrix game

−1, 1	2, −2	1, −1
4, −4	−3, 3	−1, 1

.

Then we compute the following: $E(P_1, Q) = (4/7)(-1) + (2/7)2 + (1/7)1 = 1/7$, $E(P_2, Q) = (4/7)4 + (2/7)(-3) + (1/7)(-1) = 9/7$, and $E(P,Q) = (1/5)(1/7) + (4/5)(9/7) = 37/35$. Now, $a_1 = \max(0, 1/7 - 37/35) = 0$ and $a_2 = \max(0, 9/7 - 37/35) = 8/35$, so $1 + a_1 + a_2 = 43/35$. Thus $p_1' = (1/5+0)/(43/35) = 7/43$ and $p_2' = 36/43$.

Similarly

$$\begin{aligned} F(P,Q_1) &= (1/5)1 + (4/5)(-4) = -3, \\ F(P,Q_2) &= (1/5)(-2) + (4/5)3 = 2, \\ F(P,Q_3) &= (1/5)(-1) + (4/5)1 = 3/5, \text{ and} \\ F(P,Q) &= (4/7)(-3) + (2/7)2 + (1/7)(3/5) = -37/35. \end{aligned}$$

Thus $b_1 = \max(0, -3 - (-37/35)) = 0$, $b_2 = \max(0, 2 - (-37/35)) = 107/35$, and $b_3 = \max(0, 3/5 - (-37/35)) = 58/35$ so that $1 + b_1 + b_2 + b_3 = 40/7$. Thus $q_1' = (4/7+0) \cdot (7/40) = 1/10$, $q_2' = (2/7 + 107/35) \cdot (7/40) = 117/200$ and $q_3' = (1/7 + 58/35) \cdot (7/40) = 63/200$.

Chapter 18

18.1. The four outcomes for each candidate can be ordered, from best to worst, as follows: campaign negatively while the opponent campaigns positively, campaign positively while the opponent campaigns positively, campaign negatively while the opponent campaigns negatively, and campaign positively while the opponent campaigns negatively. Hence an ordinal bimatrix game that captures this ranking is

	Negative	Positive
Negative	3rd, 3rd	1st, 4th
Positive	4th, 1st	2nd, 2nd

We have chosen to write this as an ordinal game, since it isn't easy to assign utilities to the four alternatives. Negative campaigning dominates positive for both candidates, and so the outcome (Negative, Negative)

is a saddle point outcome. Yet both candidates would be better off if they could cooperate and mutually campaign positively, playing their dominated and imprudent strategies. This is the Prisoner's Dilemma paradox.

It is difficult to give good advice to a candidate facing this famously inscrutable game. One is tempted to recommend that the candidate attempt to reach an agreement with the opponent to cooperate and campaign positively. To help enforce this agreement, it might be reasonable to threaten retaliation against the candidate if they dare to initiate a negative campaign. Perhaps your campaign could dig up some dirt on the opponent to keep in the back pocket in case the agreement falls through. And perhaps you'd want to let your opponent know that you have this damaging information, so that he can see that your negative campaign would be very damaging to him. In this way, you are effectively altering the outcomes of the game and it becomes more like this.

	Negative	Positive
Negative	$-9, -9$	$1, -10$
Positive	$-10, 1$	$0, 0$

Now the Prisoner's Dilemma is even more striking. The incentive for cooperation is enormous. Yet the saddle point strategy, the prudent strategy, and the equilibrium strategy continue to be negative campaigning for both candidates.

18.3. Suppose the high price is $22,000 and the low price is $18,000. We will set up the game so that the payoff for the purchaser, Ms. Rowe, is the amount below the high price she pays, and for the dealer, Mr. Collier, is the amount above the low price he manages to get. However, if the sale falls through, each gets a score of –$100 (for time and energy wasted).

Rowe	Collier	Price	Rowe's score	Collier's score
High	High	$22,000	0	$4000
High	Low	$20,000	$2000	$2000
Low	High	No sale	–$100	–$100
Low	Low	$18,000	$4000	0

The bimatrix is the following

	High	Low
High	$0, 4000$	$2000, 2000$
Low	$-100, -100$	$4000, 0$

For Ms. Rowe, High is prudent and Low is naive. For the dealer it is the opposite. There is no dominance, but two saddle points, which occur when both agree on the price. This is not the Prisoner's Dilemma but rather Chicken.

18.5. Each candidate can vow to raise or lower taxes. If exactly one candidate vows to lower taxes, that candidate is swept into office while the opponent is forever a political pariah and never seen again. These are the best and worst outcomes, respectively. If the two candidates make the same vow, then it will be a close election, a 50-50 toss-up, but a president taking office is better off if they vowed to raise taxes (the responsible position) rather than lower taxes (the irresponsible position). This is a Prisoner's Dilemma, with the dominant, prudent, and equilibrium position to lower taxes, leading to an irresponsible (if familiar) outcome of a president forced to lower taxes against national interest. The cooperative move in this setting is to vow to raise taxes. When both candidates cooperate, the result is a close election and a president who addresses a national crisis.

18.7. The outcomes can be put in an ordinal game as follows.

	Extreme	Moderate
Extreme	3rd, 3rd	1st, 4th
Moderate	4th, 1st	2nd, 2nd

This is clearly an instance of the Prisoner's Dilemma. The model predicts that while, in principle, political parties would like to be moderate, the risk is too great. Instead they will both end up moving toward the extreme position, even though it would be better for everyone if they could both stay moderate.

18.9. This method is a version of the equilibrium method as described in Theorem 18.9, so for matrix games it satisfies the dominance, security, stability, and Pareto criteria. For non-zero sum games, this satisfies the dominance criterion (Proposition 18.3), however it does not satisfy the security or stability criteria (as shown by Chicken) or the Pareto criterion (as shown by the Prisoner's Dilemma).

18.11. To find her prudent pure and prudent mixed strategies, Row simply ignores Column's payoffs, views her game as a zero-sum game, and finds her zero-sum prudent pure and prudent mixed strategies. Let r and $\overline{r}$ (respectively) be the guarantees for these strategies. We know $r \leq \overline{r}$ by Theorem 15.11 or Problem **15.14**. In a similar way, $c \leq \overline{c}$ (where we have written c in the bimatrix style, as explained in Figure 16.1). This shows that the prudent mixed strategy always has a guarantee at least as good as the prudent pure strategy. Thus the prudent mixed method satisfies the security criterion.

Chapter 19

19.1. There are just three voters in this system; we call them A, B, and C, in decreasing order of number of votes. The winning coalitions are AB, AC, and ABC. Voter A is critical in all three of these coalitions, so her voting power is 3. Voters B and C are critical only in AB and AC, respectively, so each of them has a voting power of 1. What is surprising is that voter C, with 1 vote, is precisely as powerful as voter B, who has 19 votes.

19.3. We call the voters A, B, C, D, and E in decreasing order of weight. The winning coalitions, with critical members identified by boldface font, are ABCDE, ABCD, **A**BC**E**, **AB**DE, **AC**DE, **BCDE**, **ABC**, **ABD**, **AB**E, **ACD**, **ACE**, **BCD** and **AB**. Counting the number of appearances of each letter in boldface gives the Banzhaf power of each voter: A, 9; B, 7; C, 5; D, 3; and E, 1. It is striking and uncharacteristic that these numbers are equal to the weights of the respective voters.

19.5. A distribution of weights that represents this voting system is $V(22; 10, 5, 5, 4, 4, 4, 4, 4, 4, 4)$. Clearly any coalition of 6 voters wins in this system, any coalition of 5 voters that includes B and C wins, and any coalition of 4 voters that includes A wins. Moreover, no coalition that is not among these wins, because no coalition of 3 voters has enough weight, no coalition of 4 voters has enough weight unless A is included, and no coalition of 5 voters has enough weight unless it includes either B and C or else A.

19.7. Suppose that the old system is $V(q; a_1, a_2, \ldots, a_n)$ and suppose that the first voter is voter A. Let $L = a_2+a_3+\cdots+a_n-q+1$. (This exact formula is unimportant. Any value of L larger than this would suffice.) The new system is $V(q+L; a_1+L, a_2, \ldots, a_n)$. (The first voter is given L extra votes, but the quota is raised by the same amount.) If a coalition includes voter A, then it is a winning coalition in the new system exactly when it was a winning coalition in the old system. That is because the coalition will weigh exactly L votes more in the new system than in the old, but the quota is set correspondingly L votes higher, so there is no difference between the systems for such coalitions. If a coalition does not include voter A, then in the new system, it cannot win, because the weight of such a coalition is at most $a_2 + a_3 + \cdots + a_n$, while the quota is $q + L = q + (a_2 + a_3 + \cdots + a_n - q + 1) = a_2 + a_3 + \cdots + a_n + 1$. Hence no such coalition can win, and voter A has veto power.

19.9. This is the weighted system that results from giving every vowel a weight of 20 and every consonant a weight of 1 and setting the quota q

equal to 70. Notice that 4 vowels alone weigh enough to reach the quota, while 3 vowels and 10 consonants do so also. Meanwhile, 2 vowels, even if accompanied by all 21 consonants, do not reach the quota. So these weights properly represent the voting system described.

19.11. (a) It is a weighted voting system, equivalent to the weighted voting system $V(5; 2, 1, 1, 1, 1, 1, 1)$. (b) Of the $2^7 = 128$ coalitions in this voting system, the winning ones in which the mayor is critical are those that consist of the mayor and either 3 or 4 council members. There are $C(6,3) + C(6,4) = 20 + 15 = 35$ such coalitions, so the Banzhaf power of the mayor is 35. (The notation $C(n,k)$ here anticipates its introduction in Chapter 20.) The coalitions in which a council member is critical are those that consist of that council member along with 4 others or that council member along with 2 others and the mayor. There are $C(5,4) + C(5,2) = 5 + 10 = 15$ such coalitions, so the Banzhaf power of a council member is 15. The ratio of these powers is $35/15 = 7/3$.

19.13. Suppose that voter i has smaller weight than voter j. If we have a winning coalition in which voter i is critical and in which voter j does not appear, then the coalition formed by removing voter i and substituting voter j is a winning coalition with voter j critical. Moreover if we have a winning coalition in which voter i is critical and in which voter j *does* appear, then voter j is critical in this coalition as well. We therefore find that, for every winning coalition in which voter i is critical, there is a corresponding coalition in which j is critical. Therefore the Banzhaf power of voter j is at least as large as the Banzhaf power of voter i.

19.15. We abbreviate the provinces AB, BC, MB, NB, NL, NS, ON, PE, QC, and SK, as is customary. If the system for amending the Canadian constitution were weighted, then there would be a quota q and weights, which we abbreviate with the two-letter names ab, bc, mb, nb, nl, ns, on, pe, qc, and sk, such that a coalition can pass an amendment if and only if it weighs q or more.

The coalition consisting of the 6 provinces with the largest populations is not a winning coalition, because it falls short of the 7 provinces required. Therefore

$$ab + bc + mb + on + qc + sk < q.$$

Moreover, the coalition consisting of the 8 smallest provinces is not a winning coalition, because the total population of these state falls short of 50% of the population of all 10 provinces. Therefore

$$ab + bc + mb + nb + nl + ns + pe + sk < q.$$

Adding these two inequalities yields

$$2ab + 2bc + 2mb + nb + nl + ns + on + qc + pe + 2sk < 2q. \qquad (*)$$

The coalition consisting of AB, BC, MB, NL, ON, PE, and SK meets both branches of the 7-50 formula, and so

$$ab + bc + mb + nl + on + pe + sk \geq q.$$

Also, the coalition consisting of AB, BC, MB, NB, NS, QC, and SK meets both branches of the 7-50 formula, and so

$$ab + bc + mb + nb + ns + qc + sk \geq q.$$

Adding these two inequalities yields

$$2ab + 2bc + 2mb + nb + nl + ns + on + qc + pe + 2sk \geq 2q.$$

This contradicts $(*)$, and so the assumption that this voting system can be weighted is false.

19.17. Assume it is a weighted voting system with quota q and weights a, b, c, and d for the four voters. Then $a + b \geq q$ and $c + d \geq q$, and so $a + b + c + d \geq 2q$. On the other hand, neither AC nor BD are winning coalitions, from which it follows that $a + c < q$ and $b + d < q$, and so $a + b + c + d < 2q$. This is a contradiction.

19.19. Let system 1 be $V(2; 1, 1, 0, 0)$ and let system 2 be $V(2; 0, 0, 1, 1)$. The winning coalitions in system 1 are AB, ABC, ABD, and ABCD, and the winning coalitions in system 2 are CD, ACD, BCD, and ABCD. The coalitions that win in one or the other of these systems are AB, CD, ABC, ABD, ACD, BCD, and ABCD. This voting system is the one considered in problem **19.17**, where it was shown that it is not a weighted system.

Chapter 20

20.1. We compute that $C(4, 2) = 4!/(2! \cdot 2!) = 24/(2 \cdot 2) = 6$. The 6 profiles, using the numbers 1, 2, 3, and 4 to stand for the 4 voters and listing just the voters who vote for candidate A, are 12, 13, 14, 23, 24, and 34.

20.3. There are $2^4 = 16$ equally likely possible patterns of boys and girls. Of these, exactly $C(4, 2) = 6$ of them consist of two boys and two girls. Hence the probability is $6/16 = 3/8$.

20.5. According to the 2010 census, Wyoming has apportionment population 568,300. Substituting this for p in Equation (20.4) yields a probability of approximately 1/84,000.

20.7. Because these 8 jurisdictions have the same voice in the Electoral College, they have precisely equal Banzhaf power. Therefore the only difference between voters in one of these jurisdictions and voters in another is associated with the probability that the voter will be critical within the state. This probability increases as the population of the state decreases, and so the greater power resides with the voters in the smaller state.

20.9. (a) It is the voters in the smallest state that are most powerful, because any two states carry the election and voters in the smallest state have the most influence over the outcome in their state. (b) For a voter in the smallest state, the probability is $1/2$ that the other two voters disagree, in which case the voter casts the deciding vote. There is then a probability of $1/2$ that the other two states disagree. Hence the probability that a voter in the smallest state casts a deciding vote in the national election is $1/4$. For a voter in the middle-sized state, the probability of casting the deciding vote within the state is $C(4,2)/2^4 = 6/16 = 3/8$. The probability is again $1/2$ that the other two states disagree, and so the final probability of casting a deciding vote is $3/16$. For a voter in the largest state, the two probabilities are $C(6,3)/2^6 = 20/64 = 5/16$ and $1/2$, so the probability of casting a deciding vote is $5/32$. (c) The smaller two states would be dummies, and the probabilities of their voters casting deciding votes plunges to 0. In the largest state, the probability increases to $C(8,4)/2^8 = 70/256 = 35/128$, because once a voter casts the deciding vote in his state, the national election is over. (d) Again the probabilities in the two smaller states become 0. The probability in the largest state increases from $5/32$ to $5/16$.

20.11. For $n = 2$ and $p = 5$, the exact value is $C(4,2)/2^4 = 6/16 = 3/8 = 0.375$, while the approximate value is $\sqrt{2/\pi} \cdot \frac{1}{\sqrt{5-1}} = 1/\sqrt{2\pi} \approx 0.399$. For $n = 7$ and $p = 15$, the exact value is $C(14,7)/2^{14} = 3432/16384 \approx 0.209$, while the approximate value is $\sqrt{2/\pi} \cdot \frac{1}{\sqrt{15-1}} = 1/\sqrt{7\pi} \approx 0.213$. For $n = 1000$, the exact value may be computed with symbolic algebra software and the result is $C(2000,1000)/2^{2000} \approx 0.01784$, while the approximate value is $\sqrt{2/\pi} \cdot \frac{1}{\sqrt{2001-1}} = 1/\sqrt{1000\pi} \approx 0.01784$. Note that the approximation becomes accurate to four significant figures once the population reaches even the modest value 2001.

20.13. The function $f(p) = K(p+1{,}400{,}000)/\sqrt{p}$ has derivative equal to $(p/2 - 700{,}000)/p^{3/2}$, which is 0 when $p = 1{,}400{,}000$. This is the value of p that yields the smallest value $f(p)$. The state with population closest to this is Hawaii, although they do not appear at the bottom of Table 20.1, because the states at the very bottom are those with unfavorable apportionment rounding.

Bibliography

[1] R. A. Agnew, Optimal congressional apportionment, *American Mathematical Monthly* **115** (2008), no. 4, 297–303.

[2] K. J. Arrow, *Social Choice and Individual Values*, John Wiley & Sons Inc.: New York, 1951.

[3] R. J. Aumann and M. Maschler, Game-theoretic analysis of a bankruptcy problem from the Talmud, *Journal of Economic Theory* **36** (1985), no. 4, 195–213.

[4] R. Axelrod, *The Evolution of Cooperation*, Basic Books: New York, 1984.

[5] M. L. Balinski and R. Laraki, *Majority Judgment, Measuring, Ranking, and Electing*, MIT Press: Cambridge, 2011.

[6] M. L. Balinski and H. P. Young, *Fair Representation: Meeting the Ideal of One Man, One Vote*, second ed., Brookings Institution Press: Washington, 2001.

[7] J. F. Banzhaf, Weighted voting doesn't work: A mathematical analysis, *Rutgers Law Review* **19** (1965), 317–343.

[8] ———, One man, 3312 votes: A mathematical analysis of the Electoral College, *Villanova Law Review* **13** (1968), 304–332.

[9] J. B. Barbanel, *The Geometry of Efficient Fair Division*, Cambridge University Press: Cambridge, 2005.

[10] J. Borda, *Mémoire sur les élections au scrutin*, l'Académie Royale des Sciences: Paris, 1781.

[11] S. J. Brams, *Game Theory and Politics*, Dover Publications Inc.: New York, 2004.

[12] S. J. Brams and A. D. Taylor, *Fair Division: From Cake-Cutting to Dispute Resolution*, Cambridge University Press: Cambridge, 1996.

[13] The Consortium for Mathematics and its Applications (COMAP), *For All Practical Purposes*, eighth ed., W. H. Freeman: New York, 2009.

[14] M. D. Davis, *Game Theory: A Non-technical Introduction*, Dover Publications, Inc.: New York, 1997.

[15] M. J. A. N. de Caritat, marquis de Condorcet, *Essai sur l'application de l'analyse à la probabilité des décisions rendues à la pluralité des voix*, De L'Imprimerie royale: Paris, 1785.

[16] J. Deegan and E. Packel, A new index for simple n-person games, *International Journal of Game Theory* **7** (1978), 113–123.

[17] K. Devlin, *The Unfinished Game: Pascal, Fermat, and the Seventeenth Century Letter that Made the World Modern*, Basic Books: Englewood Cliffs, NJ, 2008.

[18] C. Dodgson, A method of taking votes on more than two issues (1876), *The Theory of Committees and Elections* (D. Black, ed.), Cambridge University Press, Cambridge, 1958.

[19] M. Dresher, *Games of Strategy: Theory and Applications*, Prentice-Hall Inc.: Englewood Cliffs, NJ, 1961.

[20] M. Drton, G. Hägele, D. Haneberg, F. Pukelsheim, and W. Reif, A rediscovered Llull tract and the Augsburg Web Edition of Llull's electoral writings, *Le Médiéviste et l'ordinateur* **43** (2004).

[21] P. Edelman, Getting the math right: Why California has too many seats in the house of representatives, *Vanderbilt Law Review* **59** (2006), 297.

[22] D. S. Felsenthal and M. Machover, Postulates and paradoxes of relative voting power—a critical re-appraisal, *Theory and Decision* **38** (1995), no. 2, 195–229.

[23] P. C. Fishburn, *The Theory of Social Choice*, Princeton University Press: Princeton, NJ, 1973.

[24] V. Fon, Integral proportional system: Aligning electoral votes more closely with state popular votes, *The Supreme Court Economic Review* **16** (2008), 99–129.

[25] A. Gelman, J. N. Katz, and F. Tuerlinckx, The mathematics and statistics of voting power, *Statistical Science* **17** (2002), no. 4, 420–435.

[26] E. J. Gilbert and J. A. Schatz, An ill-conceived proposal for apportionment of the U. S. house of representatives, *Operations Research* **12** (1964), no. 4, 768–769.

[27] O. G. Haywood, Military decision and game theory, *Operations Research* **2** (1954), no. 4, 365–385.

[28] J. K. Hodge and R. E. Klima, *The Mathematics of Voting and Elections: A Hands-on Approach*, American Mathematical Society: Providence, RI, 2005.

[29] E. V. Huntington, The apportionment of representatives in Congress, *Transactions of the American Mathematical Society* **30** (1928), 85–110.

[30] R. J. Johnston, On the measurement of power: some reactions to Laver, *Environment and Planning* **10A** (1978), 907–914.

[31] J. S. Kelly, *Social Choice Theory: An Introduction*, Springer: New York, 1987.

[32] H. Margolis, The Banzhaf fallacy, *American Journal of Political Science* **27** (1983), no. 1, 321–326.

[33] K. O. May, A set of independent necessary and sufficient conditions for simple majority decisions, *Econometrica* **20** (1952), no. 4, 680–684.

[34] J. Maynard Smith, *Evolution and the Theory of Games*, Cambridge University Press: Cambridge, 1982.

[35] S. Nasar, *A Beautiful Mind*, Simon and Schuster: New York, 1998.

[36] J. F. Nash, Non-cooperative games, *Annals of Mathematics* **54** (1951), 286–295.

[37] L. Penrose, The elementary statistics of majority voting, *Journal of the Royal Statistical Society* **109** (1946), no. 1, 53–57.

[38] W. Poundstone, *The Prisoner's Dilemma*, Anchor Books: New York, 1992.

[39] D. G. Saari, *Chaotic Elections! A Mathematician Looks at Voting*, American Mathematical Society: Providence, RI, 2001.

[40] ______, *Decisions and Elections; Explaining the Unexpected*, Cambridge University Press: Cambridge, 2001.

[41] L. Shapley and M. Shubik, A method for evaluating the distribution of power in a committee system, *American Political Science Review* **48** (1954), 787–792.

[42] R. J. Sickles, The power index and the Electoral College: A challenge to Banzhaf's analysis, *Villanova Law Review* **14** (1968), no. 1, 92–96.

[43] P. D. Straffin, *Game Theory and Strategy*, New Mathematical Library, vol. 36, Mathematical Association of America: Washington, 1993.

[44] P. Tannenbaum, *Excursions in Modern Mathematics*, sixth ed., Pearson Prentice Hall: Upper Saddle River, NJ, 2007.

[45] A. D. Taylor, *Social Choice and the Mathematics of Manipulation*, Cambridge University Press: Cambridge, 2005.

[46] A. D. Taylor and A. M. Pacelli, *Mathematics and Politics: Strategy, Voting, Power and Proof*, second ed., Springer: New York, 2008.

[47] J. von Neumann, Zur Theorie der Gesellschaftsspiele, *Mathematische Annalen* **100** (1928), 295–300.

[48] J. von Neumann and O. Morgenstern, *Theory of Games and Economic Behavior*, 60th anniversary ed., Princeton University Press: Princeton, NJ, 2007.

[49] J. D. Williams, *The Compleat Strategyst: Being a Primer on the Theory of Games of Strategy*, Dover: New York, 1954.

[50] H. P. Young, Extending Condorcet's rule, *Journal of Economic Theory* **16** (1977), 335–353.

Index